THE SOVIET MANNED SPACE PROGRAMME

THE SOVIET MANNED SPACE PROGRAMME

An illustrated history of the men, the missions, and the spacecraft

Phillip Clark

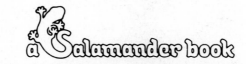

a Salamander book

Published by Salamander Books Limited
LONDON • NEW YORK

a Salamander book

Published by Salamander Books Ltd,
52 Bedford Row,
London WC1R 4LR,
United Kingdom

© Salamander Books Ltd 1988

ISBN 0 86101 369 7

Distributed in the UK by
Hodder & Stoughton Services,
P.O. Box 6,
Mill Road,
Dunton Green,
Sevenoaks,
Kent TN3 2XX

The Author

Phillip Clark, FBIS, is one of the world's
leading authorities on the Soviet space
programme. He was born and educated in
Bradford, and holds an Honours Degree in Mathematics and Computing from the Open Univer-
sity. He has written numerous detailed analyses
of Soviet space operations which have appeared
in the *Journal of the British Interplanetary
Society* and *Spaceflight*, and he is a regular
speaker at the British Interplanetary Society's
Soviet Space Forums. His other written work
includes a contribution on spaceflight published
in *Le Grand Atlas de l'Espace*, and he is currently preparing features for inclusion in the
Magill Survey of Science: Space Exploration
series, and *The Joy of Knowledge*. In 1988 he presented a special edition of the BBC television
science programme "Antenna" which was
filmed in Moscow, and which was devoted to the
Soviet Union's plans for exploration of the Martian system. He currently works for Commercial
Space Technologies Ltd in London, where he
is responsible for producing detailed reports
on the status of the Soviet and Chinese space
programmes.

257507

Dedication

*To the memory of Anthony Kenden
(1947-1987)*

Credits

Editor: Philip de Ste. Croix
Designer: Mike Jolley
Colour artwork: Mike Badrocke
© Salamander Books Ltd
Proof reading: Roseanne Eckart; Harry
Coussins
Filmset: SX Composing Ltd, England
Colour and monochrome reproduction:
Kentscan Ltd, England

Printed in Italy

Contents

Introduction

The Soviet space programme is the largest space enterprise currently operating, and it will remain so for the foreseeable future. In its early years the Soviet programme was shrouded in secrecy, with only scraps of information being officially released. As a result, it was the subject of much speculation in the West: some observers viewed it as an enterprise geared simply towards beating the Americans at all costs, while others saw a slow, stumbling programme based upon a greatly inferior technology.

Although the first satellite was launched by the Soviet Union in 1957, it was ten years before the launch vehicle was seen publicly. The Proton booster was introduced in 1965 but nearly twenty years elapsed before the vehicle was clearly shown and described in any detail. However, the climate is changing and the launch of the Energiya shuttle booster in May 1987 was shown on Moscow Television (and transmitted worldwide) only a day after the event. The watershed seems to have been the forced openness of the Apollo-Soyuz programme of the mid-1970s, followed by the guest cosmonaut programme which began in 1978. When non-Soviet Bloc cosmonauts (France, India, Syria) were training, it was impossible to keep details of the programme secret.

Apart from the collaborative aspect of the Soviet space programme, during the 1980s – and particularly in the wake of the loss of the American *Challenger* Shuttle Orbiter in 1986 – the Soviet Union has offered elements of its space programme on the commercial market. In order that potential users might take the offers seriously, and have sufficient information to make commercial decisions, a more open space programme was essential.

In January 1980 the British Interplanetary Society held its first Technical Forum meeting in London, the topic being the Soviet space programme. Since then, the meetings have become annual events (in May or June), encouraging open discussions on any aspect of Soviet space activity. The meeting of May 1981 received a surprise when Claude Wachtel arrived from France and presented a paper based upon Soviet data describing unflown variants of the Vostok and Soyuz spacecraft; although the information had been published the previous year in a Soviet book, no one in Britain was aware of it.

In many ways, this new information meant that the history of the Vostok and Soyuz programmes needed to be reassessed and two years later, together with Ralph Gibbons, I prepared two papers for the June 1983 Technical Forum meeting, looking anew at the Vostok, Voskhod and Soyuz programmes. In the context of papers which the British Interplanetary Society could publish, as much information as possible about the histories of these programmes was included, but as time

Above: *Ready to go! The Soyuz-T 6 crew on their way to the pad; (l to r) Chretien, Dzhanibekov and Ivanchenkov.*

ORBITAL DATA

Two sources of orbital data are used in this book. In the text and for the Vostok and Voskhod missions the orbits announced by the Soviet Union are usually quoted. However, for the Soyuz, Salyut and Mir programmes the tables of orbital data are derived from the Goddard Space Flight Center's Two-Line Orbital Elements. Issued three times a week, these include virtually daily orbital measurements of Soviet manned spacecraft, and therefore all manoeuvres of the spacecraft can be detected, whether announced by the Soviet Union or not. The data derived from the Two-Lines assume that the Earth is a sphere with a radius of 6,378km; different "models" of the Earth could be used, but they would not alter the arguments presented here.

In the tables of Two-Line derived orbital data the "Epoch" is quoted. This is the date and time at which the orbit was measured. The time is given as decimals of a day, GMT: for example, 16.25 June is 06.00 GMT on 16 June.

The text for the Soyuz, Salyut and Mir programmes normally uses Soviet announced orbits (unless the Two-Lines are specified) and therefore there should be no confusion over the source of the information being used. The Two-Line orbits and the Soviet orbits give different orbital altitudes (although the orbital periods are generally almost identical); this is because the Soviets use a different model of the Earth for their data.

TIMES

The Soviet Union announces all mission events as Moscow Time; during the 1980s a complication was introduced when Moscow Summer Time was introduced during April-October, but the word "Summer" was omitted when times were announced! Here, unless times are specified as Moscow Time, they are quoted as Greenwich Mean Time (GMT).

---

went on more and more information was published by the Soviets about the development of the manned programme. To take a specific example, for more than a decade it had been known that a significant number of cosmonauts depicted in Soviet photographs had never flown in space and who were otherwise unidentified; as a result these men were the subject of much speculation. At a stroke, in a series of newspaper articles to celebrate the 25th anniversary of Yuri Gagarin's first manned space flight, the Soviet authorities openly identified and discussed the unflown men who had been selected with Gagarin. Of course, there are still some "missing" men, but the openness connected with the 1960 selection of cosmonauts and more recently the 1962 selection of women cosmonauts suggests that similar information may be released about other unflown cosmonaut-trainees.

This book, therefore, is the logical development of the two British Interplanetary Society papers of 1983, being expanded to include not simply an overview of the spacecraft design changes and mission types but a review of each flight, showing it as part of a continually evolving programme. The abandoned manned lunar programme of the late 1960s is discussed in depth and the Salyut orbital stations are fully detailed. The most recent developments in the Mir orbital complex are discussed and a forward look provides a study of the new launch vehicles being developed in the late 1980s and the possible manned exploration of Mars before the end of this century.

Detailed specific references are not given for the information provided in this book, but most chapters feature a list of "Further Reading", generally including the books and research papers which have provided the information for the Chapter. The publications *Spaceflight* and *JBIS* (Journal of the British Interplanetary Society) are widely quoted, and they are available from the British Interplanetary Society (27/29 South Lambeth Road, London SW8 1SZ). During the mid-1970s the Society was the world-leader in its publication of discussions relating to the Soviet space programme.

This book is aimed at the general reader, but it also contains information which will interest the experienced observer of the Soviet space scene. A particular example of this are the detailed orbital data quoted for the Salyut and Mir orbital stations, showing all of the manoeuvres completed by these vehicles; this has never before been detailed in a single book. The more general reader will be able to follow the story without having to understand all the details of rocket technology (specific impulse, etc.).

The study of the Soviet space programme is like a detective story – an aspect which attracted this writer to the subject in 1969. Soviet authorities were releasing information slowly, and only careful analysis could fill in the apparent gaps in Soviet state-

ments. Correspondence with Geoffrey Perry of the Kettering Group allowed analytical skills to be developed, and over the years a worldwide correspondence and exchange of views and information has evolved. No-one in the West knows the full story of the Soviet space programme, but all analysts have something to contribute.

In a work of this size, a large number of acknowledgements is essential. While the manuscript of the book was being prepared, it was reviewed, corrected and updated with the help of Rex and Lynne Hall. Rex Hall is perhaps the world's foremost authority on the Soviet cosmonaut team. When he began his studies, the "missing cosmonaut" stories were a quagmire of fact and fiction, with a dash of unwarranted speculation included for good measure, but over the years he has developed his studies into an exact science. Rex also supplied all

Above: *The first Energiya booster waits on the pad in May 1987; this vehicle heralds a new age of Soviet spaceflight.*

the information which has been used in Appendix 4, dealing with spacecraft designers, etc.

Nicholas L. Johnson, in Colorado Springs, is to be thanked for supplying back issues of the Two-Line Orbital Elements, these being essential for the listings of Soyuz and Salyut orbits which are quoted here. Additionally, my exchanges of views and information with him for more than a decade has proved invaluable. The Goddard Space Flight Center is to be thanked for the continued mailings of Two-Line Orbital Elements. Of all my correspondents whose ideas have contributed to this book, Alan Bond, Ralph F. Gibbons, John A. Parfitt and David R. Woods (whose drawings

have been used as essential reference material for some of our artwork) need a special mention. Discussions within Commercial Space Technologies Ltd have been essential for the development of the chapters dealing with the lunar programme and the future programmes; however, the ideas expressed here are those of *this* author and not necessarily those of Commercial Space Technologies Ltd. As far as the illustrations in this book are concerned, I would like to express particular thanks to Mike Badrocke, who prepared all the artwork, and to Theo Pirard of the Space Information Center in Pepinster, Belgium, who very generously supplied original photographs from his own incomparable collection. Last, but by no means least, I would like to thank Dr Diane Holmes for her work in managing to help check the proofs while over-burdened with other work.

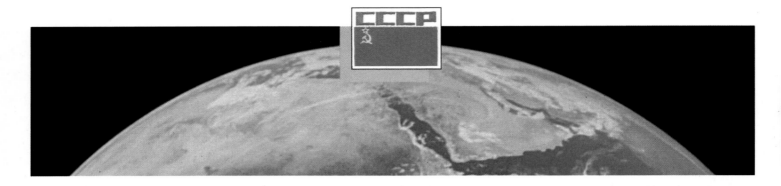

Chapter 1: **Laying the Foundations**

On 4 October 1957 the Western world was stunned by the news that the Soviet Union had launched the world's first artificial satellite, even though the Soviets had long been indicating that such a launch was imminent. Sputnik 1 itself was a simple payload, but it was quickly followed by the biological payload Sputnik 2 and, following a launch failure, the major geophysical research satellite Sputnik 3.

Soviet publications reveal that a great deal of research had been conducted in the Soviet Union from the beginning of the present century into various problems concerned with spaceflight, before the first actual launches occurred. An extensive review of this research cannot be conducted here, and the interested reader is referred to other sources listed at the end of this chapter which examine this early history in detail.

Preparations for Sputnik 1

At the end of World War II the Soviet Union viewed the outside world with trepidation. The Soviet defeat of the invading German forces on the Eastern Front and their subsequent occupation of the eastern part of Germany (now the German Democratic Republic) gave strength to the faltering regime which had been brought to power by the Revolution in 1917. However, the Communist World saw itself threatened by the United States and Europe, and feared a further confrontation.

By the early 1950s Soviet scientists had succeeded in building an atomic bomb, although they still lacked the means to deliver such a weapon to the United States. The United States could, however, penetrate Soviet defences because its bomber forces had access to overseas bases for refuelling; the Soviet forces facing the West did not have this capability. Therefore, the Soviets turned to rocketry to develop the means to deliver nuclear weapons that could strike the United States.

Unlike the Americans, who found that they were able to miniaturise their atomic weapons when they turned to developing a missile carrier system, the payload requirement that the Soviets had to achieve to ensure an intercontinental capacity was formidably large.

Early in the 1950s Soviet scientists, led by Sergei P. Korolyov – who would later become the "Chief Designer of Spacecraft" – were given the task of designing a missile

Above: *The original SS-6 Sapwood missile sits on the launch pad in 1957. The first flight was in May but it was three months before a full-range flight was successfully accomplished. Presumably the long nose-cone is a test object.*

Left: *Konstantin Tsiolkovski (1857-1935) is now the revered "Father of Cosmonautics" in the Soviet Union. A school teacher in the town of Kaluga, he derived the theoretical mathematics of rocketry which inspired others like Korolyov actually to design and fly rockets. Although he died virtually unknown, he is now considered a hero.*

Left: A photograph taken by an American U-2 reconnaissance plane of the launch complex for the SS-6 and later Sputnik and Vostok boosters. The boosters are brought to the pad by rail from the bottom of the picture. The flame pit extends from beneath the launch pad to the top of the picture.

Below: A rare photograph showing the launch of an SS-6 missile. Ignition complete, the four petals of the "tulip" fall away from the booster as it leaves the launch pad. The launch complex and support equipment are designed to allow launches to take place from the same pad within 2-3 days.

recent test firings having been completed on 3 August.

Thus, in the third quarter of 1957 the Soviet Union had a vehicle available which was capable of launching an atomic weapon and also orbiting a satellite. However, no suggestion of an impending Soviet launch was voiced in the United States, and when it came the size of the payload was far greater than anything which the Americans could achieve.

Sputnik 1

The period from mid-1957 to the end of 1958 was designated the International Geophysical Year by the world's scientific community, and both the Soviet Union and the United States had indicated that they would attempt to launch scientific satellites within that period. While the American preparations continued in the full glare of publicity, Soviet preparations were hidden from the world.

On 16 April 1955 an article in the Soviet evening paper *Vechernaya Moskva* announced that a spaceflight commission had been set up in the Soviet Union, and no secret was made that one aim of the commission was to ensure the launching of a Soviet satellite. 29 July 1955 saw the Americans announce that they would launch a number of small scientific satellites during the impending International Geophysical Year. Only three days later, the Soviet scientist Leonid Sedov revealed Soviet plans for launching a satellite in the same time-scale.

Further concrete signs that the Soviets would soon launch a satellite became apparent in 1957. The President of the Soviet Academy of Sciences announced on 9 July that Soviet scientists had solved the problem of orbiting an artificial satellite, and the Academy's *Astronomical Journal* indicated that the transmission frequencies of the first satellite were to be 20 megacycles and 40 megacycles. The public record shows that these announcements made no impact in the West whatsoever.

It is uncertain when the final "go-ahead" was given for the first satellite to be prepared. It was rumoured to be scheduled for 15 September 1957, the hundredth anniversary of the birth of Konstantin Tsiolkovsky, the Soviet space pioneer whose theoretical work had inspired Korolyov and many other rocket scientists around the world. Whatever was planned, Korolyov decided that the first satellite to be launched would be of the simplest characteristics: it would be a sphere with a mass of about 80kg.

Because of its historical importance, information of a very detailed nature for the first Sputnik follows here. According to Academician Tikhonravov, a senior member of the Korolyov design bureau, in 1973, the guidelines decided upon for the design of the first satellite were:

1 The artificial Earth satellite was to be of maximum simplicity and reliability, so that it would have the best possible chance of completing the tasks assigned to it. At the same time, technical solutions of the problems of ensuring its pressurisation and temperature control were to be of a kind

system which would be capable of carrying a heavy atomic warhead as far as the United States. The vehicle with which Korolyov attained this capability became known in the West as the SS-6 or Sapwood rocket (later to become the basis of the world's most-used launch vehicle); the Soviet designation for the vehicle is "R-7". One may wonder how much of Korolyov's design for the SS-6 was influenced by its potential as a satellite carrier.

The SS-6 consisted of a long central core booster stage, atop which sat the weapon: the core stage was surrounded by four strap-on boosters. The first testing of the SS-6 seems to have taken place in 1956;

and it is reported that there were some 40 to 50 tests in all (many resulting in failure) which involved the staging of the strap-on boosters and operations near the peak altitude. In the summer of 1957 an American U-2 reconnaissance aircraft is said to have overflown the test base at Tyuratam, and apparently it photographed an SS-6 ICBM on the launch pad. The long-range testing of the SS-6 began in May 1957, and continued until August that year. Eight launches were made, culminating in two flights in August which reached a range of 6,500km. On 26 August 1957 the Soviet Union announced to the world that it possessed an operating ICBM, with the most

that could be used in the planning of a more complex spacecraft.

2 The body of the satellite was to be spherical in order to determine, as accurately as possible, the density of the atmosphere along the orbital path of the unguided satellite.

3 The artificial satellite was to carry radio equipment working without interruption on at least two wavelengths, and with sufficient power to ensure the reception of its signals over great distances by a large number of radio tracking stations and amateur observers, in order to obtain statistical data on the propagation of radio waves through the ionosphere under varying conditions.

4 The antenna system was to be designed to exclude the effects of the spinning of the satellite on the intensity of the received signals. In order to work out a method for using radar to measure the orbit by means of passive reflections of the signals, angle reflectors were installed on the body of the carrier rocket [ie, the core stage of the SS-6]. The rocket was also to carry equipment for recording cosmic radiation.

5 The equipment on board the satellite was to obtain its power from batteries with a high energy output capable of ensuring the functioning of the equipment for at least two or three weeks.

6 The positioning of the satellite on its carrier rocket and the method of separation was to be such that it would ensure that there would be no failure at separation or at the moment when the antennas deployed.

7 The radio transmitting equipment of the satellite was designed to ensure: One watt transmissions on frequencies of about 20 megacycles and 40 megacycles; transmission of two telemetry signals, recording pressure and temperature inside the satellite by means of a form of frequency modulation; power supply from silver-zinc accumulators; an uninterrupted life of 14 days and nights.

The first Sputnik was revealed after launch to have been a sphere with a

MASS BREAKDOWN OF SPUTNIK 1

1.	**THE BODY OF THE SPUTNIK**	
	Upper half of casing	5.8kg
	Lower half of casing	5.9kg
	Shield	1.6kg
	Other parts	0.6kg
	Total	**13.9kg**
2.	**EQUIPMENT, WIRING & POWER SOURCES**	
	Power source unit	51.0kg
	Radio transmitters	3.5kg
	Remote control switch	1.6kg
	Ventilator	0.2kg
	Diffuser	1.2kg
	Commutator unit	0.7kg
	Pressurisation ducts and pyrotechnic switch	0.4kg
	Reinforcements	0.8kg
	Total	**59.4kg**
3.	**ANTENNAS**	8.4kg
4.	**OTHER PARTS**	0.3kg
	Total mass of Sputnik 1	**82.0kg**

Notes: The data are based upon the Tikhonravov paper listed in the Further Reading notes at the end of this chapter: the total mass of Sputnik 1 is normally given as 83.6kg.

diameter of 0.58m with four antennas; the mass of the satellite was said to have been 83.6kg, a figure which differs slightly from that given in the table, which is based upon Tikhonravov's summary. The objectives of the Sputnik 1 mission were five-fold, according to Soviet reports:

1 To test the method of placing artificial Earth satellites into orbit and their separation from carrier rockets.

2 To supply data on the density of the upper atmosphere, essential for the calculation of the lifetime of a satellite.

3 To test radio and optical methods of measuring orbits.

4 To find out how radio signals of varying frequencies passed through the atmosphere.

5 To check the principles of pressurisation control under the conditions of spaceflight and the efficiency of the seals used.

The Sputnik 1 satellite was well below the maximum payload capacity which the SS-6 booster could place in orbit, but even so it was an order of magnitude in excess of what the Americans planned. By 4 October 1957 the satellite had been mounted in the nose-cone of the SS-6 launch vehicle and it stood on the launch pad at Tyuratam. The mass of the vehicle was about 260 tonnes

Below: *An exploded view of Sputnik 1, the first satellite to be successfully launched. Inside the simple spherical container were a chemical battery system, radio transmitter and other electronic equipment. The attachment points for three of the four whip antennas are visible on the right-hand outer casing of the satellite.*

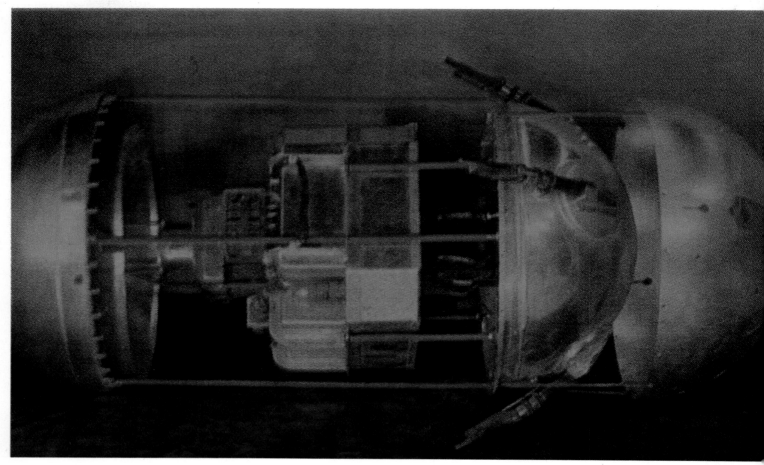

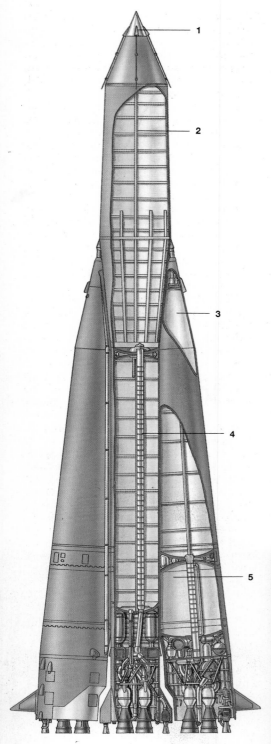

The Sputnik 1 Launch Vehicle

1 Sputnik 1 inside a protective conical shroud.
2 Liquid oxygen tank for the core booster.
3 Liquid oxygen tank for a strap-on booster.
4 Kerosene tank for the core booster.
5 Kerosene tank for a strap-on booster.

Although used as the Sputnik booster in 1957, the basic SS-6 missile with upper stages added and with some slight up-rating is still in use as the launch vehicle for the manned Soyuz-TM missions. The use of non-storable propellants on the SS-6 meant that it could never have been seriously considered as an operational ICBM, because propellant would need loading just before launch. More modern ICBMs use storable propellants which allow the booster to sit on a pad or (more usually) in a silo for many weeks or months ready for immediate launch. Today operational ICBMs use solid propellant which has an even longer lifetime inside the vehicle. The Sputnik booster was launched on four space missions: the first, second and fourth orbited Sputniks 1-3, while the third failed to reach orbit. According to unofficial western reports, even as the fourth Sputnik launch was taking place, a three stage version of the booster was being prepared.

and when lift-off came at 22h 28m 04s Moscow Time (19.28.04 GMT) a total of five engines generating nearly 500 tonnes of thrust came to life. It took about 300 seconds to reach orbital velocity and altitude, and after injection the Sputnik separated from the core of the SS-6 booster (the booster's empty mass would be about 7.5 tonnes), and its four antennas were extended. The first man-made satellite had entered orbit around the Earth, a success at its first launch attempt. The space age had arrived.

The resulting world reaction to the launch of Sputnik and the concern among the American military hierarchy has been reviewed in detail elsewhere (see Further Reading at the end of the chapter), and will not be repeated at length here. The American satellite plans hinged on their small Vanguard programme, and an orbital launch was not scheduled for 1957. The Vanguard programme therefore came under pressure and in early December an attempt was made to launch a small test satellite: the booster exploded only seconds after launch. The first American satellite, Explorer 1, was not orbited until 31 January 1958 (1 February GMT), and by this time the Soviets had launched Sputnik 2 and gained an impressive lead over the Americans in what was to become popularly known as the "Space Race".

The cause of the furore continued in its solitary orbit. The initial orbit of Sputnik 1 was tracked in the West 65.1°, 215-939km with an orbital period of 96.2 minutes. Its batteries long since silent, Sputnik 1 remained to circle the Earth until it decayed on 4 January 1958. The more massive carrier rocket body which had placed Sputnik 1 in orbit had re-entered the atmosphere a month before the satellite, the decay occurring on 1 December 1957.

A Dog in Orbit

After the success of Sputnik 1, the Soviets had no intention of resting on their laurels. A far more ambitious flight was planned less than a month after Sputnik 1. There are some reports that the Soviets had not originally planned to launch a biological satellite until the third Sputnik, but in fact Sputnik 2 carried a dog; presumably, the original second Sputnik was to have been a scientific mission.

Since they were using an ICBM for orbital flights which had been designed to carry a heavier atomic weapon than the Americans planned, the payload mass was not a critical limiting factor to the Soviets in the early days of spaceflight. With little warning, Sputnik 2 took to space, carrying a cabin and equipment with a mass of 508.3kg (plus the mass of the attached core booster, 7.5 tonnes). Although the mass was a surprise, the payload was even more so: it was a dog named Laika carried in a pressurised cabin. During the first week of the flight, biological data were returned to Earth and these would prove invaluable in the planning of the embryonic manned space programme. The launch came so early in the space programme that the problem of a safe re-entry had not been solved, and consequently Laika was put to sleep after about a week in orbit.

Unlike Sputnik 1 and Sputnik 3 which followed, Sputnik 2 remained attached to the booster stage of the launch vehicle. The payload was about 4m long and about 2m in diameter at the base (overall, taking the

Above: *Laika, the first animal to be launched into orbit around the Earth. This husky-type bitch was launched inside a pressurised cabin as part of the Sputnik 2 payload, but no recovery was possible at the end of the mission.*

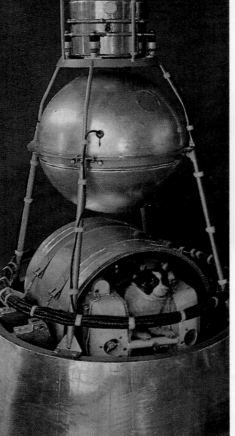

Left: *The Sputnik 2 payload had a mass of about half a tonne. This remained attached to the core of the Sputnik 2 booster after launch. The cabin in which Laika rode in space forms the lower part of the satellite proper.*

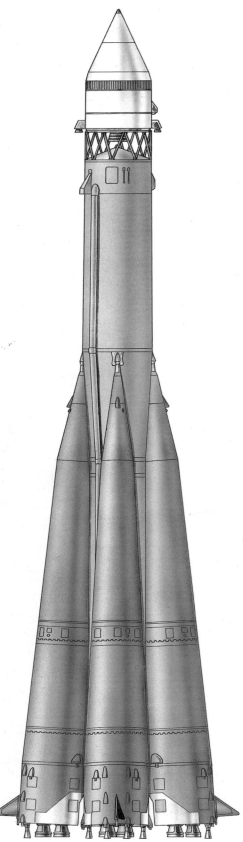

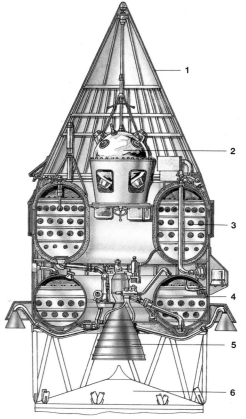

shape of a cone), but the total length of the object in orbit must have been about 32m. Sputnik 2 remained in orbit for 162 days, re-entry coming on 14 April 1958.

There were two further launches in the Sputnik programme before the Soviets switched their attention to the Moon. The first, on 3 February 1958, was unsuccessful when an intended Sputnik 3 failed to reach orbit. The actual Sputnik 3 was launched on 15 May 1958 and entered an even more eccentric orbit than that of Sputnik 2 (see table). The satellite had a mass of 1,327kg of which 968kg comprised scientific equipment. Sputnik 3 was an advanced geophysical laboratory, which explored the Earth's ionosphere, magnetic field, radiation belt, cosmic rays etc. and it returned scientific data until its re-entry into the atmosphere on 6 April 1960.

The Lunar Programme

After the launching of the first artificial Earth satellites, the Soviet Union's next goal was a flight towards the Earth's near-est neighbour, the Moon. To launch a spacecraft to the Moon, the Soviets added a small third stage to the SS-6 missile. The Sputnik booster had been designated the SL-1 and SL-2 vehicles in Western intelligence circles ("SL" indicating Satellite Launcher; the system is explained in Appendix 1). The difference between the SL-1 and SL-2, however, is none too clear from open sources. The lunar variant with its third stage became the SL-3, and in an uprated version this was to become the booster for manned space missions during 1961. On the lunar missions, the booster placed the payload on a direct ascent trajectory towards the Moon, without entering an initial Earth parking orbit.

A series of launch failures dogged the lunar probe programme in 1958, but January 1959 saw the first acknowledged Soviet mission launched towards the Moon. Luna 1 was intended to hit the Moon, but in the event it flew past its target. Following further failures, Luna 2 was launched in September 1959 and this successfully crashed onto the lunar surface at a point east of the Marc Serenitatis – the first man-made object to reach another world. A third announced Luna probe followed the next month, and this looped around the Moon, taking photographs of parts of the far side of the lunar surface which are never seen from the Earth. Like its predecessors, it was also instrumented to study the physical and radiation environment of cislunar space.

After the three acknowledged missions in 1959, the two launches in 1960 were an anti-climax since they failed to enter a correct trans-lunar trajectory and were not announced by the Soviets. Thus ended the first series of lunar exploration probes. Now the stage was being readied for a manned spacecraft.

Luna 1 Launch Vehicle Third Stage Assembly
1 Protective payload shroud, separated while the third stage RO-7 engine is burning.
2 Luna 1 capsule sitting in its container. The SL-3 booster placed this on a direct trajectory to the Moon.
3 Torus tank, containing liquid oxygen.
4 Torus tank, containing the kerosene.
5 The RO-7 engine assembly, with a main combustion chamber and four vernier engines (two are visible).

6 Top of the second stage core booster with open truss network connecting the two upper stages of the launch vehicle.

A modification of this upper stage was later used for the manned Vostok flights (the engine performance was improved and a protective cylindrical shroud which covered the Vostok retro-rocket was added to the rocket stage), and is still in use for the occasional launch of unmanned non-recoverable satellites in the Earth resources programme.

SL-3 Launch Vehicle
This is the original version of the three stage SL-3 Vostok booster which was used for the three successful launches of Luna missions in 1959. The third stage detailed above was added to a somewhat modified version of the original Sputnik launch vehicle. This variant of the booster had a mass of 279 tonnes, compared with the Sputnik 3 booster's 267 tonnes. It seems probable that this particular variant of the SL-3 vehicle was only used for the Luna missions in 1958-1960 and that all the unmanned flights –

starting with the Vostok tests – which took place with military and meteorological satellites used the actual Vostok variant of the upper stage. The failure rate of this vehicle was high, according to western reports: of the eleven launches during 1958-1960, only three in 1959 were acknowledged by the Soviets. The remaining missions either failed to reach orbit or the orbits were so inaccurate that the Soviets ignored the missions (the Soviets were using a direct launch to the Moon which called for great launch accuracy).

SPUTNIK PROGRAMME LAUNCHES					
LAUNCH DATE	**SATELLITE**	**MASS** (kg)	**INCL** (deg)	**ORBITAL PERIOD** (min)	**ORBITAL ALTITUDE** (km)
1957 4 Oct	Sputnik 1	83.6	65.1	96.17	228-947
3 Nov	Sputnik 2	508.3*	65.3	103.75	225-1.671
1958 3 Feb	Sputnik	1,250 ?		failed to reach orbit	
15 May	Sputnik 3	1,327	65.2	105.8	230-1,880

Notes: The masses and orbital data are as announced by the Soviet Union. The mass of Sputnik 2 (marked *) excludes the second stage core booster which remained attached to the payload.

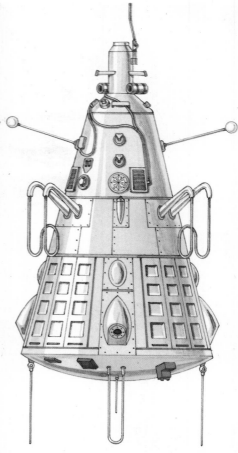

Sputnik 3
Sputnik 3 may originally have been scheduled to be the first Soviet satellite. If it had been launched instead of Sputnik 1 in 1957 the effect on American morale to have had a 1.3 tonne Soviet satellite in orbit before the small (less than 10kg) Vanguard would have proved devastating. It appears that Korolyov erred on the side of caution and decided to fly the simple Sputnik 1 instead of the complex Sputnik 3 payload in 1957. As a sophisticated (for its time) geophysical observatory, Sputnik 3 returned data until it was destroyed on entering the Earth's atmosphere in 1960. The launch of Sputnik 3 marked the fourth and final use of the basic "core + strap-on" Sputnik booster; the third launch had failed to orbit its payload. Of the missions launched by the Soviets during 1957-1959, Sputnik 3 was by far the most successful.

FURTHER READING

Daniloff, N., *The Kremlin and the Cosmos*, Alfred J. Knopf, New York, 1972

Govorchin, G.G., "The Soviets In Space – An Historical Survey" in *Spaceflight*, vol 7 (1965), pp74-82.

Peebles, C., "Tests of the SS-6 Sapwood ICBM" in *Spaceflight*, vol 22 (1980), pp340-342.

Peebles, C., "A Traveller In The Night" in *JBIS*, vol 33 (1980), pp282-286, 311.

Peebles, C., "Soviet Launch Losses" in *Spaceflight*, vol 29 (1987), pp163-166. (This paper deals with the failures disclosed in a de-classified CIA document in the L.B. Johnson Library).

Tikhonravov, M.E., "The Creation of the First Artificial Earth Satellite: Some Historical Details", paper given to the IAF Congress, 1973.

Tokaty, G.A., "Soviet Space Technology" in *Spaceflight*, vol 5 (1963), pp58-64.

Tokaty, G.A., "Foundations of Soviet Cosmonautics" in *Spaceflight*, vol 10 (1968), pp335-346.

LAUNCHES OF THE FIRST LUNAR PROBES			
LAUNCH DATE	**SATELLITE**	**MASS (kg)**	**INTENDED MISSION**
1958 1 May ?	Luna	350 ?	Lunar fly-by/impact?
25 Jun ?	Luna	350 ?	Lunar fly-by/impact ?
22 Sep ?	Luna	350 ?	Lunar fly-by/impact ?
15 Nov ?	Luna	350 ?	Lunar fly-by/impact ?
1959 2 Jan	Luna 1	361.3	Lunar fly-by: impact failed
9 Jan ?	Luna	375 ?	Lunar impact ?
16 Jun ?	Luna	375 ?	Lunar impact ?
12 Sep	Luna 2	390.2	Lunar impact
4 Oct	Luna 3	278.5	Photography of lunar far side
1960 12 Apr ?	Luna	300 ?	Lunar photography ?
18 Apr ?	Luna	300 ?	Lunar photography ?

Notes: The missions with questioned launch dates were not announced by the Soviet Union and failed to attain the intended trans-lunar trajectories. The failures are usually derived from a de-classified CIA document recovered from the Lyndon B. Johnson Library by Curtis Peebles in 1985. In these cases, the masses can only be estimated and the probable missions suggested. The masses exclude the final stage of the SL-3 launch vehicle, which may or may not have included supplemental experiments.

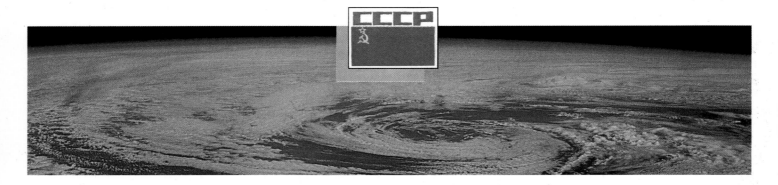

Chapter 2: **The Development of Vostok**

Even before their first satellite had been launched, Soviet space scientists were planning ahead; seriously thinking not only of placing a man in space but also of landing a man on the Moon. Tikhonravov's review of Sputnik 1, noted in Chapter 1, reveals that in November 1956 Soviet scientists began work on preliminary plans for the design of a manned spacecraft, and that they were also looking at the requirements for a manned lunar landing. This was at the time of Korolyov's first deliberations which led to the development of Sputnik 1.

With the launch vehicle introduced for the Luna series of flights – the SL-3 variant of the Sapwood – the Soviets had a booster capable of placing well over four tonnes into a low Earth orbit, a mass sufficient for a manned spacecraft. Within a year of Sputnik 1 being launched, the Soviets had begun work on a manned space programme.

Alternative Goals

In the United States the debate which led to the eventual decision to launch a manned spacecraft was conducted in public, but the equivalent Soviet process was hidden from the public record. From comments made over the last two decades, it has become clear that a major debate relating to the advantages of a manned space programme over unmanned exploration raged in the Soviet Union, but only recently, in a booklet which celebrates the 25th anniversary of the Gagarin mission, have the alternative proposals been revealed.

During 1958 Korolyov was actively supporting the proponents of a manned orbital mission as the next goal (after the scheduled Luna missions), but there were opponents to this scheme who championed rival space plans. One school recommended the development of a craft which could carry a man along a suborbital ballistic trajectory. This would allow limited experience of space conditions to be gained before a man was fully committed to orbit. In the end, it was decided that although this project would be easier to achieve than manned orbital flight, it would slow down the overall programme and result in a dissipation of scientific effort and design work. It will be recalled, however, that just such a sub-orbital flight programme was undertaken by the United States, with two such manned missions

taking place in 1961 before John Glenn was placed in orbit in February 1962.

The unmanned space programme lobby took a different line and supported plans for an advanced automatic satellite (no other details of this have been released). Since this programme would add nothing to the goal of a manned space mission at some date, and would absorb the resources of design bureaux and assembly shops, it also was turned down.

During August 1958 reports and feasibility studies of the alternative proposals were submitted for consideration, and in November the council of the chiefs of the design bureaux met to discuss the alternatives. The council decided to begin work immediately on the preparation of a manned orbital spacecraft - the programme which would become Vostok.

Following the authorisation for Vostok being given at the end of 1958, the actual design of the manned spacecraft began in early 1959. Work on the design and some of the manufacturing processes was conducted in parallel, an unusual approach for such a technical project. The initial details of the hull design were available in March 1959, and the advanced design was

ready in May. This contained the initial data required for the design of the spacecraft's circuitry, attitude control, communications and other systems.

What is officially described as the "electrical analogue" of Vostok was completed by the end of 1959. In its outward appearance, it seemed identical to the manned Vostok: it included all the instrumentation which a manned craft required, but did not include the thermal (heat) shield. During the early part of 1960 the first landing tests were conducted with the Vostok descent capsule.

The Design of Vostok

At the time of the initial manned flights, some brief glimpses of what was supposed to be Vostok were revealed in Soviet documentary films, although it is now known that all that was being shown was the cone-cylinder of the payload shroud being mated to the booster's orbital stage. What

Below: *The production line for the Vostok spacecraft re-entry vehicles. Even for the Vostok missions in 1961 the Soviets had established their standard procedure of assembling the vehicles in bulk.*

these early clips did disclose, however, was that the same orbital stage which had been displayed with the 1959 Luna probes was being used to place Vostok into orbit. Even the Moscow Parades showed the Vostok as being the shroud-plus-stage (plus a stabilizing annulus), carried underneath a helicopter, and to complicate matters further the film released to celebrate Vostok 2 showed the same reconstruction but with stubby wings added!

When the true Vostok design was publicly revealed in 1965, it was shown to be a two-part spacecraft. The descent module, in which the cosmonaut sat, was spherical in shape, and below that was the double-cone of an instrument module which contained the retro-rocket system. To this day, when Vostok is put on display it is usually with the third stage of the launch vehicle added to cover the inverted cone leading to the retro-rocket system: the only place which seems to display the Vostok without its final rocket stage is the Tsiolkovsky Museum in Kaluga. However, full cutaway details of Vostok are available from Soviet literature.

The overall mass of Vostok was about 4.73 tonnes: the length of the complete craft was 4.4m and the maximum diameter was 2.43m. On manned missions it was planned that the launch would be made into an orbit low enough to ensure natural decay into the atmosphere within ten days of launch, and therefore the Vostok was launched with supplies sufficient to last the cosmonaut for ten days, to give him a chance of surviving the mission if the retro-rocket were to fail to fire when the planned de-orbit manoeuvres were scheduled to occur.

The spherical descent module which housed the cosmonaut had a mass of 2.46 tonnes and a diameter of 2.3m. To soft-land such a heavy capsule with a cosmonaut inside would have required an enormous parachute and probably an additional soft-landing retro-rocket system — and the inclusion of these would certainly have pushed the mass of Vostok beyond the payload capacity of the SL-3 booster which was scheduled to carry out the launches. Therefore, the Soviet designers incorporated an escape system which would allow the cosmonaut to eject from his craft during the final descent to Earth at an altitude of about 7km. After an initial single parachute system had slowed the descent module, the cosmonaut would eject and his ejector seat would first deploy a small drogue parachute and then a main parachute. On the way down, the cosmonaut would separate from his seat, and land at a velocity of about 5m/sec. In the meantime, the Vostok descent module itself would land nearby at a speed of about 10m/sec - enough (at least) seriously to injure the cosmonaut had he remained inside, since the deceleration overloads could have reached about 100g (100 times the acceleration due to gravity).

The double cone of the instrument module beneath the descent module was 2.25m long and 2.43m at its maximum diameter. The mass of the module was about 2.27 tonnes. At the base of the instrument module was the retro-rocket system,

The Vostok Manned Spacecraft
1 Command control antenna.
2 Communications antenna.
3 Housing for the umbilical connectors (mounted on the Vostok exterior).
4 Entry/exit hatch above cosmonaut's head.
5 Food locker.
6 Tensioning bands, strapping the descent and instrument modules together.
7 Whip antennas.
8 TDU-1 retro-rocket system.
9 Communications antennas.
10 Access hatch to instrument module interior.
11 Instrument (or equipment) module.
12 Electrical harness.
13 Oxygen and nitrogen gas containers for life support system.
14 Ejector seat with cosmonaut.
15 Radio antenna.
16 Porthole with "Vzor" optical orientation device.
17 Technological hatch.
18 TV camera.
19 Ablative heat shield.
20 Electronics package.

The Vostok spacecraft was designed to support a single person in orbit for up to ten days, although the longest manned flight was actually just under five days. The drawing above clearly shows the two parts of the spacecraft: the upper descent module housed the cosmonaut, his instruments and escape system, while the lower instrument module contained propellant and the engine system. The basic vehicle was later modified for the multi-manned Voskhod programme and has seen wide use in the unmanned Cosmos photo-reconnaissance programme. Additionally, variants are now being flown as biological satellites and as the Foton materials processing satellite.

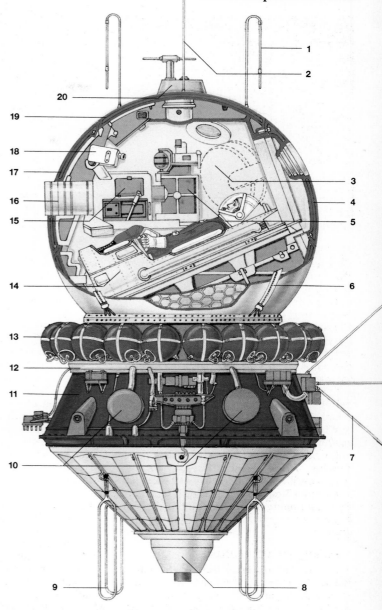

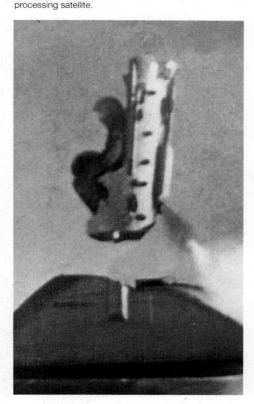

Vostok Descent Module
This is a cosmonaut's view of the Vostok descent module as he is being lowered into it. On the left hand side is the parachute container (1) and lid. In front of the cosmonaut is the panel with his limited controls and instruments (2), while the porthole and "Vzor" are close to his feet (3). The apparently empty space (4) is where the seat is located during the flight. Although not roomy, the Vostok capsule was larger than the American Mercury; in fact, there was enough space for the cosmonaut to be allowed to float "free" in the capsule.

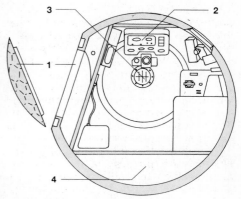

Left: *A cosmonaut in a Vostok ejector seat being launched from an aircraft, simulating the ejection from the descending Vostok sphere.*

LAUNCHES IN THE KORABL-SPUTNIK PROGRAMME							
LAUNCH DATE	SPACECRAFT	DOG(S)	MASS (kg)	INCLINATION (deg)	PERIOD (min)	ALTITUDE (km)	LIFETIME (d.hh.mm)
1960 15 May	KS 1	None	4,540	65 65	91.2 94.25	312-369 307-690	(4. — . —)
23 Jul	KS 2-1	2 Dogs ?	4,600 ?	Failed to reach orbit			
19 Aug	KS 2	Strelka & Belka	4,600	64.95	90.7	306-340	1.03. —
1 Dec	KS 3	Pchelka & Mushka	4,563	64.97	88.47	180-249	1. — . —
22 Dec ?	KS 4-1	2 Dogs ?	4,575 ?	Failed to reach orbit			
1961 9 Mar	KS 4	Chemushka	4,700	64.93		183.5-248.8	0.01.46
25 Mar	KS 5	Zvezdochka	4,695	64.9	88.42	178.1-247	0.01-45

Notes: The two launch failures have been confirmed by the Soviet Union, although the second was described simply as being three weeks after Korabl-Sputnik 3. Orbit data are those announced by the Soviet Union – in a case of two after initial data was refined. Mission durations are given as accurately as Soviet data allows: Korabl-Sputnik 1 should have been de-orbited after four days, but instead it was pushed into a much higher orbit.

designated "TDU-1" (Braking Engine Installation) by the Soviets. The TDU-1 was designed by the bureau headed by A.M. Isayev during 1959, and it used a self-igniting propellant consisting of a nitrous oxide oxidiser and an amine-based fuel. Developing a thrust of 1.614 tonnes and having a specific impulse (see Glossary) of 2,610m/sec (equivalent to 266 sec), the total ignition time of 45 seconds meant that the fuel load was about 275kg. In turn, this implies that the TDU-1 was able to reduce the velocity of a Vostok by about 155m/sec for re-entry into the atmosphere.

Vostok Tests

During 1960 and 1961, in preparation for the manned phase of the Vostok programme, the Soviet Union launched a series of satellites under the name "Korabl-Sputnik" (normally translated as Spaceship-Satellite) which were unmanned tests of the Vostok. The first flight carried no biological specimens, but the remaining flights carried dogs and other animals to provide data on how living organisms fared in space.

As the testing of Vostok landing characteristics was being undertaken, the "electrical analogue" of Vostok, noted above, was readied for an orbital flight. The Soviets have now revealed that there were three variants of the Vostok assembled during this period:

Vostok-A, Non-recoverable variant.
Vostok-B, Recoverable biological variant.
Vostok-V, Man-rated Vostok variant.
("A", "B" and "V" are the first three letters of the Cyrillic alphabet.)

The first launch of a Korabl-Sputnik came on 15 May 1960, and clearly this was the "electrical analogue" without a heat shield because no recovery attempt was to be made. Since many analysts consider that the retro-fire attempt at the end of the mission was indicative of a recovery attempt, part of the launch announcement from *Pravda* on 16 May is reproduced below to clarify this matter:

"After the necessary data have been obtained from the spaceship-satellite the pressurised cabin weighing approximately 2.5 tonnes will be separated from it. In this launching return of the pressurised cabin is not intended, and after the reliability of its functioning has been tested and it has been separated from the spaceship satellite, the cabin, as well as the

spaceship-satellite itself, will at a command from the Earth begin to descend and will terminate its existence as it enters the dense layers of the atmosphere."

On 18 May at 23.52 GMT the Soviets tried to orient the Korabl-Sputnik for re-entry, but it would seem to have been oriented almost exactly 180 degrees from the planned orbital altitude. When retro-fire came, the Korabl-Sputnik was placed not on a descent trajectory but into a higher orbit. The velocity change attained was about 90m/sec, implying a burn time of about 26 seconds: clearly, this would have been insufficient for a de-orbit leading to a recovery.

According to Soviet accounts, the next Korabl-Sputnik launch attempt was on 23 July 1960 but a failure of the booster occurred during the orbital injection phase, and the payload did not reach orbit. It is uncertain whether dogs were carried on this mission: their presence would imply a recovery attempt.

Whatever had been planned for the July mission, the next flight which began on 19 August was clearly intended for recovery. The main experimental payloads were the two dogs, Strelka and Belka, although

other biological specimens were also carried: rats, white mice, flies, plant seeds, fungi, chlorella algae and a plant "spiderwort". Korabl-Sputnik 2 remained in orbit for about a day, after which the retro-rocket was fired. The instrument module separated and burned up in the atmosphere, while the descent craft continued for a successful recovery. During the descent, the experimental animals were ejected from the main module in what would become the cosmonaut's ejector seat, and they were recovered close to the main descent craft. The landing was accomplished only 10km away from the planned site.

Following this successful mission there were two failures in the Korabl-Sputnik programme at the end of 1960. Korabl-Sputnik 3 was launched on 1 December with a payload of two dogs, Pchelka and Mushka: the orbit used was lower than the previous Korabl-Sputniks, and was more similar to those employed for the manned missions. Korabl-Sputnik 3 introduced a new solar orientation system, rather than the previously-used infra-red attitude control system. Whether this new system caused the failure has not been reported, but after retro-fire the satellite re-entered the atmosphere at the incorrect angle and burned up.

The second flight in December was launched three weeks later. Again failure ensued when the third stage of the launch vehicle failed to operate properly, and prevented the Korabl-Sputnik from attaining orbit; however, the Soviets state that the descent craft was recovered safely with its biological payload.

At the beginning of 1961 three more Vostok spacecraft were delivered to the Tyuratam launch site, and these were the first of the man-rated vehicles. The first two would be launched with dogs on board, while the third would carry a man. Korabl-Sputnik 4 was launched on 9 March with a single dog, Chernushka, together with other biological specimens, and was successfully recovered after one orbit. The final Korabl-Sputnik appeared on 25 March with the dog Zvezdochka, and again a successful recovery followed a single Earth orbit. Thus the way was clear for a man to fly into space.

The First Cosmonaut Team

In October 1959 the Soviet magazine "Ogonyok" carried photographs of men apparently training for space missions – quite a scoop. In fact they were simply ground test engineers and pilots undertaking the basic testing of equipment prior to the real cosmonaut-trainees beginning their work. The first team of actual cosmonauts will be detailed in full here for the historical record. Although training began

Left: The 1960 cosmonaut team. Along the front row are (left to right) Popovich, Gorbatko, Khrunov, Gagarin, Korolyov, Korolyov's wife with Popovich's daughter, Karpov (training director), Nikitin (parachute trainer), Fedorov (doctor); second row are Leonov, Nikolayev, Rafikov, Zaikin, Volynov, Titov, Nelyubov, Bykovsky and Shonin; third row are Filatyev, Anikeyev and Belyayev.

Above: *Gagarin in the bus, on his way to the Vostok 1 launch. Behind him in a pressure suit is Titov, his back-up, while behind Gagarin are Nelyubov (support cosmonaut) and Nikolayev.*

Left: *The ejector seat from a Korabl-Sputnik mission in 1961 at the landing site, with a dummy cosmonaut. The dog which completed the flight was also ejected with the suit.*

THE 1960 COSMONAUT GROUP	
NAME	**MISSION ASSIGNMENTS/REMARKS**
I. N. Anikeyev	Dismissed, 1961.
P. I. Belyayev	Joined Voskhod 2 training group, April 1964; commander, Voskhod 2; died 1970.
V. V. Bondorenko	Killed 1961.
V. F. Bykovsky	*Joined Vostok training group, 1960; support, Vostok 2; back-up, Vostok 3; flew Vostok 5.
V. I. Filatyev	Dismissed, 1961.
Y. A. Gagarin	*Joined Vostok training group, 1960; flew Vostok 1; killed 1968.
V. V. Gorbatko	*Vostok support, 1963; joined Voskhod 2 training group, 1964; back-up commander, Voskhod 2 (stood down).
A. Y. Kartaskov	Left group, 1960.
Y. V. Khrunov	*Joined Voskhod 2 training group, April 1964; back-up pilot, Voskhod 2.
V. M. Komarov	*Joined Vostok training group, 1961; back-up commander, Vostok 4 (stood down); support, Vostok 5; joined Voskhod 1 training group, March 1964; commander, Voskhod 1; died 1967.
A. A. Leonov	*Joined Voskhod 2 training group, April 1964; pilot, Voskhod 2.
G. G. Nelyubov	Joined Vostok training group, 1960; support, Vostok 1; dismissed 1961; died 1966
A. G. Nikolayev	*Joined Vostok training group, 1960; back-up, Vostok 2; flew Vostok 3.
P. R. Popovich	*Joined Vostok training group, 1960; flew Vostok 4.
M. A. Rafikov	Retired, 1962.
G. S. Shonin	*Joined Voskhod 3 training group, April 1965; pilot, Voskhod 3 (cancelled).
G. S. Titov	Joined Vostok training group, 1960; back-up, Vostok 1; flew Vostok 2.
V. S. Varlamov	Joined Vostok training group, 1960; left team, 1960; died 1980
B. V. Volynov	*Joined Vostok training group, 1962; support then back-up, Vostok 4; back-up, Vostok 5; joined Voskhod 1 training group, March 1964; back-up commander, Voskhod 1; joined Voskhod 3 training group, April 1965; commander, Voskhod 3 (cancelled).
D. A. Zaikin	*Joined Voskhod 2 training group, 1964-1965; back-up commander, Voskhod 2 (replacing Gorbatko).
Notes: This table lists the twenty men selected for training in 1960, together with their assignments within the Vostok and Voskhod programmes. Cosmonauts who were assigned to missions after Voskhod are noted with an asterisk (*). More details of the cosmonauts who did not fly are given in the text, while a complete list of mission assignments is given in Appendix 3.	

in 1960, many details of the training group were not revealed by the Soviet authorities until 1986, the twenty-fifth anniversary of Gagarin's spaceflight.

In 1959 the Soviet authorities began to recruit military pilots for possible cosmonaut training, with the initial screening of candidates taking place in October. During January-February 1960 the final selection of twenty pilots was made, and the first training session began on 14 March. The head of the cosmonaut training programme was Colonel-General N.P. Kamanin and his deputy was Colonel E.A. Karpov, a specialist in aviation medicine. The twenty men selected for training are listed in the accompanying table. Of these, twelve made orbital flights (not all of which were within the Vostok programme, of course). The details of the remaining eight unflown cosmonauts are given below.

Ivan Nikolayevich Anikeyev was dismissed from cosmonaut training in late 1961 following an incident when he was found drunk at a railway station which resulted in a clash with a military patrol; also involved in this incident were Filatyev and Nelyubov (see below).

Valentin Vasilyevich Bondarenko (also known as "Valentin Junior" and "Tinkerbell") was the youngest member of the cosmonaut team. He died on 23 March 1961 in an accident at the end of a ten day isolation test. He had been in an oxygen atmosphere and after he discarded an alcohol-dipped swab onto a hot ring, a fire began. Although he tried to extinguish the fire himself, he was unable to control it before the outside scientists equalised the cabin pressure with normal atmospheric pressure. He was pulled from the fire alive, but died soon afterwards in hospital. As an aside, one must wonder why Bondorenko was training in a pure oxygen atmosphere, since this was not to be used on the Vostok spacecraft.

Valentin Ignatyevich Filatyev (known as "Valentin the old one") was dismissed along with Anikeyev and Nelyubov in 1961.

Anatoli Yakovlevich Kartashov left the cosmonaut team in summer 1960, following haemorrhages caused by a centrifuge test: the overloads reached eight times the acceleration due to gravity.

Grigori Grigoryevich Nelyubov could have been one of the top cosmonauts: he was an egotist, always wanting to be, but never quite being, the number one. He was dismissed from the cosmonaut group in 1961, together with Anikeyev and Filatyev, and he was posted to the Far East. Apparently he could not accept his dismissal, and while undergoing "a serious spiritual crisis" and in a state of intoxication he killed himself on 22 February 1966 by throwing himself under a train.

Mars Zakirovich Rafikov left the cosmonaut team in 1962 for unspecified reasons.

Valentin Stepanovich Varlamov fractured a vertebra in a swimming pool accident on 24 July 1960 and left the team soon afterwards. He died in October 1980 from a brain haemorrhage.

Dmitri Alekseyevich Zaikin left the cosmonaut group in April 1968, when an ulcer was diagnosed while he was undergoing Soyuz training.

Of the twelve cosmonauts from the 1960 selection who completed space missions, five completed Vostok flights, three completed Voskhod flights and four made their first flights on Soyuz spacecraft. Some were active long enough even to make flights to the Salyut orbital stations.

Countdown to Vostok 1

The trainee cosmonauts were housed at first in a building at the Khodynskoye Field within the Frunze Central Airfield, and when lectures began in March 1960 the curriculum was almost totally given over to aviation medicine. The cosmonauts became quickly disenchanted with their lot and upon hearing this Korolyov assigned a group of his designers to begin lecturing the trainees on the theory of spacecraft design and orbital dynamics. In the early months the training facilities were limited, and therefore six of the trainee cosmonauts were selected to form a group for advanced training, with the intention that this group would supply the crews for the Vostok missions: this group comprised Gagarin, Kartashov, Nikolayev, Popovich, Titov and Varlamov. Kartashov was soon dropped (following the centrifuge acci-

dent) and he was replaced by Nelyubov. Varlamov then suffered his swimming accident and he was replaced by Bykovsky.

With the premier group of six cosmonauts established, Korolyov began to visit the training centre more frequently and became involved with their training. Later, he invited them to his design bureau and showed them the Vostok spacecraft.

The day after the death of Bondarenko, 24 March 1961, the group of six cosmonauts left for their first trip to Tyuratam, to witness the launch of Korabl-Sputnik 5 which was to be the final dress rehearsal for the manned space mission. On 28 March a press conference was held at the Academy of Sciences, during which the results of the two 1961 Korabl-Sputnik missions were discussed. The final preparations for the launch of a manned Vostok were now underway, and on 3 April the Soviet Government officially approved the launch of a manned spacecraft. At this time, the crew had not been finalised, but there were considered to be two "leaders": Yuri Alexeyevich Gagarin and Gherman Stepanovich Titov.

A Man in Space

The six cosmonauts in the Vostok training group, together with the leaders of the training programme, flew to the Tyuratam launch site again on 5 April and the following day K.N. Rudnev, the chairman of the State Commission, arrived. On 7 April, Korolyov and O.G. Ivanovskiy, the design project leader for the Vostok, agreed for Gagarin and Titov to undergo a familiarisation session with the spacecraft that would actually be launched. 8 April saw the State Commission meeting to decide

Left: *The Vostok 1 launch vehicle being raised to the vertical at the launch pad after being transported from the assembly building by rail. The booster was moved to the pad the day before launch.*

VOSTOK 1: MISSION SUMMARY		
DATE (1961)	**TIME** (GMT)	**EVENT**
8 Apr		State Commission agrees to flight: planned orbit is 180-230km, duration, 90 minutes.
10 Apr	Evening	Gagarin named as prime cosmonaut, with Titov as his back-up.
11 Apr	02.00-04.00	Vostok 1, atop its launch vehicle, is transported to the launch pad.
12 Apr	02.30	Gagarin and Titov woken up.
	03.50	Bus containing Gagarin and Titov (and others) arrives at the launch pad.
	04.10	Gagarin is inside Vostok 1: switches on the radio transmitter.
	04.50-05.10	Hatch problem noticed: hatch re-opened and re-sealed.
	06.07	Launch of Vostok 1.
	06.09	First stage strap-on boosters fall away.
	06.10	Separation of payload shroud.
	06.12	Separation of second stage (core) booster: ignition of third stage.
	06.21	Separation of third stage in orbit.
	06.49	Vostok 1 enters the Earth's shadow.
	06.57	Vostok passing over America.
	07.02	Radio Moscow announces the launch of Vostok 1.
	07.09	Vostok leaves the Earth's shadow.
	07.25	Ignition of retro-rocket engine.
	07.35	Descent module Parachute opens at 7km; Gagarin ejects and descends separately.
	07.55	Gagarin lands 26km south-west of Engels, in Saratov region.
Notes: This chronology is generally based upon data given in the Soviet book *Our Gagarin*. The landing time of the descent craft (as opposed to Gagarin's) has not been released. The planned orbit (see 8 April entry) is lower than the one actually attained and is more akin to those found on later Vostok missions: possibly Vostok 1 entered a higher-than-planned orbit?		

who would make the flight and on 10 April Kamanin summoned Gagarin and Titov and informed them that the choice for the flight was to be Gagarin, with Titov as his back-up. Gagarin and Titov continued training on 11 April, with both men experiencing their first trials with the actual Vostok 1 spacecraft.

12 April was the historic day that a man first entered space. After entering the spacecraft, Gagarin immediately established radio communications with Korolyov and the launch team. An hour after he entered the craft, a problem was discovered with the main spacecraft hatch: an electrical contact designated KP-3 was showing that the hatch was improperly closed. As a result, the hatch was removed and then replaced; the alarm was over.

Launch from Tyuratam came at 09.07 Moscow Time (06.07 GMT) and the mission of Vostok 1 lasted for 108 minutes; landing came 26km south-west of Engels in the Saratov region of the Soviet Union. Little information was given about the work which Gagarin undertook while in space, and in fact his experimental programme was almost zero. He was simply a passenger in his spacecraft. There were worries that if Gagarin experienced an adverse reaction to weightlessness he might endanger the mission if he had actual control of his spacecraft. Therefore, the manual controls were locked before the launch; on the instructions of General Kamanin (head of cosmonaut training),

Left: *The launch of Vostok 1, beginning the first manned space mission. The four arms of the "tulip" booster support structure fall back as the booster leaves the pad, while the gantry is retracted.*

Below: *A painting by Leonov and Sokolov, showing Gagarin's descent (right) after ejecting from the main Vostok sphere (left). All of the cosmonauts ejected on Vostok flights.*

Below: *Gagarin in his pressure suit, sitting inside Vostok 1. It is apparent from this picture how much room Gagarin had in the descent module, although he remained strapped in his ejector seat.*

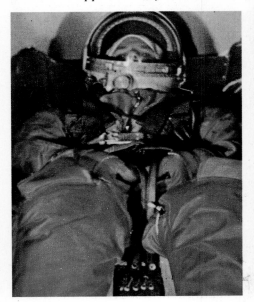

engineer Oleg Ivanovsky and training manager Col Mark Gallai a three-digit code was set on the control's "logic clock", and Gagarin was not told of the code (1-2-5). However, if Vostok 1 were to lose its ground command link which would control the retrofire manoeuvre, then Gagarin could unseal an envelope which contained the code and thus unlock the controls. Since there was no call for this action, Gagarin never took control of his spacecraft.

Live television pictures of Gagarin were sent back to Earth, and while in orbit Gagarin simply reported on his condition and sent "fraternal greetings" to the countries over which he was passing. A small supply of food was carried in the form of paste in tubes, but for such a short flight it was not really needed.

In order to establish Gagarin's flight as a record Gagarin was supposed to land in his spacecraft; if he landed separately, people would assume that he had abandoned Vostok due to problems and therefore the flight would not be recognised as a successful "first". As a result, the Soviets were very evasive about describing Gagarin's landing, hinting that he landed while still inside the descent craft. However, Gagarin *did* land separately from the descent craft using the standard Vostok landing system illustrated. The first open admission of this does not seem to have appeared until the

Vostok Launch and Landing Sequence
1 Launch from the "Baikonur Cosmodrome" at Tyuratam.
2 Separation of the four strap-on boosters.
3 Separation of the payload shroud.
4 The second stage core booster separates and the third stage ignites.
5 Vostok enters orbit and the third stage separates and falls away.
6 Vostok orients itself and retro-fire takes place.
7 The descent and instrument modules separate.

8 Sphere orients itself for re-entry.
9 The hatch above the cosmonaut is automatically blown off.
10 Cosmonaut ejects as sphere parachute container hatch separates.
11 Pilot parachute deploys on ejector seat.
12 Pilot parachute separates.
13 Cosmonaut separates from ejector seat at about 41km altitude and descends and lands with his own parachute.
14, 15, 16 Descent module parachute deploys and module lands.

Vostok Descent Procedure
This shows an "exploded view" of the descending Vostok capsule as the cosmonaut ejects. On the left hand side of the sphere, the descent module's parachute container hatch has separated. The cosmonaut's own parachute can be seen stowed at the top of the seat. On Gagarin's flight the Soviets initially did not acknowledge that he had ejected from the capsule — such an admission might have prevented the flight from being recognised as a

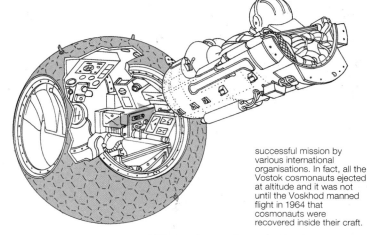

successful mission by various international organisations. In fact, all the Vostok cosmonauts ejected at altitude and it was not until the Voskhod manned flight in 1964 that cosmonauts were recovered inside their craft.

Above: *Making an interesting comparison with the drawing of the cosmonaut ejecting, this photograph shows a Vostok sphere on display, with the top of an ejector seat and cosmonaut mock-up clearly shown inside.*

Left: *Reputedly Gagarin's capsule, this is a recovery photograph of a Vostok sphere. Vostok 1 was apparently in a bad condition after landing and had to be heavily repaired prior to being displayed.*

publication of the 1978 book *Our Gagarin*. This book describes the pilot parachute on the descent craft opening and Gagarin's ejection from the descent craft some 20 minutes before the landing when the descent rate was 200m/s. At an altitude of 4km the ejector seat dropped away from Gagarin and he continued his parachute descent sustained only by his life support system. At about the same altitude the main Vostok parachute opened.

Thus ended the first manned space mission. Gagarin immediately became a world personality. American hopes of a space "first" were dashed. They had only the prospect of a sub-orbital manned space mission in the foreseeable future, while the Soviets were already planning a manned orbital flight of longer duration. President Kennedy's declared ambition in May 1961 of landing an American on the Moon before the end of the decade must have seemed a remote dream.

Vostok 2 to Vostok 6

It was intended to follow the Gagarin flight with a manned mission which would last for three orbits, but political pressure was put upon Korolyov to prepare a mission lasting a day. Thus, was Vostok 2 planned. The second Vostok was launched on 6 August 1961 with Gherman S. Titov on board. It was on this flight that the spectre of space sickness first raised its head. Titov, according to most accounts, was unwell for most of his 25 hour mission, and as a result he was unable to perform many significant experiments. It is believed that he conducted Earth photography and he was given some limited control of his spacecraft: during the first and seventh circuits he was allowed to fire the spacecraft's attitude control thrusters. A supply of paste food was carried in tubes (as with Vostok 1), but how much was consumed because of his illness was not stated. After spending more than a day in space, Titov returned safely to Earth, but, as a result of his space sickness, he was grounded since the illness was not understood at the time.

By the end of 1961, three of the six cosmonauts from the main training group were no longer available for flights: Gagarin and Titov, of course, had already flown, and Nelyubov had left after being involved in trouble at a railway station. During the month of Titov's flight, Komarov joined the main training group: after being assigned as the back-up cosmonaut for Vostok 4, Komarov fell ill and was removed from flight status, being replaced by Volynov.

Also in late 1961, it was decided that one of the later flights in the Vostok programme would carry a woman on board. As a result, on 14 March 1962 five women joined the cosmonaut team. V.V. Tereshkova made the actual flight, while the names of the other four women (announced in 1987) were T.D. Kuznetsova, V.L. Ponomareva, I.B. Soloveva and Z.D. Yorkina.

There was a break of more than a year between Vostok 2 and the next manned Soviet space missions. On 11 August 1962 Vostok 3 was launched, with Andrian G.

Above: Titov (left) training for Vostok 2. Some of the early cosmonaut training equipment like this has now been donated for the use of the "Young Pioneers" group in Moscow.

Left: Titov (left) and Gagarin (right) prior to the Vostok 2 launch. After acting as Gagarin's back-up, Titov suffered from space-sickness during his mission and never made a second flight.

Below: Popovich in the bus taking him to the launch pad for the fourth Soviet manned launch. Already in orbit, Nikolayev was waiting to begin the first Vostok "group flight".

Nikolayev on board. The following day, from the same launch pad, Vostok 4 ascended with Pavel R. Popovich. The two craft were launched into similar orbits, and they approached to within 6.5km of one another; however, since the craft could not manoeuvre, an actual rendezvous could not be considered as a realistic goal. Both missions were destined to be long duration flights by 1962 standards: Nikolayev remained in orbit for four days while Popovich was in space for three days. For these missions the cosmonauts had packed meals rather than the tubes of paste: about 700 grammes was allowed for each man's daily consumption. Nikolayev and Popovich were both permitted to unstrap themselves from their ejector seats and actually "float" weightless in the cramped Vostok interior. Medical monitoring was completed by on-board equipment, following the experiences of Titov.

When Nikolayev and Popovich were safely returned to Earth with no ill-effects during their missions, the Soviets began to feel happier about longer flights, having realised that space sickness was something which depended upon the individual cosmonaut rather than being a direct result of long flights.

Once more, there was a long break before another manned space mission. On 14 June

1963 Valery F. Bykovsky was launched inside Vostok 5, and he completed a mission lasting for five days; this is still the record for the longest one-manned space flight. However, his mission was totally overshadowed by Vostok 6.

Because of the Vostok 3/Vostok 4 dual mission, a further launch was expected to join Vostok 5 in orbit. What was not expected was that the cosmonaut would be a woman. Rumours of illness have surrounded reports of Valentina Tereshkova's

Above: *Valentina Tereshkova (right) and Tatiana Kuznetsova (left) during Vostok landing tests. Kuznetsova was selected with Tereshkova in March 1962, but was not actually assigned to the mission.*

Below: *Tereshkova gives a farewell atop the Vostok 6 booster's gantry, before her launch to become the first woman in space. Already in orbit was Bykovsky aboard Vostok 5.*

mission. It has been suggested that Tereshkova was originally the back-up cosmonaut, and the prime candidate became ill on the day of the launch. This may explain why the Soviets have been somewhat reticent about naming Tereshkova's back-up (now known to have been Irina Soloveva) since she may have originally been assigned to the mission in only a support role.

Once Vostok 6 had been launched with Tereshkova, it is further rumoured that she became ill in orbit. Against this one can weigh another rumour that Tereshkova had originally been slated for a 24 hour mission, but because everything was going so well, the mission was extended to three days. What is clear is that Vostok 6 was launched into an orbit that closely matched that of Vostok 5: it has been claimed by the Soviets that on the first orbit, the minimum distance between the Vostoks was about 5km.

With the successful recoveries of Vostok

5 and Vostok 6 on 19 June, the public record showed that the Vostok programme ended. However, 1963-1964 was a period of revising plans and changing goals in the Soviet manned space programme, and details of these changes have only recently been hinted at in the Soviet literature.

Plans for a Vostok 7

It is believed that during 1963 two new trainees joined the cosmonaut group, separately from the military pilot and engineer selection of January 1963. Possibly after the Vostok 5/Vostok 6 missions had been completed, B.B. Yegorov and V.G. Lazarev began training for a Vostok 7 mission. Both men had extensive backgrounds in medicine and Vostok 7 was due to be a one person mission lasting for a week. The cosmonauts trained for about six months for Vostok 7, before the mission was cancelled. No reason for the cancellation was made publicly. Vostok 7 was probably scheduled for launch in the summer of 1964. There seem to have been no plans finalised for Vostok missions beyond Vostok 7, although as the next section will describe, a follow-on Vostok-Zh programme was studied by Korolyov.

Post-Vostok Studies

At the end of 1961, Korolyov was already looking beyond the simple orbital missions of the Vostok spacecraft to plan greater things. During the period 1962-1964 the Soviet space administrators considered the missions that might follow Vostok: two programmes were studied (Vostok-Zh and the Soyuz complex), but then dropped in their original concepts. However, the studies which were undertaken were important for the future of the manned space programme, and these are the programmes to be discussed here. Further details are to be found in the book edited by M.V. Keldysh which catalogues the design studies undertaken by Korolyov (some of which were not launched), and which is listed in the Further Reading notes.

Vostok-Zh Studies

Although there were dual flights of Vostok in 1962 (Vostok 3/Vostok 4) and 1963 (Vostok 5/Vostok 6), none of the craft was able either to sustain long-duration rendezvous and station-keeping, or attempt an actual docking. In 1962 Korolyov undertook a preliminary study of an improved

MISSIONS WITHIN THE VOSTOK PROGRAMME								
LAUNCH DATE & TIME	SPACECRAFT	COSMONAUT	BACK-UP	MASS (kg)	INCLINATION (deg)	PERIOD (min)	ALTITUDE (km)	LIFETIME (d.hh.mm.ss)
1961 12 Apr 06.07	Vostok 1	Gagarin	Titov	4,725	65.07	89.1	175-302	0.01.48.—
6 Aug 06.00	Vostok 2	Titov	Nikolayev	4,731	64.93	88.6	178-257	1.01.11.— (1.01.18.—)
1962 11 Aug 08.30	Vostok 3	Nikolayev	Bykovsky	4,722	65	88.5	183-251	3.22.09.59 (3.22.22.—)
12 Aug 08.02	Vostok 4	Popovich	Komarov/Volynov	4,728	65	88.5	180-254	2.22.44.— (2.22.57.—)
1963 14 Jun 12.00	Vostol 5	Bykovsky	Volynov	4,720	65 64.97	88.4 88.27	181-235 175-222	4.22.56.41 (4.23.06.—)
16 Jun 09.30	Vostok 6	Tereshkova	Soloveva	4,713	65	88.3	183-233	2.22.40.48 (2.22.50.—)
1964 Summer ?	Vostok 7	Yegorov	Lazarev ?	4,750 ?	Mission cancelled			Scheduled 1 week

Notes: The launch times noted above are GMT. The orbital data is that announced by the Soviet Union: two sets of data are given for Vostok 5 because it seems to have failed to attain the intended (first) orbital altitude. Mission durations are shown as cosmonaut flight time with spacecraft. Mission durations are shown as cosmonaut flight time with spacecraft flight time in parentheses. The Soviets have not publicly given the two differing times for Vostok 1. The Vostok 7 mission was cancelled after the crew had been training for six months.

Vostok, which was designated Vostok-Zh ("Zh" being the eighth letter of the Cyrillic alphabet). The Vostok-Zh mission would require three basic building blocks:

1 Manned Vostok-Zh spacecraft.
2 A number of rocket modules.
3 A payload module.

Korolyov's plan envisaged the launch of the manned Vostok-Zh spacecraft, followed a day later by a rocket module. The manned Vostok would rendezvous and dock with the rocket module: a small manoeuvring engine would then be discarded from the rocket block. Further rocket modules – the number depending upon the mission – would be launched at daily intervals and these would dock with the manned craft. Once the required number of rocket modules had been docked in orbit, the payload module would be launched, and that in turn would dock with the complex: the Vostok-Zh would be at one end of the complex, the payload module at the other end and between them the docked rocket modules.

As the rocket modules were launched, it was planned that they should reach an orbit about 5-10km away from the Vostok-Zh. The initial Vostok-Zh approach would be automatic, but when the rocket module was about 100-200m away the crew (number unspecified) would take over control and dock manually with the rocket module. After docking, electrical and mechanical latches would be secured. The mass of the completed Vostok-Zh assembly would be about 12 tonnes.

Once the payload module was docked, and the final "go-ahead" was given, the Vostok-Zh would undock and return to Earth with its crew. Then, one-by-one, the

Below: *Gagarin in his orange pressure suit saying farewell to Korolyov (right) at the bottom of the Vostok 1 launch vehicle. For Korolyov – with an involvement in rocketry since the 1930s – this would be the peak of his career.*

rocket blocks would ignite to take the payload module to its required orbit. The published mission description unfortunately does not specify the actual payload which would be launched using this method. The mission description suggests that the Vostok-Zh complex would have been assembled from sections launched using the tried-and-tested SL-3 "Vostok" booster. While the rocket blocks and payload could possibly have been launched atop the SL-3, the launch of a manoeuvrable Vostok derivative – possibly with a multi-man crew – would probably have required the use of the more powerful SL-4 "Soyuz" booster.

Only the design study of Vostok-Zh is discussed in Soviet sources, and it is uncertain whether this was ever a programme backed by actual funding and hardware development. One may speculate that some of the changes planned to turn a Vostok into a Vostok-Zh may have been incorporated in the Voskhod programme when that began in 1964.

Early Apollo Studies

In the early part of 1962, Yuri Gagarin and other cosmonauts attended a lecture dealing with a new manned spacecraft: the Soyuz. Rather than being a rival to the Vostok-Zh design studies, the Soyuz seems to have been regarded even then as the main Post-Vostok spacecraft. The collection of Korolyov's design studies includes a description dated 1964 of a three spacecraft Soyuz complex. Presumably this was the version which was shown to this cosmonaut team two years earlier.

Before looking at the Soyuz complex as envisaged in 1964, it is worth examining an early concept for an Apollo craft that was generated by the General Electric Company in the USA. According to some reports, Korolyov and the other space designers had full access to the Western press which detailed the spaceflight planning, and so it is possible that contemporary US

design studies would have some influence over Soviet thinking.

In 1960, NASA called for design submissions for a spacecraft destined to become Apollo, which would be used for a manned lunar mission – either lunar orbit or lunar landing. One of the designs was submitted by General Electric, and this envisaged a three part spacecraft:

1 Propulsion module: length 4.6m, diameter 3.0m, flaring to 5.4m at the interstage which would mate the craft with a Saturn booster.
2 Descent module: length about 2.5m, diameter 2.8m.
3 Mission module: pear-shaped, length 3.1m, maximum diameter 1.7m.

The total mass of the GE Apollo was to be 7,473kg, with the descent module being 2,184kg. It is uncertain whether these figures refer to a vehicle capable of simply flying around the Moon or of going into lunar orbit; if the latter were to be the case, the dry masses of the mission and propulsion modules total only about 1,250kg which is very low. It would seem more likely that the figures refer to a circumlunar, rather than a lunar orbit, variant.

For re-entry, it was intended that the GE Apollo would split into its three component modules. The descent module would make a parachute landing on water: just before touch-down the heat shield would be separated (like Soyuz), and then a Mercury-type floatation bag to cushion the landing would be deployed (unlike Soyuz). It is interesting to see how this design study parallels Korolyov's plans for the development of a Soyuz complex.

The Soyuz Complex

The reason for briefly considering the GE Apollo concept here is that when details of the Soyuz craft were released, they showed more than a passing resemblance to the US scheme. Certainly, the three module design is common, as is the beehive shape of the descent module. During 1962-1964 Korolyov was investigating a complex which could be assembled in orbit, using post-Vostok technology; this became Soyuz. As originally conceived, the Soyuz complex would have used three spacecraft:

1 Soyuz-A, the manned craft, which had three modules (orbital, descent and instrument).
2 Soyuz-B, an unmanned rocket block which would be launched "dry" but with a small rendezvous module attached.
3 Soyuz-V, an unmanned tanker to carry fuel and oxidiser to the Soyuz-B rocket block.

The three spacecraft will first be described, and then the probable use of the complex will be examined.

Soyuz-A exhibits many of the features of the Soyuz craft which actually came into use during 1966-1967. Scaling the Soviet pictures, the overall length of the craft can be estimated as approximately 7.7m, slightly longer than the Soyuz which flew. The individual dimensions for the three spacecraft modules can further be scaled as follows: orbital module: length, 3.1m; diameter, 2.3m; descent module: length,

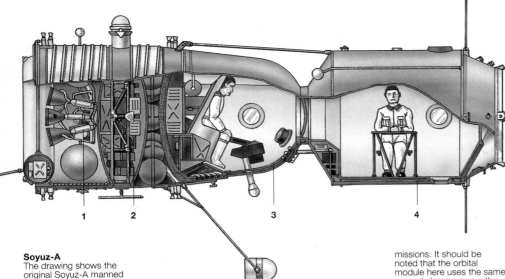

Soyuz-A

The drawing shows the original Soyuz-A manned spacecraft, the design of which was shown to cosmonauts in early 1962. This was planned to carry men into Earth orbit and then on a mission to circle the Moon.
1 The Earth orbital manoeuvring system, similar to that which actually flew on manned missions.

2 The Soyuz instrument module, containing the propellant tanks and electrical systems; on this version apparently no solar panels were planned.
3 The Soyuz descent module, capable of carrying either two or three men.

This module's design was apparently little changed for the actual manned version which flew in 1967.
4 The orbital module for Soyuz-A was distinctly cylindrical, compared with the later ellipsoid shape which was used on actual

missions. It should be noted that the orbital module here uses the same up-and-down axes as the descent module, but the later version had the "down" axis pointing towards the descent module.

It is interesting to note that the design of the docking system for the spacecraft is not shown, and was possibly not finalised.

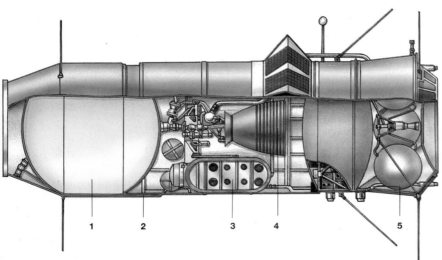

Soyuz-B Rocket Block (above)
1 The Soyuz-B oxygen tank.
2 The outer structure of the rocket block stage.
3 The torus tank which would have carried kerosene fuel.

4 Main propulsion system for the Soyuz-B rocket stage.
5 The in-orbit manoeuvring system retained for the rendezvous manoeuvres with tanker modules. Similar to the Soyuz-A propulsion system, this

would have been separated in Earth orbit, prior to the launch towards the Moon.

Of the three spacecraft which formed the Soyuz complex, the Soyuz-B would have been the first to be launched.

Soyuz-V Tanker Module (below)
1 The in-orbit propulsion system, which once more would apparently have been based upon the Soyuz-A propulsion unit.
2 The propellant tank inside the tanker, which would have carried either kerosene or liquid oxygen into orbit for transfer to a Soyuz-B rocket block.

It seems probable that up to four Soyuz-V tankers were required to carry all of the Soyuz-B propellant into orbit.

2.0m (the curving heat shield adds a further 0.4m to the length), diameter, 2.3m; instrument module: length, 2.6m; diameter, 2.5m. The Soviet sketches of the vehicle do not indicate any docking system attached, so it may be fair to assume that the final designs were still being studied in 1963-1964 when the drawings were published.

The mass of the Soyuz-A with three cosmonauts (their mass totalling 225kg) is given as 5,800kg in one part of Korolyov's description, but elsewhere Korolyov quotes 6,450kg: The latter figure is more in line with the Soyuz which actually flew in 1966-1967. A total of 830kg of propellant was scheduled to be carried in the Soyuz-A, contained in both a toroidal tank at the rear of the instrument module and spherical tanks inside the instrument module. There is no indication of solar panels being intended for use on Soyuz-A. A single large search antenna was to be carried (hinged at the interface of the descent and instrument modules), rather than one antenna at each end of the Soyuz, as actually flew. The original orbital module looks distinctly cylindrical, while the orbital module which actually flew was almost spherical.

Soyuz-B was the rocket block module, with which the Soyuz-A would rendezvous and dock. It would have been launched with its main propellant tanks empty, but with a small module attached which would carry the rendezvous and docking propulsion system and propellant. The overall length of the Soyuz-B was to be about 7.8m, including the rendezvous block (length, about 3.1m). The diameter of the Soyuz-B was to be about 2.5m.

The total mass of the Soyuz-B at orbital injection was estimated to be 5,700kg: this included the rendezvous block with a mass of 2,400kg, containing 1,490kg of propellant. Therefore, the dry rocket block itself had a mass of about 1,800kg. The mass of the main Soyuz-B rocket engine was 140kg.

The Soyuz-V tanker module was a cylinder 4.2m long and 2.5m in diameter. At orbital injection the tanker had a mass of 6,100kg, 4,155kg of this consisting of propellant for the Soyuz-B. The propellant carried to permit rendezvous manoeuvres with the Soyuz-B rocket block was 490kg, which means that the dry mass of the Soyuz-V would have been 1,455kg.

The masses of the three Soyuz craft range from 5,700kg to 6,100kg (possibly up to 6,450kg), and this indicates that all three modules would have had to be launched using the SL-4 booster, which the Soviets now call the "Soyuz" because it was used from 1967 onwards to launch the manned Soyuz missions (it had previously launched the manned Voskhods in 1964-1965).

Deducing the Mission

Although the Soviets have not revealed the actual target for the Soyuz complex, the mission has been described in general terms. It was to begin with the launch of a Soyuz-B rocket block into a 65 degree, circular 226km orbit. At daily intervals, up to four Soyuz-V tankers would be launched to carry propellant to the Soyuz-B. As each tanker was drained of its cargo, it would have been discarded, and presumably de-orbited. Once the Soyuz-B was fully fuelled, the manned Soyuz-A was to be launched and it would dock with the Soyuz-B. The rendezvous and in-orbit manoeuvring units would then be dis-carded, and the Soyuz-B rocket block would be ignited to lift the complex out of low orbit. This is as far as the Soviet description goes, but knowledge of the propellant combination and the various masses allows us to make some informed deductions and identify the ultimate target as the Moon.

If we assume that there were three Soyuz-V missions to a Soyuz-B rocket block, then the fuelled mass of the Soyuz-B becomes 1,490kg (dry mass) + 12,465kg (propellant mass delivered in three missions), resulting in a total mass of 13,955kg. Looking at the cutaway of the Soyuz-B rocket block suggests that the propellants were kerosene and liquid oxygen, and in turn this would allow the specific impulse of the rocket block engine to be about 3,140m/sec (equivalent to about 320 seconds). Adding on the Soyuz-A space-craft mass (initially 5,800kg), which will be reduced by about 155kg in consumed ren-dezvous propellant, gives an initial mass for the Soyuz-A/Soyuz-B of 19,600kg: this implies a velocity change of 3,170m/sec. This makes an interesting comparison with the velocity change required to place Apollo 11 on its trans-lunar trajectory: 3,039m/sec. It would therefore seem that the Soyuz complex was designed to allow a Soviet flight around the Moon.

It must be stressed at this point that apart

Below: *The ultimate goal of the "space race" during the 1960s, a race which the Americans would win with the Apollo 11 flight. The Soviet lunar programme was dogged by development problems, particularly with the giant SL-15 booster which never made a successful flight and was scrapped after its third failure.*

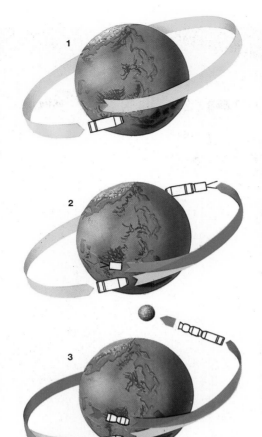

Soyuz Complex Mission Profile
1 Launch of a Soyuz-B into low Earth orbit.
2 Launches of up to four Soyuz-V tankers (at daily intervals) which dock with the Soyuz-B in turn, transfer their propellant and are de-orbited.
3 The Soyuz-B block fully fuelled, the manned Soyuz-A is launched and docks with the Soyuz-B. According to the Soviet account, the Earth orbit would be 65°, 225km and it is stated simply that the Soyuz-B discards its in-orbit manoeuvring system and carries the Soyuz-A out of its low orbit. Calculations show that the propellant load on the Soyuz-B would be sufficient to carry the manned complex around the Moon.

Admittedly ambitious, this Soyuz mission was cancelled as being too complex. The programme was re-structured in about 1964, delaying the first flight for about two years. Although Soyuz was still intended for a lunar flight, it would fail.

from the rocket engine specific impulse (3,140m/sec) which has been estimated by the author, all the numerical data given here are available from Soviet sources: even the specific impulse is derived from known Soviet rocket engines which use the deduced propellant of kerosene and liquid oxygen.

A possible up-rated Soyuz-A circumlu-nar mission could be supported (the manned craft having a mass of 8,040kg) if four flights of the Soyuz-V tanker were used to fuel the rocket block. This number of tanker flights would allow a velocity change of 3,780m/sec to be given to the Soyuz-A variant which the Soviets describe. However, even with this propel-lant load added to the Soyuz-A propellant there would not have been the capability to place the manned spacecraft in lunar orbit and bring it back to Earth. Should such a mission be planned, a larger version of the Soyuz-B would be required, because the lunar orbit injection and trans-Earth injec-tion manoeuvres totalled 1,945m/sec.

Therefore, the evidence seems to indi-cate that the SoyuzA/Soyuz-B/Soyuz-V complex was designed to allow a trans-lunar mission to be flown using the orbital assembly and refuelling of payloads which could be launched by the SL-4 booster.

Soviet Goals Re-Directed

History shows that neither the Vos-tok-Zh nor the Soyuz complex spacecraft were launched on missions. So, what hap-pened? As already suggested, Vostok-Zh does not seem to have been more than a paper study, but the details published for the Soyuz complex have more substance.

With the benefit of hindsight, we can see that technical aspects of the Soyuz pro-gramme seem to have been re-thought, with the Soyuz-B and Soyuz-V craft being dropped, thus delaying the introduction of the Soyuz spacecraft. A Soyuz in a stripped-down version could be flown around the Moon, not using the orbital assembly technique just detailed but by using the far more powerful Proton booster which was being developed by the Chelo-mei design bureau. The outcome of these deliberations and plans for a Soviet manned lunar programme will be dis-cussed in as much detail as Soviet data and western analyses allow in Chapter 4.

Because Soyuz was being re-thought, after the Vostok flights, the manned pro-gramme seemed to be in for a long delay unless some interim missions could be prepared. Soyuz was originally scheduled to begin manned flights in 1964-1965; its delay meant that they could not now be ex-pected until late 1966. Out of this setback, the interim Voskhod programme was born.

FURTHER READING

Golovanov, Y., *Our Gagarin*, Progress Publishers, Moscow, 1978.
Hall, R., "The Soviet Cosmonaut Team (1960-1971)" in *JBIS*, vol 36 (1983), pp468-473.
Hall, R., "The Soviet Cosmonaut Team (1956-1967)" in *JBIS*, vol 41 (1988), pp107-110.
Hooper, G.R., *The Soviet Cosmonaut Team*, GRH Publications, Woodbridge, Suffolk, 1986.
Keldysh, M.V. (editor), *The Creative Legacy of Academician Sergei Pavlovich Korolyov* (in Russian), Nauka Publish-ers, Moscow, 1980.
Shayler, D.J. and Hall, R., *The Soviet Cos-monaut Detachment: 1960-1985*, Astro Info Service, Halesowen, 1985.
Tsygankov, V., *Yuri Gagarin*, Nauka Publishers, Moscow, 1986.
Tsymbal, N. (Editor), *First Man In Space – The Life and Achievement of Yuri Gaga-rin*, Progress Publishers, Moscow, 1984.
All of the new details of the unflown cos-monaut trainees from the 1960 selection were first given in a series of articles in *Izvestia*, 2-6 April 1986, with a further article appearing on 20 May 1986.
The Vostok-Zh and Soyuz data in the Kel-dysh book are summarised in:
Wachtel, C, "Design Studies of the Vos-tok-J And Soyuz Spacecraft" in *JBIS*, vol 35 (1982), pp92-94.

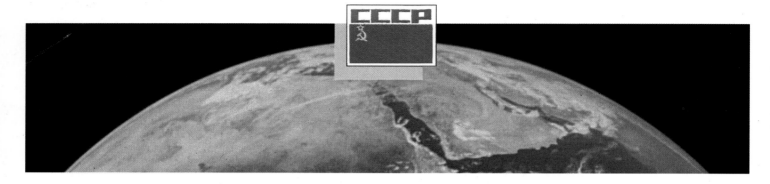

Chapter 3: **The Voskhod Missions**

Because of changing plans and priorities and the slower-than-expected development of Soyuz, the Soviet Union required an interim programme to plug the gap between the final Vostok missions in the summer of 1963 and the first manned missions in the Soyuz programme which were probably re-scheduled for late 1966. It was in this climate that the Voskhod programme was born. Little has been revealed about Voskhod as far as official Soviet data are concerned. Although the second and last manned Voskhod was launched in 1965, it was fifteen years before a clear photograph of the Voskhod in assembly was released. This confirmed what Western analysts had long suspected: it was simply an up-rated Vostok. Despite its "stop-gap" nature the two flights of Voskhod did give the Soviets two significant space "firsts": the first multi-manned spacecraft and the first "walk" in space.

The Birth of Voskhod

With the Soyuz programme delayed, it seems likely that political pressure from Khrushchev resulted in Korolyov's team dusting off their Vostok-Zh studies and wondering if and how those plans could be adapted for a few missions in the period 1964-1966. It was expected that the first American Gemini launch would take place in late 1964 (actually it was delayed until March 1965), and the Soviets wanted a craft which could at least beat Gemini to some of its goals. When it flew, Gemini would be capable of:

1 Two-man space missions.
2 Manoeuvring in orbit.
3 Rendezvous and docking.
4 Extra-vehicular activity ("space-walks").
5 Extended duration missions.

Using Vostok/Vostok-Zh technology, which of these capabilities could the Soviets hope to attain? With modifications, more than one man could be squeezed into a Vostok capsule, and a long duration mission could be attempted. Also, with the addition of an airlock, a man could possibly emerge into space. However, at the time it was presumably difficult to modify the Vostok to allow in-orbit manoeuvres, and therefore in-flight manoeuvring was not planned for Voskhod.

The end of 1963 and the beginning of 1964 saw major changes in the Soviet

Above: *A rare photograph of a Voskhod spacecraft in assembly. The instrument module is surrounded by the adapter at the base; atop the modified Vostok sphere is the back-up retro-rocket.*

space programme: the manned Vostok 7 was cancelled. The Soyuz complex was cancelled, and the Soyuz programme redirected. Unmanned Soyuz craft could be launched on trans-lunar trajectories using the new Proton launch vehicle, while Earth orbital versions could be launched on the SL-4 booster. The Voskhod programme was initiated to maintain a Soviet manned presence in space while the Soyuz was being readied to fly.

Since it was an interim programme, Korolyov apparently planned only three missions:

1 Late 1964 – Voskhod 1. Three men in orbit for at least a day.
2 Early 1965 – Voskhod 2. Two men in orbit for a day, with one completing a "spacewalk".
3 Late 1965 – Voskhod 3. Two men in orbit for two weeks.

Two training groups for Voskhod missions were put together in 1964; the first, formed in March, consisted of eight men to train for Voskhod 1, and the second formed in April comprised four men to train for Voskhod 2. The Voskhod 3 training group was not formed until 1965.

After the success of Voskhod 1, and possibly after the Voskhod 2 flight, Korolyov seems to have considered some further Voskhod missions, bringing the total number of manned Voskhods to six; these are considered in connection with Voskhod 3, since the missions were not actually flown.

The Voskhod Spacecraft

The Voskhod spacecraft was based upon the Vostok, which was described in Chapter 2. There were two variants of Voskhod launched on manned missions, and outwardly they differed from a Vostok as follows: Voskhod 1 carried a back-up retro-rocket; in addition Voskhod 2 carried

MASS BREAKDOWN OF VOSKHOD AND VOSTOK			
	VOSKHOD 1	**VOSKHOD 2**	**VOSTOK**
Back-up retro-rocket (tonnes)	0.145*	0.145*	–
Airlock (tonnes)	–	0.250*	–
Descent module (tonnes)	2.9*	3.1*	2.46*
Instrument module (tonnes)	2.28	2.19	2.27*
Total mass (tonnes)	5.325	5.685	4.73
Announced mass (kilogrammes)	5,320	5,682	4,730

Notes: The figures marked with an asterisk are from Soviet sources, with the Voskhod instrument module masses obtained using repeated subtraction from the announced masses; the figures are given as accurately as the original Soviet data. For comparison, the actual Voskhod masses announced (to the nearest kilogramme) are also listed, together with comparison data for Vostok (mean values).

Left: *Voskhod 2 on the launch pad, atop its SL-4 booster. The "blister" on the side of the payload shroud houses the retracted airlock which would allow Leonov to perform the first spacewalk.*

Below: *The launch of Voskhod 1, carrying three men in a modified Vostok spacecraft. This was the first manned flight of the Soyuz SL-4 booster.*

back-up if the prime TDU-1 retro-pack failed to ignite.

The mass of the solid propellant rocket system which formed the back-up retro-rocket was 143kg, including 87kg of (an unspecified) propellant. The total impulse (burn time multiplied by thrust) of the engine was 19.6 tonnes. Since the thrust of the engine was about 12 tonnes, this allows the burn time to be calculated as 1.55-1.70 seconds. In turn, the specific impulse of the engine can be calculated (using standard rocket-engine equations) as 2,197m/sec (or 224 seconds).

The inclusion of extra crew members inside the Vostok sphere meant that the ejector seats could not be used on Voskhod, and therefore couches were to be used for the crew. To squeeze the two or three man crew in Voskhod, the seats had to be re-aligned through 90° compared with the Vostok orientation. The elimination of the ejector seats posed another problem for Korolyov, because the landing deceleration would be too great for a crew to survive without injury. However, with Voskhod it was possible to add a retro-rocket system to the parachute pack because the use of the SL-4 launch vehicle did not impose the payload mass restrictions that Vostok atop its SL-3 had suffered. The soft-landing retro-rocket was located underneath the main parachute and when deployed it was situated above the descent module; this would fire as the descent module approached touchdown, reducing the terminal velocity from 8-10m/sec to 0-2m/sec.

Considering that the Soviet Union has always stressed that safety has always been a major consideration in its manned space

an extendable airlock to allow a spacewalk to be completed. If it had been launched, Voskhod 3 would probably have looked like Voskhod 1. The total length of Voskhod was 5m and the maximum diameter was 2.43m. The mass breakdown of the Voskhod craft is given in the accompanying table, together with Vostok data to allow easy comparison.

The masses of the Voskhods were up to a tonne greater than that of Vostok, and therefore the SL-4 booster, intended for Soyuz, was slated to launch them. However, the Voskhod masses were lower than

the payload masses eventually intended to be launched on the SL-4, and the only way to prevent excessive accelerations because of the lower fuel requirement to a Vostok orbit, was to launch Voskhods into orbits with apogees 200-300km higher than the Vostoks had used. However, this posed Korolyov a problem: if such a high orbit were to be used, the Voskhod would not naturally decay out of orbit within ten days (as would a Vostok) if the retro-rocket failed to ignite. Therefore, Korolyov added a solid propellant retro-rocket to the top of the Voskhod; this would be used as a

Voskhod Landing Sequence
1 After retro-fire, re-entry begins.
2 Hatch of the parachute container separates.
3 Drogue parachute deploys.
4 Descent under drogue parachute.
5 Drogue parachute separates.
6 Voskhod descends under its two main parachutes. The parachute system includes a small solid

propellant rocket which fires just before landing to cushion the ground impact.

Voskhod was the first Soviet spacecraft inside which the cosmonauts were recovered. Without the rockets to cushion the final landing the crew would have had a very rough landing and could have been badly injured. Voskhod 2 landed far off course and had to wait for a day before rescue came.

Above: *The three Voskhod 1 cosmonauts, squeezed inside what had been designed as the single-man Vostok capsule. The crew did not wear pressure suits because with them they would not fit inside!*

Left: *The Voskhod 1 cosmonauts walking towards their launch vehicle. Left to right, they are Feoktistov, Komarov and Yegorov. They reportedly had to diet so they could squeeze into the capsule.*

programme, it is surprising to learn that there was no emergency escape system for Voskhod. Vostok had carried the ejector seat which was actually used as part of the standard cosmonaut recovery procedure. On Soyuz which followed, the rocket system atop the payload shroud doubled as an escape system, but on Voskhod there was none. When asked about Voskhod escape systems, the Soviets (when they have responded) have claimed that the launch vehicle was so reliable that an escape system was not required. However, in tacit refutation of this argument, Soyuz used the same booster, nearly three years later, and an escape system was deemed necessary then.

Voskhod 2 was about 360kg heavier than Voskhod 1, and most of this extra weight was contributed by the addition of a 250kg airlock which was to permit a crew member to exit the spacecraft and perform a spacewalk. At launch, this was mounted almost flush with the Voskhod sphere, and

in this form its depth was 0.7m. Designed like a concertina, the airlock extended to a height of 2.5m. Its inner diameter was 1m and the outer diameter was 1.2m. The hatch which opened to allow the final exit into space was 0.7m in diameter. After the spacewalk was over, the airlock was designed to be discarded.

The projected Voskhod 3 craft for a long duration mission would probably have appeared more like Voskhod 1; but its mass might well have been closer to that of Voskhod 2 because of the extra supplies required for a two week mission.

Three Men in Orbit

Of the Voskhod missions planned, Voskhod 1 was the simplest: the launch and recovery of a multi-manned spacecraft. To supplement the pilots and engineers already training at Zvezdny Gorodok (where the cosmonaut training centre is located, near Moscow), five or six new men were recruited specifically for Voskhod 1.

Three were physicians while either two or three (Soviet reports differ on this) were "science workers" (the Soviet phrase). The three doctors were B. B. Yegorov, V. G. Lazarev and A. Sorokin, and the two science workers, or "engineers", were K. P. Feoktistov and G. Katys.

One can speculate that a third engineer might have been selected for Voskhod 1 training, but for some reason he left the programme so early that his rôle has never been acknowledged; (in the West, O. G. Makarov, who later became an engineer cosmonaut, and the spacecraft designer B. V. Rausenbach have been suggested as cosmonaut candidates for this mission, but without Soviet confirmation).

The training group for Voskhod 1 was put together in March 1964, probably very soon after the Voskhod programme had been defined. The new men selected for the Voskhod 1 mission trained with two 1960 cosmonauts, and although a three-manned mission was planned the men trained in the following two groups:

V.M. Komarov	B.V. Volynov
K.P. Feoktistov	G. Katys
V.G. Lazarev	B.B. Yegorov
	A. Sorokin

If there were only two engineers chosen for Voskhod 1 training, it is uncertain with whom Katys trained; logically, he would have been part of the Volynov team, although this has not been confirmed by Soviet sources. One curiosity is that Yegorov was training with men who would

Above: *The two cosmonauts who flew on Voskhod 2 during their training. Both wearing EVA (extra-vehicular activity) suits, Belyayev is on the left and Leonov, who completed the first spacewalk, is on the right.*

eventually form the back-up crew, while he flew the Voskhod 1 mission with Komarov and Feoktistov.

Because Voskhod 1 was derived from the proven Vostok craft, only a single precursor mission was required: Cosmos 47 was launched on 6 October 1964 and remained in orbit for a day. The Soviets did not announce the recovery, and in fact they simply treated the mission as a standard Cosmos flight.

Voskhod 1 itself was launched on 12 October and remained in orbit for just over a day (see table of mission data). The cosmonauts flew for the first time without spacesuits: officially this was because the Soviets were confident about Voskhod, but it must have happened because there simply was not enough room for three cosmonauts in spacesuits inside the Voskhod sphere. Because there were three men in what was intended to be a one-manned spacecraft, the in-flight experiments were reduced to a basic minimum. Yegorov is

thought to have performed some medical experiments, but they have never been detailed. Presumably some Earth photography was undertaken. There have been persistent reports that a longer mission was planned (despite the duration of Cosmos 47). One report states that when Komarov was instructed to end the Voskhod mission after a day in space and he protested, Korolyov quoting *Hamlet*, told him that "there are more things in Heaven and Earth, Horatio, than are dreamt of in your philosophy", an allusion to the change in leadership at the Kremlin. When Voskhod 1 was launched, Krushchev ruled the Soviet Union, but when it returned Brezhnev and Kosygin had taken control. The first major public appearance of the new leadership was at the ceremony to greet the Voskhod 1 crew on their return to Moscow.

A Walk in Space

The training group for the second manned Voskhod mission was formed in April 1964, and consisted of four men, all of whom were from the original 1960 selection: P.I. Belyayev, V.V. Gorbatko, Y.V. Khrunov and A.A. Leonov. Gorbatko stood down due to ill health, and was replaced by D.A. Zaikin.

As with Voskhod 1, the Voskhod 2 mission had a single precursor test mission, Cosmos 57. This was launched on 22 February 1965, but the mission was less than a total success. Presumably the Cosmos was intended to test the extension, pressurisation, de-pressurisation and separation of the airlock, with the descent craft returning after 24 hours. However, within hours of launch, a cloud of debris was being tracked in orbit: 142 additional objects were catalogued from the mission. Probably, there was a problem with the de-pressurisation/re-pressurisation system, or maybe the airlock module simply disintegrated at separation? Catalogue listings show that the main object from the mission (1965-012A - presumably the Voskhod spacecraft itself) re-entered the atmosphere on the day of launch. If it was still intact, it is unknown whether this was due to simple orbital decay (unlikely, because of the high orbit) or an emergency return and recovery after the unexpected appearance of the cloud of debris.

Whatever the problems were on the Cosmos 57 test flight, the manned Voskhod followed within a month (the interval between Cosmos 47 and Voskhod 1 was a week). Launched on 18 March, the craft carried two men: Pavel Belyayev and Alexei Leonov. Leonov was destined to become the first man to "walk" in space. Launch was at 07.00 GMT and immediately after attaining orbit the cosmonauts prepared for Leonov's spacewalk. Both were dressed in spacesuits at launch (unlike the Voskhod 1 crewmen), but Leonov had to be helped to put on his life support system for use outside, as well as to connect the umbilical cord which would attach him to Voskhod. The sequence of pressurising and de-pressurising the airlock which allowed Leonov access to space is shown in the accompanying diagram overleaf.

As the first orbit was being completed, at 08.30 the outer hatch of the airlock was opened and Leonov climbed into open space. Attached to the 5m long umbilical which supplied him with air (an emergency supply was in a back-pack) and a radio-telephone link, Leonov floated free of the Voskhod for over ten minutes. As the umbilical twisted, Leonov also gyrated because he had no control system to counter the torque of the umbilical. According to Leonov, if anything had gone wrong at this stage, it would have been possible for Belyayev to leave Voskhod

MISSIONS WITHIN THE VOSKHOD PROGRAMME								
LAUNCH DATE & TIME	**SPACECRAFT**	**PRIME CREW**	**BACK-UP CREW**	**MASS** (kg)	**INCLINATION** (deg)	**PERIOD** (min)	**ALTITUDE** (km)	**LIFETIME** (d.hh.mm.ss)
1964 6 Oct 07.15 ?	Cosmos 47	Unmanned	Unmanned	5,200 ?	64.77	90	177-413	1.—.—.—
12 Oct 07.30	Voskhod 1	Komarov Feoktistov Yegorov	Volynov Katys Lazarev	5,320	65	90.1	178-409	1.00.17.03
1965 22 Feb 07.40 ?	Cosmos 57	Unmanned	Unmanned	5,500 ?	64.77	91.1	175-512	<1.—.—.—
18 Mar 07.00	Voskhod 2	Belyayev Leonov	Zaikin/Gorbatko Khrunov	5,682	65	90.9	173-495	1.02.02.17
2 Jul	Cosmos	2 Dogs ?	Unmanned	5,700 ?	Failed to reach orbit			
1965-1966	Voskhod 3	Volynov Shonin	Shatalov Beregovoi	6,000 ?	Mission cancelled			Two weeks ?
1966 22 Feb 20.10 ?	Cosmos 110	2 Dogs Veterok, Ugolek	Unmanned	5,700 ?	51.90	95.3	187-904	21.18.—.—

Notes: All of the launches within the Voskhod programme are listed here. The launch failure in July 1965 has been rumoured, and may have carried two dogs on a Voskhod 3 precursor mission. It is uncertain who backed-up who on the projected Voskhod 3 mission: Beregovoi may have been Volynov's back-up and Shatalov Shonin's back-up. The orbital data is that announced by the Soviet Union. Mission durations are as accurate as Soviet and other data allow.

and rescue him, although whether this course of action would have been permitted if the occasion had arisen is uncertain.

A serious problem did arise when Leonov tried to get back into the airlock. During the spacewalk, the lack of external pressure on the spacesuit in the near vacuum of space had caused it to balloon more than anticipated, and Leonov had to struggle for his life before he could get back into the airlock, and repressurise it. His spacewalk had lasted for 12 minutes 9 seconds, although he had been in a vacuum environment for about 20 minutes.

The problems with Voskhod 2 were not over yet. When the time came for re-entry the day after launch, the prime TDU-1 retro-rocket system failed to operate at the first recovery attempt, and Voskhod completed an extra orbit. During this time, the ground controllers decided that they would attempt a return using the back-up solid propellant retro-rocket system, and this was successfully fired at the next landing opportunity. Voskhod 2 was safely returned to Earth, far off course in a snow-covered forest, but with the crew alive and well. It was not until the next day that rescue teams reached the crew and returned them to civilisation proper. Details of this emergency landing were not actually released until March 1966 after the emergency return of the American Gemini 8 mission.

One interesting point is not covered in official reports detailing the use of the back-up retro-rocket system. If the complete Voskhod were to have been de-orbited, then the back-up could only reduce the velocity by about 35-40m/sec – just enough to de-orbit Voskhod, but probably not enough to ensure a controlled re-entry. It is, therefore, possible that before the back-up engine fired, the whole of the Voskhod 2 instrument module was separated, so allowing the back-up system to achieve a velocity reduction of 60-65m/sec – enough for a controlled re-entry.

Voskhod 2 marked the end of the Voskhod programme as far as the public flight record was concerned, although further flights were being studied.

Cancelled Voskhod Missions

According to the autobiography of V.A. Shatalov, he was chosen along with G.T. Beregovoi, G.S. Shonin and B.V. Volynov to train for a Voskhod 3 mission. The training group was probably formed in April 1965 with a view to a flight in late 1965 or early 1966. In view of Volynov's back-up experience, he and Shonin were presumably scheduled to fly the mission, with Shatalov and Beregovoi as back-ups. Photographs are available of these pairings training in Voskhod simulators. Voskhod 3 was planned as a long duration flight, lasting about two weeks.

During 1988 evidence appeared which suggested that further manned Voskhod flights through to Voskhod 6 may have been considered for the 1965-1966 period. Although cosmonauts and cosmonaut-candidates were selected, actual training groups may not have been formed. One rumoured mission was a repeat of Voskhod 2, but with a woman performing a spacewalk. Some evidence, apparently confirming that this mission was considered, surfaced when a photograph was published in 1987 of Tereskhova training in a Voskhod 2 EVA suit.

Korolyov also seems to have decided that the pilot cosmonauts were not giving a picture of space missions which actually reflected the flight experiences, and so in July 1965 two journalists, Y. Golovankov and Y. Letunov joined the training programme. Korolyov hoped that their ability to communicate a story would allow an accurate description of the flight to be written in terms which ordinary men could appreciate.

A further use of Voskhod as a manned biological research satellite is a possibility. It has been rumoured that Professor Y.A. Illyin was selected as a cosmonaut candidate in the group of Voskhod 1 physicians,

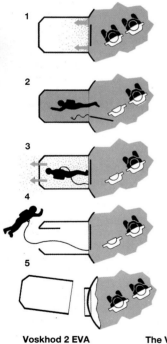

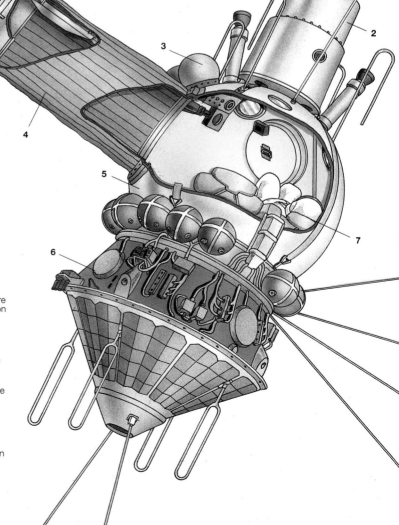

Voskhod 2 EVA Sequence
1 Leonov (left) and Belyayev inside Voskhod with the airlock expanded.
2 Leonov opens the inner airlock hatch and floats inside.
3 The inner airlock hatch is shut (keeping the Voskhod sphere pressurised) and the atmosphere is vented out of the airlock. For safety, the airlock control panel in the cabin is duplicated inside the airlock itself.
4 Leonov opens the outer airlock hatch and floats outside to perform the EVA.
5 The above procedure is reversed after the EVA until Leonov is back inside Voskhod with Belyayev; the airlock is then discarded into space.

The Voskhod 2 Spacecraft
1 Television and movie cameras to record Leonov's historic spacewalk.
2 Back-up solid propellant retro-rocket which had to be used on Voskhod 2 for de-orbit when the main retro-rocket failed to operate.
3 Spheres of gas for use with the airlock.
4 The fully extended airlock, with Leonov outside, attached to Voskhod by his umbilical. This was discarded in orbit after the spacewalk.
5 The modified Vostok descent module sphere in which the cosmonauts rode into space and returned to Earth.

6 Modified Vostok instrument module, incorporating a TDU-1 engine. Voskhod 2 represented the only failure of the engine to operate on a manned flight.
7 Belyayev inside the Voskhod sphere, wearing an EVA spacesuit.

Apart from the addition of the back-up retro-rocket and, in the case of Voskhod 2, the airlock, the exterior of the Voskhod craft was little changed from the original Vostok design. The Soviets claim that Belyayev could have gone outside to rescue Leonov if necessary, but in reality such an operation actually taking place is difficult to believe.

but in a 1988 interview with this writer Illyin said that this was incorrect. He had been selected in 1965 – together with two other physicians (Y.A. Senkevich and A.S. Kisilev) – for a Voskhod mission, which would have been a two-manned, two week mission with biological experiments conducted in orbit on rabbits and other animals. Illyin, Senkevich and Kisilev completed the initial selection procedure and undertook a five-day flight simulation in a Voskhod trainer, but the mission was cancelled before the physicians were paired with pilot cosmonauts.

At some time during 1965, Voskhod 3 was cancelled, much to the chagrin of the cosmonaut team. At the end of 1965 Titov said that the Voskhod had completed its manned programme, and therefore the mission which involved a woman performing an EVA, the journalist mission and the bio-Voskhod mission were also cancelled during 1965. Illyin, who in 1988 was in charge of the unmanned biological Cosmos satellite programme, said that the bio-Cosmos missions were authorised in part to replace his cancelled flight and also because the Soyuz missions would have no place for such research.

The cosmonauts selected for Voskhod 3 were transferred to the Soyuz programme, although the first training groups for Soyuz were not to be formed for about another six months. The two journalists continued as cosmonaut candidates until shortly after Korolyov's death (suggesting that Korolyov may still have considered their flying on a Soyuz mission, but that such plans died with him).

Cosmos 110

On 22 February 1966 the last flight in the Voskhod programme was launched, but since it was unmanned it was designated as part of the Cosmos programme. Cosmos 110 carried two dogs in orbit, Veterok and Ugolek, and the purpose of the mission was to study the dogs' reaction to a long period of weightlessness. The spacecraft was recovered after about 22 days in orbit, with both dogs apparently having survived the flight in good health.

At the end of 1965 the American Gemini 7 had completed a two-week mission in orbit with two astronauts, and the most that Voskhod 3 could do would be to match this achievement, rather than exceed it. In addition, Gemini 7 had been involved in a rendezvous and station-keeping test with the manned Gemini 6, which was something that Voskhod could not emulate. Therefore, one may wonder whether the three-week mission of Cosmos 110 was a replacement for the cancelled Voskhod 3 mission, which had effectively been upstaged by the American Gemini flights.

Voskhod in Retrospect

As has been stressed, the Voskhod programme came into existence as a stop-gap measure between Vostok and the delayed Soyuz programme. As such, it allowed the Soviets to gain two major space "firsts": the orbiting of the first multi-manned spacecraft and the first walk in space. In this role, it was a success, although the near loss of Voskhod 2 when the main retrorocket failed must have seriously worried the Soviets; possibly that failure was partly responsible for the re-think and cancellation of Voskhod 3 ? However, whatever its limitations, it allowed the Soviets a continued presence in space, although the Voskhod 3 cancellation and delays in Soyuz meant that a gap of more than two years did open up between Voskhod 2 and Soyuz 1, the next manned mission actually to be flown.

Below: *The prime crew training for the proposed Voskhod 3 mission. Volynov (rear) had trained for Voskhod 1 previously, while this was Shonin's first assignment. The planned two-week mission was cancelled in 1965.*

Above: *A television picture of Leonov performing his spacewalk. Good quality pictures of this historic event are not available since Leonov could not retrieve the camera from outside Voskhod. The next Soviet EVA would not be until 1969.*

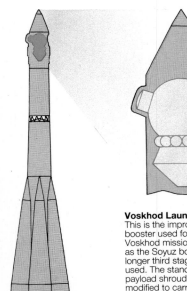

Voskhod Launch Vehicle
This is the improved Vostok booster used for the Voskhod missions. Known as the Soyuz booster, a longer third stage was used. The standard Vostok payload shroud was modified to carry the expandable airlock on the side of Voskhod 2.

FURTHER READING

Leonov, A. A., "My First Walk In Space" in *The Unesco Courier*, June 1965, pp4-11.

The political role of Voskhod, as well as the rest of the Soviet space programme, is reviewed in:
Vladimirov, L, *The Russian Space Bluff*, Tom Stacey Ltd, London, 1971.

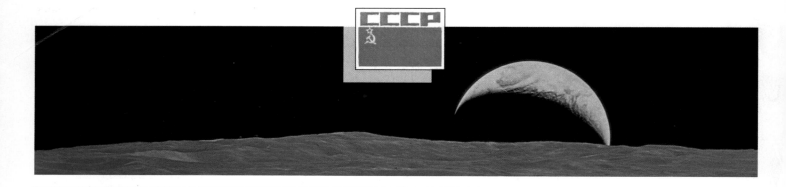

Chapter 4: **Soyuz and the Manned Lunar Programme**

During 1964-1965 Western observers anticipated a long series of manned flights in the Voskhod programme, not realising that it was only a limited capability, interim version of Vostok. In reality, the next major series of manned space missions would be part of the Soyuz programme. Soyuz has evolved through a number of distinct phases, as can be seen from the accompanying historical family tree.

It seems clear in retrospect that Soyuz was originally planned as the craft with which the Soviets would attempt a manned lunar landing programme. Manned engineering flights took place in Earth orbit during 1967-1970, but after Apollo 11's successful landing on the Moon in 1969 Soyuz was adapted as a ferry for space station missions. One ferry type flew in 1971 and a second type was introduced following the Soyuz 11 accident which resulted in the deaths of three cosmonauts. In 1980 a new Soyuz ferry variant, Soyuz-T, began manned operations, lasting until 1986. In that year the latest modification, Soyuz-TM, began to fly.

There were also some "solo" Soyuz missions in the mid-1970s, but these were not within the mainstream Soyuz programme and flew using another Soyuz variant.

The Soviet Union has never discussed the lunar role of Soyuz, although in a two module version it was flown on unmanned circumlunar missions within the Zond programme (1968-1970). These will be examined in detail later in the chapter. Initially, however, we shall look in detail at the original Soyuz variant, as well as the launch and landing techniques. The various Soyuz flights themselves will be discussed throughout the remaining chapters of this volume.

Soyuz Spacecraft Design

The version of Soyuz which began to fly in 1966 was based upon the Soyuz-A design from the Soyuz complex, which was described in Chapter 2. The three module concept was retained, but with some differences. From top to bottom the three Soyuz modules are designated Orbital, Descent and Instrument by the Soviets. However, it is clear that this is not the only configuration which was considered: sketches in a book written by the cosmonaut Konstantin P. Feoktistov show two alternative configurations with the orbital and descent modules interchanged.

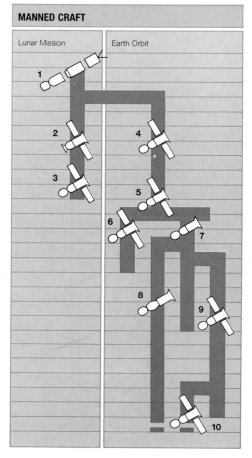

MANNED CRAFT

Lunar Mission | Earth Orbit

"Family Tree" Showing the Major Variants of the Soyuz Spacecraft
1 The original Soyuz complex (1962-1963).
2 The Zond lunar spacecraft (1967-1970).
3 Propulsion tests of the lunar Soyuz (1967-1971).
4 Research vehicle for testing in Earth orbit (1966-1970).
5 Original Soyuz space station ferry (1971).
6 Soyuz variant configured for "solo" missions (1973-1976).
7 Specialised Soyuz space station ferry (1972-1981).
8 Progress unmanned cargo freighter (1978-the present day).
9 Soyuz-T ferry (1979-1986).
10 Soyuz-TM ferry (1986-the present day).

The diagram illustrates how the Soyuz craft have evolved over 25 years.

Soyuz was capable of carrying one, two or three men which suggests that there might have been a trade-off between the crew size and either mission duration or experimental payload carried. The masses of the original Soyuz ranged from 6,450kg (Soyuz 1) to 6,646kg (Soyuz 8). Unlike Soyuz-A, the Soyuz 1-9 series carried a pair of solar cell "wings" which generated electrical power; these had a total surface area of nearly 14m² each being 3.6m by 1.9m.

The dimensions of the three Soyuz modules are listed in the relevant table; the only module to differ significantly from its Soyuz-A counterpart in shape is the orbital module, which was now more spherical, rather than being slightly cylindrical.

Soyuz Propulsion

The Soyuz propulsion system was designated KTDU-35, and it was designed by the Isayev Bureau during 1962-1966. The system consisted of two individual engines operating from the same propellant supply: a prime system and a back-up system. The prime engine had a specific impulse of 2,750m/sec (equivalent to 280.4 seconds) and a thrust of 417kg: the back-up engine had the parameters 2,650m/sec (270.2 seconds) and 411kg respectively. The total burn time of the system was 500 seconds, which implied a propellant load of 755kg. In fact, the Soviets claim a propellant load of 500-900kg for Soyuz (the latter mass suggesting a total burn time of 595 seconds).

Initially, the Soviets gave no details of the Soyuz propulsion system, but as a guide to performance they said that the Soyuz could manoeuvre up to an altitude of 1,300km. This can be interpreted in various ways, but the following explanation seems the most logical. If orbits in the same orbital plane with altitudes 190-230km (initial orbit), 190-1,300km (final orbit) and 50-1,300km (recovery

TYPICAL SOYUZ SPACECRAFT DATA				
	DOCKING MODULE	**ORBITAL MODULE**	**DESCENT MODULE**	**INSTRUMENT MODULE**
Shape	Cylinder	Spheroid	Bee-Hive	Cylinder
Length (m)	1.2	2.65	2.2	2.3
Diameter (m)	1.7	2.25	2.3	2.3
Dry mass (kg)	325	1,000	2,675	2,075
Propellant mass (kg)	–	–	–	500

Notes: This data is taken for a typical Soyuz in the initial series which flew during 1966-1970; individual missions will differ from these figures. For example, Soyuz 6 and Soyuz 9 did not carry docking units and the propellant load could apparently vary from 500kg to 900kg.

1970 lasted for 17.7 days with a crew of two cosmonauts, so that a capability for slightly more than a 35 "man-day" mission was demonstrated.

Launching a Soyuz

The Soyuz spacecraft and its SL-4 launch vehicle components are delivered to the Tyuratam Cosmodrome from their separate assembly shops. It would seem that all booster assembly work is completed with the vehicle horizontal. The four strap-ons are attached to the second stage core booster while the Soyuz is undergoing separate check-out procedure. The Soyuz is fuelled before being mated to the third stage of the booster and then the Soyuz payload shroud is slipped over the spacecraft. The base of the third stage is finally mated to the top of the central core of the lower assembly.

The Soyuz and its booster are then transported to the launch pad horizontally on a railway trailer. On arrival at the launch pad the train complex is backed onto the pad, to allow the base of the booster to be positioned above a flame deflector pit. A cradle on the railway transporter raises the booster to the vertical and four stabilizing launch pad arms are deployed to cradle the vehicle. The train then returns to the assembly shop area.

The four arms of the stabilizer ("tulip" as the Soviets call it) are then joined by two service towers around the booster. The launch of the mission normally comes 1-2 days after the booster has been transported to the launch pad. The propellant for the launch vehicle is delivered to it – again using a railway system – when it is on the pad, only a few hours before the planned launch. The cosmonauts enter the Soyuz by first climbing up into the orbital module and then dropping down into their seats in descent module: clearly, on a multi-manned mission the last cosmonaut to enter must be the one in the centre seat!

Prior to the launch, the two side service structures are retracted, so that the booster is suspended over the flame deflector pit by only the four tulip arms. The five engines of the four strap-on boosters and the central core (which doubles as the second stage) are ignited simultaneously,

---

Below: *The original Soyuz propulsion system (designated KTDU-35), with the torus propellant tank used on the Soyuz 1-9 missions. The two solar cell wings extend from either side of the craft.*

Above: *A night-time launch of the Soyuz spacecraft in 1977, using the three stage SL-4 booster. The launch facilities are built to allow the pad to be re-used within a day of a launch.*

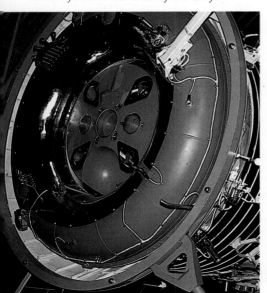

orbit) are assumed, then the propellant requirement is 735-760kg, depending upon the specific impulse of the engine used (prime or back-up ?). This would seem to agree with the previously-calculated propellant load of 755kg.

The propellant of UDMH (fuel) and nitric acid (oxidiser) was carried in four spherical tanks within the instrument module and a toroidal tank mounted externally at the rear of the instrument module. In all probability the full propellant load was not carried on the Soyuz 1-9 series; about 500kg would easily account for all the observed orbital manoeuvres.

The Soviets state that Soyuz was capable of supporting manned missions of up to 30 days duration, although no "solo" Soyuz flew such a mission (later versions accomplished longer manned missions, but on such flights they were docked with orbital stations). The extended Soyuz 9 mission in

Early Soyuz Designs
Below we see early design concepts of the Soyuz spacecraft, showing how the arrangement of the orbital and descent modules changed. On the left is the arrangement which actually flew; in the centre the orbital module is below the descent module with connecting hatches. A slightly different design is shown on the right.

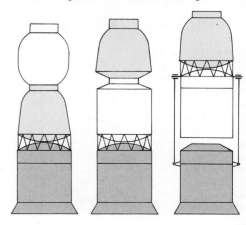

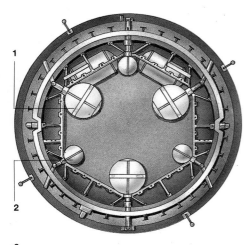

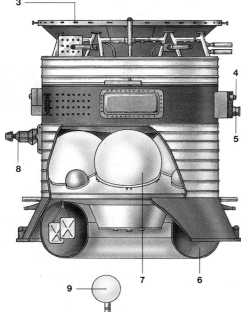

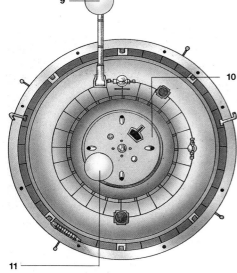

and when their thrust is sufficient to overcome the force of gravity the booster begins to lift-off. This is unlike many Western launch vehicles which are held onto the launch pad until first stage thrust has built up and are then released. As the Soyuz booster begins to ascend the four tulip arms fall back (through the action of counterweights) to their retracted positions.

The four strap-on boosters fire for up to 120 seconds after lift-off, and about 40 seconds later the payload shroud surrounding the Soyuz spacecraft and the tower atop the shroud are separated. The tower fires first and separates, and then the shroud splits into two sections down its side and falls away from the booster. The second stage core continues to fire until 270-280 seconds into the mission and when it separates, the third stage has already begun to fire. The third stage shuts down 530 seconds after launch, by which time the spacecraft is in orbit. (These figures are derived from the Soyuz 19 (ASTP) and Soyuz-T 6 missions, and should be broadly representative of typical Soyuz flights.)

Of course, things can go wrong with a launch, and the Soviets have included launch escape operations into the Soyuz design. The escape system is often misunderstood by Western analysts and is usually considered to be the sole reason for the tower being carried atop the payload shroud. However, if this were to be true, then the tower would not need to be carried on the launches of the unmanned Progress cargo missions to orbital stations. However, the Progress also includes the tower atop its shroud. In fact, the equipment which relates solely to the escape sequence necessary for manned flights is a set of four flaps arranged at 90° intervals around the Soyuz shroud, about a third of the way up; these are not to be found on the unmanned Progress shroud.

If an abort is required before the payload shroud has separated, either on the launch pad or immediately after launch, the shroud splits in two about half way up its

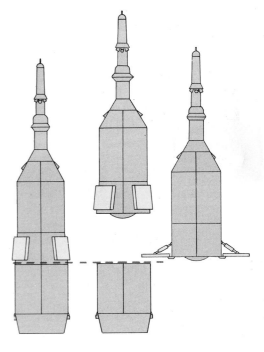

The Soyuz Abort System
Left, the Soyuz inside its shroud sits atop the launch vehicle. In the case of an abort, the rockets on the shroud tower fire as the shroud splits across the plane of the Soyuz instrument module and descent module interface (centre). The manned capsule is carried to altitude and four aerodynamic flaps are lowered to stabilise the craft (right). Subsequently, the Soyuz descent module slips out of the shroud and parachutes back to Earth after reaching a peak altitude of about one kilometre. This system was built in to the Soyuz launch vehicle from the start, but was used operationally for the first time in September 1983 when the Soyuz-T 10 mission aborted.

length, explosive bolts separate the Soyuz descent and instrument modules and the shroud tower's solid propellant rocket motors fire to haul the upper shroud containing the Soyuz descent and orbital modules away from the booster. The four aerodynamic flaps, now at the base of the shorter shroud, are deployed to stabilise the assembly. The descent module separates from the assembly, deploys a parachute, and lands down-range from the launch site.

The abort system has only been required once up to the end of 1987, when the booster for the intended Soyuz-T 10 spacecraft caught fire during the final stages of the countdown in September 1983 (see Chapter 12). During the launch pad abort, the Soyuz-T descent module was lifted to a peak altitude of 950m (3,120ft) and landed 2.5km (1.5 miles) away from the launch pad. An earlier launch failure of a Soyuz (in April 1975, see Chapter 8) occurred later in the flight sequence after the shroud had separated, so that the Soyuz propulsion system was used to take the craft away from its errant booster. Nominally, the Soyuz abort system can carry the descent module up to 1,200m altitude and it can land up to 3km from the launch pad.

Soyuz Recoveries

The recovery procedure adopted for the original Soyuz spacecraft was modified for the later Soyuz-T and Soyuz-TM missions; the original sequence will be described here, while later chapters will deal with the revised procedures. The actual timings can vary depending upon the altitude of the spacecraft in orbit. Here, the Soyuz 19 mission (ASTP) is used as a guide.

Nominal Separation Procedures
Below shows the normal separation of the Soyuz tower and shroud during the flight to orbit. The tower fires about 160 seconds after launch and separates from the vehicle; a split-second later, the shroud splits into two halves along its longitudinal axis, and falls away, exposing the Soyuz spacecraft inside.

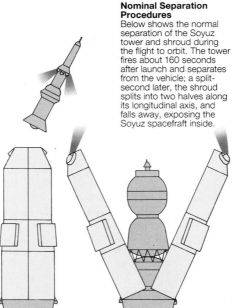

Three View of the Soyuz Instrument Module
1 Hydrogen peroxide tanks.
2 Pressurisation tanks.
3 Ring which supports the descent module.
4 Beacon.
5 Sun sensor.
6 Toroidal propellant tank (only until Soyuz 9).
7 Main engine propellant tanks.
8 Infra-red sensor.
9 Rendezvous radar beacon.
10 KTDU-35 propulsion system.
11 Cut-out for folded beacon, 9.

The general design of the Soyuz instrument module remained almost constant throughout the programme, major modifications only coming with the Soyuz-T and Soyuz-TM variants. Externally the main changes were the deletion of the rear toroidal tank after Soyuz 9, and the deletion of the solar panels for the space station ferry variant following Soyuz 11 (the "solo" Soyuz retained the panels). Soyuz-T carried a new engine system, and, once more, solar panels.

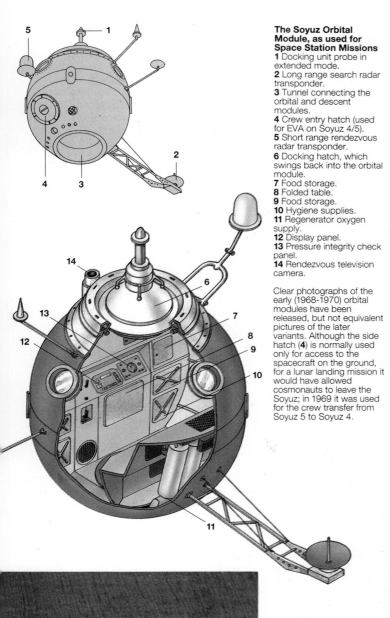

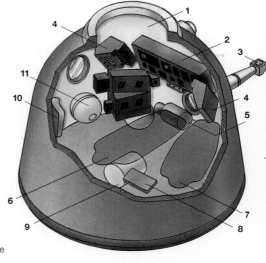

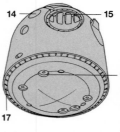

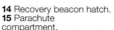

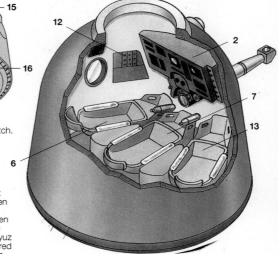

The Soyuz Orbital Module, as used for Space Station Missions

1 Docking unit probe in extended mode.
2 Long range search radar transponder.
3 Tunnel connecting the orbital and descent modules.
4 Crew entry hatch (used for EVA on Soyuz 4/5).
5 Short range rendezvous radar transponder.
6 Docking hatch, which swings back into the orbital module.
7 Food storage.
8 Folded table.
9 Food storage.
10 Hygiene supplies.
11 Regenerator oxygen supply.
12 Display panel.
13 Pressure integrity check panel.
14 Rendezvous television camera.

Clear photographs of the early (1968-1970) orbital modules have been released, but not equivalent pictures of the later variants. Although the side hatch (4) is normally used only for access to the spacecraft on the ground, for a lunar landing mission it would have allowed cosmonauts to leave the Soyuz; in 1969 it was used for the crew transfer from Soyuz 5 to Soyuz 4.

The Soyuz Descent Module (right)

Upper: Soyuz ferry variant
1 Connecting hatch to orbital module.
2 Main instrument panel.
Lower: Soyuz-T variant. Items in common with the Soyuz ferry have the same numbers.
3 15-degree field of view periscope.
4 Command signal device panel.
5 Heat exchanger condenser.
6 Flight engineer's seat.
7 Commander's seat.
8 Gas analyser.
9 Regenerator oxygen.
10 Carbon dioxide absorber.
11 Gas mixture supply system.
12 Control panel.
13 Research engineer's seat.
Inset: exterior view, with the heat shield separated.

14 Recovery beacon hatch.
15 Parachute compartment.
16 Solid propellant soft-landing retro-rockets.
17 Landing radar.

The interior of the Soyuz descent module has been changed periodically. Originally, up to three men without spacesuits were carried, but after the Soyuz 11 loss, it was reconfigured for two men in flight suits

Launch and Landing Sequences of a Typical Soyuz Mission

1 Launch atop an SL-4 booster.
2 Separation of the four strap-on boosters.
3 Separation of tower and payload shroud.
4 Separation of the second stage core.
5 Orbital injection followed by third stage separation.
6 Orientation in orbit and firing of the instrument module engine.
7 Separation of the orbital module from the descent and instrument modules.
8 Separation of the descent module from the instrument module.
9 Descent module begins re-entry into the atmosphere.
10 Parachute hatch cap separates.
11 Drogue and then main parachutes are deployed to slow down the descent module.
12 Continuing descent under the main parachute, as the heat shield drops from underneath the descent module.
13 Solid propellant rockets fire in the base of the descent module to cushion the final touchdown.

Above: *The landing of a Soyuz descent module. The dust is not caused by the impact, but by the solid propellant retro-rockets firing.*

Above: *The recovery of a Soyuz descent module. The missions of the Soyuz-derived Zond 6 and Zond 7 ended on land, and the returned capsules would have looked similar.*

The spacecraft was oriented in orbit and retro-fire begun 39 minutes 50 seconds before landing. On Soyuz 19 the retro-fire lasted for 3m 14s: 12 minutes after retro-fire began the spacecraft split into its three component parts. The orbital and instrument modules were allowed to burn up in the atmosphere while the descent module was protected from the heat of atmosphere re-entry by an ablative heat shield. Some 26 minutes after retro-fire initiation, the main Soyuz parachute opened at an altitude of 9-10km. A further 4 minutes later the heat shield at the base of the descent module was separated and at an altitude of just 1.5-2m, four solid propellant rocket engines fired to cushion the landing.

The Soyuz spacecraft has been used in various forms for manned space missions for more than twenty years, and its reliability and performance characteristics have been improved with time. Variations from the design described here will be described as each variant is introduced. The next section looks at the possible role of Soyuz within the Soviet manned lunar programme plans, while the actual Soyuz missions will be covered starting in the following chapter.

Destination Moon

The evidence of cosmonaut statements and writings available in the West indicates that the Soviet Union was hoping to fly a series of manned lunar missions during the late 1960s in competition with Apollo. However, direct confirmation of this programme has not been forthcoming from the Soviet authorities. If the cosmonaut statements are based upon fact, then one must deduce that the Soyuz spacecraft was to be an integral part of the lunar programme, because the Soviets would not develop two different manned spacecraft

in the same time period. To discount Soyuz as part of the lunar programme, one would also have to assume that all the flights of the actual manned lunar craft were cancelled or failed to reach orbit; this is hardly credible.

Physical hardware evidence of a manned lunar effort is supplied by the series of Zond circumlunar craft, a group of flights within the Cosmos programme that took place between 1967 and 1970, and the (unsuccessful) development of a giant Saturn-5 class launch vehicle, designated in the West as the SL-15 or Type-G booster. (Using the latter designation system, the G-1 was a three stage variant to be used for Earth orbital missions only, while the G-1-e would have an extra stage added for the trans-lunar injection manoeuvre).

Many of the events described here are contemporary with those described in the next chapter which is devoted to the early Soyuz missions. The lunar mission is described here for convenience in order that the Soyuz missions may be presented in unbroken chronological order.

The Zond Programme

The Zond programme is the most tangible part of the manned lunar programme – especially since even the Soviets have implied that it was part of an effort to get man to the Moon. The accompanying table provides a summary of the flights which took place within the Zond programme (Zond 1 to Zond 3 were much smaller interplanetary probes launched in 1964-1965, and were completely unrelated to the flights considered here), including the rumoured launch failures.

Let us first review what we know about the physical appearance of the Zond spacecraft. No vehicle representative of the Zond 4-Zond 8 series has been put on public display; the illustrations released at the time of the Zond 5 mission suggested that the craft might be based upon the contemporary Venera spacecraft (masses 1.1-1.2 tonnes) launched by the four stage Molniya booster (SL-6). These suppositions were proved to be wrong when photographs of the Zond 5 descent module canister on its way back to Soviet Union via India were made available, and these revealed that the return craft was far larger than the small Venera craft. Later Zond illustrations indicated a close relationship with the Soyuz vehicle. Line drawings at the time of Zond 6 suggested this, and in 1971 a more detailed drawing confirmed the Soyuz relationship. Finally, a book by cosmonaut K.P. Feoktistov (*Cosmic Apparatus*) published in Moscow in 1983 includes virtually all the technical details one could want about Zond.

The Zond was based upon a two module Soyuz design: the standard instrument module was used, but with smaller solar panel "wings" and without the torus propellant tank. The standard descent module was also carried, although the heat shield was strengthened to give added protection against heat generated by the lunar mission re-entry velocities. In place of the spheroid orbital module, a smaller inverted cone was sited atop the descent module. Feoktistov describes this as

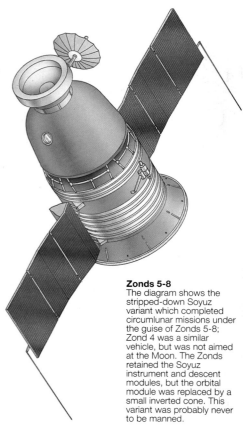

Zonds 5-8
The diagram shows the stripped-down Soyuz variant which completed circumlunar missions under the guise of Zonds 5-8; Zond 4 was a similar vehicle, but was not aimed at the Moon. The Zonds retained the Soyuz instrument and descent modules, but the orbital module was replaced by a small inverted cone. This variant was probably never to be manned.

simply a "Support Cone". The table provides a summary of the numerical data quoted by Feoktistov for Zond. The mass ranges quoted here (5.3-5.5 tonnes) differ slightly from those listed in the table of Zond launches (5.14-5.39 tonnes); the latter are taken from cumulative totals of spacecraft masses listed by V.P. Glushko in various publications. However, despite these slight discrepancies, the masses are within the same general range.

The Soviets have openly described the Zond spacecraft's propulsion system. Designated KTDU-53, it is derived from the KTDU-35, used on Soyuz. In fact, the Zond system used only the prime Soyuz engine system, with a specific impulse of 2,750m/sec (equivalent to 280.4 sec) and a thrust of 417kg. The burn time of the engine is not quoted, but Feoktistov states that the propellant load was about 400kg (implying a total burn time of about 270 seconds). The lack of a back-up propulsion

ZOND SPACECRAFT DATA	
MASS (tonnes)	
Spacecraft	5.3-5.5
Descent module	2.9-3.1
Instrument module	2.25
Support cone	0.15
LENGTH (metres)	
In Earth orbit	5.0
In lunar flight	4.5
DIAMETER (metres)	
Instrument module (max)	2.72
Descent module	2.2
Span across solar panels	9.0

Notes: This table is drawn from data in Table 3.5 (page 135) of the Feoktistov book, *Cosmic Apparatus*. The two lengths quoted suggest that the support cone atop the descent module at launch is not retained for the lunar mission. It could be separated in Earth orbit or soon after trans-lunar injection.

tested, carrying the orbital and descent modules to altitude. This would require no relocation of the flaps and the use of a standard Soyuz shroud tower. What we actually see is a shroud with the flaps placed at just the right height to bear the base when the descent module alone is being carried away, and a smaller rocket assembly on the tower, required because a far smaller mass is being carried.

The reason for the "Support Ring" becomes clear when the abort sequence is considered. On Soyuz, the orbital and descent modules are actually suspended within the shroud, the attachment points being between the two modules. Of course, Zond has no orbital module. Therefore, it would seem that the support module marks the attachment or suspension system for the Zond descent module within its shroud.

Cosmos 146 and 154

Between the initial two tests of Soyuz within the Cosmos programme (Cosmos 133 and Cosmos 140) and the Soyuz 1 mission, two other Cosmos flights occurred which are regarded as being connected with the manned programme generally and the Zond programme in particular. Cosmos 146 was launched on 10 March 1967 into a fairly low Earth orbit: on 11 March there seems to have been a manoeuvre as the orbit was changed from 184-288km to 185-339km. After 8-9 days it is believed that the spacecraft re-entered the atmosphere for a recovery. Cosmos 154 was launched four weeks later on 8 April, but exhibited no manoeuvre capability during its flight of about 11 days.

In some quarters these flights are interpreted as unsuccessful attempts to launch Zond spacecraft out towards the Moon, but this cannot be the case. Lunar launches always observe a specific relationship between the launch site and the lunar position (the lunar "Greenwich Hour Angle") and neither Cosmos 146 nor Cosmos 154

system on Zond is perhaps the best piece of evidence that Zond would not have carried a precious human cargo. To allow communications with the spacecraft at lunar distances a large dish antenna was carried and deployed at the top of the descent module.

The launch vehicle for the Zond missions was the four stage Proton SL-12 variant. The first stage consists of a large central tank containing the nitrogen tetroxide (oxidiser) and six outer tanks containing UDMH (unsymmetrical dimethylhydrazine - fuel). The second and third stages were carried in tandem and used the same propellants. These three stages originated from the Chelomei Design Bureau. The fourth stage, originally designated "Block-D", originated from the Korolyov Bureau and used kerosene (fuel) and oxygen. On a Zond mission the first two stages were sub-orbital and the third stage placed the Block-D/Zond combination into a low parking orbit. On the first orbit the Block-D

stage would ignite to take the Zond onto a deep space trajectory.

The question of whether Zond itself was ever to carry men will be discussed in detail later in this chapter, but some light is thrown on the matter by a photograph of what seems to be a Zond payload shroud that has been released by the Soviet Union. The shroud is shorter than that of a Soyuz (as one would expect for a two module payload) with a smaller rocket assembly on the shroud tower. Four aerodynamic flaps are visible near the top of the shroud, an arrangement that parallels the Soyuz escape system, suggesting that the intention was indeed to carry men and that this was an abort system.

It has been argued that this might simply be a test of concept for the abort system, but the relocation of the aerodynamic flaps implies that this is not the case. For a simple test, all that would be required is for the upper half of the Soyuz shroud to be

LAUNCHES IN THE ZOND PROGRAMME							
LAUNCH DATE & TIME	**SPACECRAFT**	**MASS** (kg)		**INCLINATION** (deg)	**PERIOD** (min)	**ALTITUDE** (km)	**LIFETIME** (d.hh.mm)
1967 10 Mar 11.31*	Cosmos 146	5017		51.5	89.2	184-290	8-9
			Then	51.5	89.7	185-339	
8 Apr 09.00*	Cosmos 154	5,020 ?		51.5	88.6	185-220	11 ?
21 Nov 18.45*	Zond	5,140 ?		Failed to reach orbit			–
1968 2 Mar 18.30*	Zond 4	5,140 ?		51.5	88.4	192-205	7 ?
			Then	highly eccentric orbit			
22 Apr ?	Zond	5,140 ?		Failed to reach orbit			–
14 Sep 21.42	Zond 5	5,140 ?		51.5	88.5	193-219	6.18.26
			Then	Barycentric orbit			
10 Nov 19.11	Zond 6	5,140 ?		51.5	88.6	186-232	6.18.59
			Then	Barycentric orbit			
1969 7 Aug 23.55*	Zond 7	5,390 ?		51.5	88.5	190-214	7 ?
			Then	Barycentric orbit			
1970 20 Oct 19.55*	Zond 8	5,390 ?		51.5	88.7	202-223	6.20 ?
			Then	Barycentric orbit			

Notes: The launch times for Zond 5 and Zond 6 were announced by the Soviet Union. The remaining launch times (marked *) are calculated using the Two-Line Orbital Elements (Cosmos 146, Cosmos 154), extracted from the *RAE Table of Earth Satellites, 1957-1982* (Zond 4, Zond 7, Zond 8) or calculated from the lunar launch window (Zond failure in November 1967). The parking orbits for Zond 4 to Zond 8 are taken from the *RAE Tables*, while the others are derived from the Two-Lines. The actual mass for Cosmos 146 has been announced, while the other masses are derived from Soviet cumulative totals which are available. Flight times are quoted as accurately as the available data allows, with only Zond 5 and Zond 6 being known to the nearest minute.

were launched close to the appropriate lunar launch window for that day.

Comments made by Glushko indicate what the actual missions for these satellites probably were, although his data only relate to Cosmos 146. He characterises this mission as one which attained "second cosmic velocity" – that is, Earth escape velocity – and he gives the mass of the spacecraft as 5,017kg. One can therefore suspect that Cosmos 146 was launched into a very eccentric Earth orbit and it was recovered after a re-entry to simulate lunar mission requirements. The mass is suggestive of a stripped-down Zond vehicle. There is a problem with this hypothesis: in an analysis of these missions, Grahn and Oslender describe that they received transmissions from the two craft while in low Earth orbit (see Further Reading), and if this is correct then the high apogee orbit cannot have been achieved within the first week of the Cosmos 146 mission. However, could the Block-D stage have retained its integrity for such a long period and completed its burn to change the Cosmos 146 orbit preparatory for a high-speed re-entry after more than a week ? It is a problematic issue.

Possibly the question just raised explains why no claim for a high-speed re-entry is made for Cosmos 154; the Block-D stage failed to ignite after eight days in low orbit, and the Zond craft simply separated and completed a re-entry at a normal Earth orbital velocity. Certainly, these two missions have not been convincingly explained in all their details.

Zond 4

The first launch of a Zond spacecraft towards the Moon seems to have been attempted in November 1967, although the spacecraft failed to reach orbit after steering problems were encountered with the second stage of the Proton booster. The next launch in the series came at the beginning of March 1968, and because of the Western speculation which has surrounded the mission the official launch announcement is reproduced below:

"In accordance with the space research programme, an automated probe Zond 4 was launched in the Soviet Union on 2 March 1968.
The automated probe was introduced into the predetermined trajectory from an intermediate orbit as an artificial Earth satellite. According to the measurement data, the automated probe Zond 4 is lying along a trajectory close to that calculated.
The aim of the flight is to explore the further regions of near-Earth space, as well as perfecting the new systems and assemblies on board.
The co-ordination and computation centre is processing the information received."

Apart from this announcement, nothing else is available from official Soviet sources about the mission, and Western analysts have debated the significance of the flight ever since. Originally it was thought that the spacecraft was being launched towards the Moon, but the lunar Greenwich Hour Angle at the time of launch shows that it was made in a direction almost directly away from the Moon. The most probable explanation of the Zond 4 flight is that it was launched onto a simulated lunar trajectory to test the deep space communications system and simulate a lunar mission re-entry, without the controllers having to worry about the disturbing gravitational influence of the Moon.

No successful re-entry was announced, and rumours have suggested that Zond 4 went into heliocentric orbit (based upon the Goddard Space Flight Center's "Satellite Situation Reports", the Zond 4 entry being inspired by all previous Zond missions having been placed into heliocentric orbits); that it remained in an eccentric Earth orbit; that it burned up on re-entry; that it crashed within the Soviet Union; or that it landed, not in the Soviet Union, but in China! Of these possibilities, it would seem logical that a recovery had been intended, but it failed for some reason.

The reason for the lack of information about Zond 4 is highlighted by a significant omission from the launch announcement. Comparison with the Zond 5 announcement shows that the two were almost identical, but that Zond 5's included an extra paragraph:

"Steady radio communications are being maintained with the probe Zond 5. According to telemetric data, all the systems and assemblies on board and the scientific equipment are functioning normally."

Clearly, it would seem that communications were lost with Zond 4 soon after it left its low Earth orbit, and possibly the Soviet authorities themselves are uncertain about the spacecraft's eventual fate.

Zond 5 and 6

The launch of Zond 5 on 14 September 1968 came at a time when the Soyuz programme was preparing to resume manned flights (Soyuz 3 flew the next month). The initial launch announcement gave no clues as to the intended mission, although the lunar Greenwich Hour Angle indicated that this time a launch towards the Moon had been made. About 67 minutes after the launch, the Block-D stage of the Proton booster ignited to place the Zond 5 craft onto a trans-lunar trajectory and the spacecraft separated from the rocket stage.

A course correction was applied on 17 September at 03.11 when the craft was 325 thousand km from Earth and the following day the spacecraft passed 1,950km behind the Moon and began its return journey to Earth. A second correction to the trajectory was applied when the probe was 143 thousand km from Earth, and this was specifically to ensure that the correct re-entry corridor would be encountered; an incorrect approach would mean that the spacecraft would either burn up on re-entry or simply "bounce off" the atmosphere into another deep space trajectory. Atmospheric entry began on 21 September at 15.54 and the descent module splashed down in the Indian Ocean (32° 63′ S, 65° 55′ E) at 16.08. When at an altitude of about 7km and travelling at 200m/sec the main parachute was deployed to slow down the craft for a safe landing. The mass at landing was 2,046kg.

Zond 5 marked the first recovery of a spacecraft from a lunar mission and its biological payload (including turtles) allowed checks to be made that living organisms would not encounter potentially harmful radiation or other problems which could adversely affect a manned mission.

Below: *A rare photograph of the Zond 5 descent module bobbing in the Indian Ocean after the first Soviet splashdown on 21 September 1968; it appears to be almost identical with the Soyuz descent module. The open hatch reveals the empty parachute compartment.*

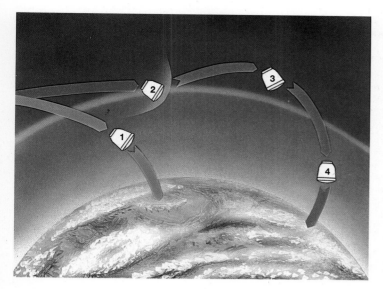

Zond Re-Entry Profiles
The two different Zond re-entry techniques after their circumlunar missions are shown here. Zond 5 (**1**) made a single high-velocity entry into the Earth's atmosphere, this involving decelerations which were higher than could comfortably be used on a manned mission. Zond 6 made an entry into the upper part of the atmosphere (**2**) and after losing some of its velocity "skipped" out again (**3**). A second entry was made (**4**) at a lower velocity, leading to recovery. This approach had lower decelerations, and could have been used on manned flights. It was thought that a manned Zond mission might have followed Zond 6, but in retrospect it seems that a manned flight was never considered.

cecraft landed at 14.10 inside the Soviet Union. The double re-entry technique was preferable for a manned mission because on Zond 5 the overloads experienced inside the capsule had been 10-15 times the acceleration due to gravity, while for Zond 6 they were only 4-7 times.

One Tass report after the Zond 6 recovery seemed to indicate that a manned Zond mission was definitely being planned:

"The aim of launching the Zond 6 probe, as well as the Zond 4 and Zond 5 probes, was to perfect the flight and construction of an automated variation of the manned spacecraft for flying to the Moon, as well as to check the functioning of systems on board under actual conditions of flight on the Earth-Moon-Earth route.."

The significance of this comment will be examined later.

Zond 7 and 8

Despite Western reports that a manned lunar mission would follow Zond 6, there was no more Zond activity until August 1969 when Zond 7 was launched. Of course, by then Apollo 11 had successfully landed two Americans on the Moon. Compared with the Zond 5 and Zond 6 missions, few details were given about the mission, almost as if the programme had now lost its importance. (From a propaganda point of view, it had.) On 9 August a

Two months later, a further circumlunar Zond mission was undertaken. Some 67 minutes after launch the Block-D stage placed the spacecraft onto a trans-lunar trajectory. On 12 November at 05.41 a correction was applied to the trajectory and two days later Zond 6 passed 2,420km from the lunar surface. On 16 November at 06.40 a second course correction was applied at a distance of 236 thousand km and a final refinement to the approach trajectory was made on 17 November at 05.36.

So far, the mission seemed to be a repeat of Zond 5.

Zond 6, however, used a radically different re-entry technique. Initially it entered the atmosphere on 17 November at 13.58 with normal lunar mission re-entry velocity — 11.2km/sec - but after losing velocity it "skipped" out of the atmosphere and completed a second re-entry at a velocity of 7.6km/sec. At a height of 7.5km and travelling at 200m/sec, the descent module's parachute opened and the spa-

LUNAR-RELATED COSMOS MISSIONS								
LAUNCH DATE & TIME	**COSMOS**	**INITIAL ORBIT**			**FINAL ORBIT**			
		Epoch	Incl (deg)	Altitude (km)	Epoch	Incl (deg)	Altitude (km)	
1967 16 May 21.45	159	**Manoeuvre 1:**					2,725m/s	
		17.34 May	51.84	200-434	31.5 May	51.60	350-60,637	
1969 16 Nov ?	—	Failed to reach orbit						
1970 6 Feb ?	—	Failed to reach orbit				·		
24 Nov 05.15	379	24.46 Nov	51.61	191-237				
		Manoeuvre 1: 25.197 Nov					263m/s	
		24.89 Nov	51.63	192-233	25.61 Nov	51.65	196-1,206	
		Manoeuvre 2: 27.666 Nov					1,518m/s	
		27.39 Nov	51.59	188-1,198	28.57 Nov	51.72	177-14,041	
2 Dec 16.44	382	2.7 Dec ?	51.6 ?	190-300 ?				
		Manoeuvre 1: 2.73 ? Dec					986m/s	
		2.7 ? Dec ?	51.6 ?	190-300	3.46 Dec	51.57	303-5,038	
		Manoeuvre 2: 5.539 Dec					288m/s	
		4.65 Dec	51.57	318-5,040	7.41 Dec	51.55	1,616-5,071	
		Manoeuvre 3: 7.663 Dec					1,311m/s	
		7.41 Dec	51.55	1,616-5,071	7.99 Dec	55.87	2,577-5,082	
1971 26 Feb 05.05	398	26.51 Feb	51.61	191-258				
		Manoeuvre 1: 28.181 Feb					252m/s	
		28.06 Feb	51.61	189-252	(6.76 Mar)	51.6	186-1,189	
		Manoeuvre 2: 28.249 Feb					1,320m/s	
		(6.76 Mar)	51.6	186-1,189	1.00 Mar	51.59	200-10,905	
12 Aug 05.30	434	12.41 Aug	51.6	189-267				
		Manoeuvre 1: 13.149 Aug					266m/s	
		12.90 Aug	51.60	188-267	13.62 Mar	51.60	190-1,261	
		Manoeuvre 2: 16.236 Aug					1,365m/s	
		15.89 Aug	51.60	188-1,262	17.55 Aug	51.54	180-11,834	

Notes: This table gives detailed orbital data for the manoeuvrable Cosmos flights directly connected with the Soviet manned lunar programme. Launch times are calculated from Two-Line Orbital Elements as GMT (hours and minutes). The orbital data and manoeuvre dates are calculated using the Two-Line Orbital Elements for the satellites. The 'Epoch' is the date and time (in decimals of a day) to which the orbit refers. In the case of Cosmos 398, the second orbit (for which the epoch is in parentheses) is that for the object discarded in the second orbit because the corresponding orbit for Cosmos 398 itself is not available from public sources. The two launch failures and Cosmos 382 were launched by Proton boosters, while the other flights used the Soyuz SL-4 booster.

course correction was applied when 260 thousand km from Earth and the spacecraft flew close to the Moon on 11 August. The spacecraft was recovered on 14 August after using the double-skip re-entry technique, the landing being near Kustani (the normal Soyuz landing area). While the two earlier missions had taken black-and-white photographs of the Moon, Zond 7 for the first time in the Soviet programme returned colour photographs of the Moon and Earth.

The final Zond launch came in October 1970, and on this occasion only the launch and landing announcements were apparently issued. On 24 October Zond 8 passed the Moon and it returned to Earth on 27 October at 13.55 in the Indian Ocean, some 760km south east of the Chagos Archipelago. This time a single re-entry was made, the probe coming back over the North Pole for the first time.

Lunar Propulsion Tests?

With the recovery of Zond 8 and the Soviets apparently shelving plans for attempting a manned lunar flight, it came as something of a surprise when a series of manoeuvring tests began within the Cosmos programme in November 1970. These can be split into two series of flights, one series using the Soyuz (SL-4) booster and the other using the three stage Proton (SL-13). Of the latter flights, two of the three launch attempts seem to have failed to reach orbit. The accompanying table provides a summary of these missions and details the manoeuvres which were completed.

While the manoeuvre tests of the satellites which attained orbit were conducted within a ten month period starting in November 1970, the very first flight in the series seems to have been the solitary Cosmos 159 mission of May 1967, which took place only a month after the Soyuz 1 accident. The rocket body orbit and the debris in the initial orbit are reminiscent of what happens when an SL-6 booster is used. This is the four stage Soyuz variant used for Molniya communications satellite missions: but the orbital inclination for a Molniya was 65° at that time, while the Cosmos 159 mission was inclined at 51.8°. It has long been believed that this was probably the first test of the fully fuelled Soyuz propulsion system, involving a single burn to depletion. Like Zond 4, the launch direction of Cosmos 159 was directly away from the Moon, although this might be just a coincidence.

The three satellites Cosmos 379, Cosmos 398 and Cosmos 434 all performed near-identical orbital manoeuvres. They were launched into initial Soyuz-type orbits using the SL-4 booster and after a period of time (which varied from one mission to another) a manoeuvre was made to an intermediate orbit, reaching out to 1,200-1,300km. In this orbit an object was discarded and then a second manoeuvre took place, resulting in the spacecraft attaining an apogee of 11,000-14,000km. It is not known with certainty whether the two manoeuvres on each flight were performed using the same propulsion system or whether the object discarded in the

Above: *A launch picture of a Zond has never been released, but this photograph shows the same Proton SL-12 variant that Zond 4-8 used. The major difference would be the use of a stubby, Soyuz-type shroud and tower atop the booster.*

Right: *The first Soviet colour photographs of the Moon and the Earth from lunar distances were those taken by Zond 7, one of which is shown here. The flight took place less than a month after the successful Apollo 11 lunar landing by American astronauts.*

intermediate orbit was the rocket block used for the first manoeuvre.

A single mission based upon the Proton booster, Cosmos 382, successfully flew during this period. There was a launch failure in November 1969 which was said to have had the same telemetry format as Cosmos 382. Additionally, the Soviets state that a Proton failed to reach orbit in February 1970 with a Cosmos satellite, and – apart from deep space failures given Cosmos numbers – Cosmos 382 was the only Cosmos mission launched on a Proton in 1968-1972.

Cosmos 382 seems to have entered a low orbit (shown with question marks in the relevant table) and at the first pass through the southern hemisphere apogee a manoeuvre took place which resulted in the first tracked orbit: 303-5,038km. A further manoeuvre took place, raising the perigee and then two objects were discarded: one might have been the rocket body and the other equivalent to the object left in the intermediate orbit for the Cosmos 379 group. The final manoeuvre was then completed by the main spacecraft.

The four flights during 1970-1971 which reached orbit have generally been considered to have involved the testing of equipment for the manned lunar mission, but whether the missions all included lunar landing craft components is unknown. The only clue which the Soviets have given was that when Cosmos 434 decayed it was said that it was "a prototype lunar cabin", a term which they normally applied to the Apollo lunar module when describing the US space programme.

The Giant Booster

For any manned lunar mission there is one further vital element required: a booster in the Saturn V class. No such booster placed a Soviet payload in orbit during 1968-1972, although there are intelligence reports that three large boosters (Type-G or SL-15 vehicles) were prepared for flight. The first is believed to have exploded on the launch pad on 3-4 July 1969 while being fuelled in readiness for a launch. The second and third were launched on 23-24 June 1971 and 24 November 1972 but they failed to attain orbit. After these failures and with the Apollo programme completed, the Soviets in all probability abandoned their manned lunar programme. An initial switch to a primarily Earth orbital programme seems to have taken place in autumn 1969 (a few months after Apollo 11's success) when the Soviets committed the future manned programme to work involving orbital stations (although it is possible that the orbital stations were originally planned as important elements in the overall lunar mission).

Was Zond to be Manned?

One question which has been debated for many years in the West is whether the Zond spacecraft as it flew would ever have been launched with a manned crew. The pro- and anti- cases are presented in summary form in the table, and the salient points will be discussed in more detail here. Let us first examine the evidence in favour of a manned Zond.

The apparent Zond launch escape system has been discussed in the description of the spacecraft earlier in this chapter. All that need be added is to say that any Soviet or American spacecraft which has used an escape system has been intended to carry men. If Zond were not to be manned, it would be the only exception. As an aside, if Zond were to be manned, how would the crew enter the craft? The shroud tower was directly above the descent module hatch – to which the support module was attached and thus the normal mode of entry would be blocked.

As far as first-hand cosmonaut testimony is concerned, cosmonaut Belyayev stated in 1968 that he was conducting helicopter training (as were American astronauts, to practice lunar landing techniques) and that he expected to make an imminent lunar flight. His back-up is often reported to have been Bykovsky.

As already noted earlier, official Soviet statements relating to Zond 6 certainly suggest that a manned variant of Zond was capable of a lunar mission, but would such a variant actually have been a Zond? If a

Lunar Boosters
A comparison of Soviet and American lunar boosters is shown here. The Saturn V (**3**) shares the same upper stage as the smaller Saturn IB (**4**), used for tests in Earth orbit. In the same way, the second and third stages of the Proton booster (**2**) were used as upper stages of the giant SL-15 (Type-G or TT-5) booster developed by the Soviets (**1**). Thus, Proton had the same role as the Saturn-IB in the lunar programme: testing components of the lunar mission before committing the giant boosters to flight.

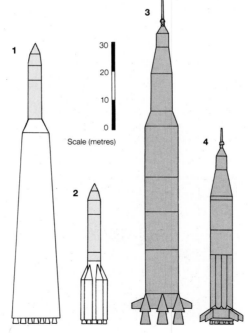

Scale (metres)

Soyuz had been launched on a manned lunar mission, the quoted statement would still hold true because Zond 4 to Zond 6 tested two of the three Soyuz components in unmanned mode on a lunar mission.

Finally, although admittedly a rather weak piece of "evidence": there have been persistent rumours of a flight. These could easily have been inspired by Western expectations of a manned Zond mission to "race" Apollo 8 to the Moon in December 1968. There is little evidence of such rumours originating in the Soviet Union, although in the West they were very widespread.

Against this evidence one can consider the launch record of the Proton booster. There had been failures in the Zond programme in November 1967 and April 1968, although there had been three consecutive successful missions: Zond 5 and Zond 6 using the SL-12 and Proton 4 using the SL-13 variant. However, the vehicle could surely not have been considered as safe for a manned flight on this record?

The duration of a manned loop around the Moon was nearly seven days, but in December 1968 the longest manned Soyuz mission had been four days on Soyuz 3 and the longest unmanned missions had been Cosmos 212 and Cosmos 213, five days each. Similarly, no Soviet manned mission had yet flown for more than five days (Vostok 5 in 1963). So, the Soviets would have been committing a Zond crew to the longest manned mission of a Soyuz-derived craft to that date, and furthermore to the longest Soviet manned mission of any type undertaken to that date.

Any Zond loop around the Moon would have been a dead-end project, because it could only have been viewed as an interim test prior to operations with the giant SL-15 booster. No more than two missions could have been flown credibly. However, it must be realised that in Voskhod the Soviets flew a dead-end craft simply to gain some space "firsts", so such a mission is not to be ruled out entirely.

Just as there is one major piece of hardware evidence in favour of Zond being intended for manned missions (the escape system), there is one major piece of conflicting evidence. As already noted, the Zond propulsion system was derived from the Soyuz system, but it did not have the back-up Soyuz engine - presumably because of mass constraints imposed by the capabilities of the Proton SL-12 booster. The lack of a back-up engine would not have mattered too much if only non-human biological missions were planned,

A MANNED ZOND: CONFLICTING EVIDENCE	
In Favour Of A Manned Flight	Against A Manned Flight
The Zond shroud apparently incorporates the equipment for an emergency abort.	The Proton booster launch record was not good, and in 1969 its success rate drastically declined.
Belyayev claimed in 1968 to be training for an imminent lunar mission.	The mission duration of 7 days was longer than any Soyuz flight to the end of 1968.
A statement from the Zond 6 press release implies a manned Zond would follow.	The duration would have been longer than any manned Soviet mission to the end of 1968.
Rumours in December 1968 of an impending manned lunar mission.	A manned Zond flight would have a dead-end project.
	The re-entry technique for a manned mission had only been tested on Zond 6.
	Soviet descriptions of the Zond propulsion system show that it incorporated no back-up system.

but it would have been a disaster if the prime system had failed (preventing the accurate course corrections required for the double-skip re-entry technique) on a manned mission.

A logical explanation for the Zond configuration actually flown is that it was being used to test in unmanned mode the main Soyuz modules which would feature in a manned lunar mission. This would allow experience to be gained in tracking and communicating with a spacecraft at lunar distances and the testing of the descent module heat shield on a lunar mission. Actual manned operations would then await the three-module Soyuz vehicle being readied for flight, this being launched in support of a lunar mission atop either a standard SL-4 booster or the giant SL-15 vehicle. If this view is accepted, it also explains the Zond 6 official statement since the Zonds did check out an automated "variant" of a manned lunar vehicle, that lunar vehicle being Soyuz.

Given that Zond was not primarily intended to be manned, could a manned flight have been mounted at short notice in response to US developments? In 1968 the Soviets were seeing delays in their manned lunar programme, as the maiden flight of the SL-15 receded into the distant future, while Apollo was overcoming its difficulties and manned lunar missions were imminent. After the two successes of Zond 5 and Zond 6, coupled with Soyuz 3, the Soviets may have felt justified in trying to launch a man in a Zond towards the Moon on about 7 December 1968 (about 17.30 for the middle of the launch window). However, no launch took place; Apollo 8 orbited the Moon later in the month, and Zond was again relegated to its original role of being an unmanned test-bed for future Soyuz lunar missions.

The Lunar Flight Plan

Since the Soviet authorities have not hinted at a possible profile for the manned lunar landing mission in 1969-1971, it is difficult for western observers to be dog-matic when approaching this subject. One can try to analyse cosmonaut statements, but these are so diverse, they imply just about every mission profile that can be imagined when taken as a whole. The Soyuz spacecraft itself gives one the impression of a rough equivalent to an Apollo command-service module, but Soyuz had no internal transfer system (at least, not until it was being used as a Salyut ferry in 1971) which would have been essential for a lunar module mission.

Two main mission profiles have been studied by Western analysts. One involves the use of more than one launch vehicle, with the Soyuz being launched after the main lunar complex has been orbited by the SL-15 booster. The other, more recent, approach suggests that a single launch would have been required with no rendezvous with a lunar module being necessary. Both these profiles are discussed below.

Rendezvous and Docking

The analysts who have done most analysis on this approach are Charles P. Vick and David R. Woods. Woods was first to have his analysis published (1976), and therefore his original ideas will be reviewed first.

Woods envisaged the launch of three boosters for the manned lunar mission. The first launch would have been the SL-15 three stage booster, with a trans-lunar stage on top. Following this, there would have been a Proton SL-13 launch, carrying the lunar module and a "habitat" module (derived from the Soyuz orbital module). The final launch would have been a manned "Heavy Zond" atop a Soyuz SL-4 booster. The "Heavy Zond" would be a fully fuelled Soyuz without an orbital module but with a docking adapter at the forward end of the descent module. The translunar rocket stage, the lunar module/habitat module and the Heavy Zond would all dock in Earth orbit, and be boosted onto a translunar trajectory.

After trans-lunar injection (TLI) and lunar orbit injection (LOI), the trans-lunar stage would be discarded, and all future course corrections would be made using the Heavy Zond propulsion system. Two of the three men in the Heavy Zond would transfer to the lunar module and this would separate (leaving the habitat module attached to the Heavy Zond). Near periselene (closest point of the orbit to the Moon) the lunar module would brake out of orbit. A single large burn would be completed, followed by a second lower-thrust burn beginning at an altitude of about 2km. The two-stage lunar module should then soft-land on the Moon.

Solar cells could be deployed from the descent stage and experiments set up on the lunar surface. After the period of exploration was complete, the lunar module ascent stage would fire to take the cosmonauts back into lunar orbit and a docking with the Heavy Zond and habitat module. After the lunar orbit part of the mission was complete, the ascent stage and habitat module would be discarded and the Heavy Zond propulsion system would fire (trans-Earth injection – TEI) to bring the cosmonauts home: re-entry would follow the double-skip profile which had been proved with the unmanned Zonds.

A modification of this profile was later preferred by Woods, and this was subsequently published in a paper by Vick (1985). This called for only two launches. The first would have been a full fuelled Soyuz, possibly with an experimental module, atop a Proton Sl-13 booster. The following day a giant SL-15 vehicle would be launched carrying a mission support module (for additional living and experimental quarters), a lunar module and a lunar braking module.

The Soyuz would dock with the SL-15 payload, still attached to the final booster stage. It was assumed that the SL-15 booster used three stages to attain Earth orbit injection and TLI: the third stage would then separate. The lunar braking module would perform LOI and in turn be cast off. The lunar landing, lunar launch, rendezvous and TEI would be performed as in the first reconstruction, with the exception that the Soyuz would retain its orbital

A Soviet Lunar Module?

Below is a concept of a possible Soviet lunar module, based upon Soyuz-type technology. The Soviets, of course, have never displayed a full-scale vehicle like this, but some photographs are available which show a small model of a similar design undergoing landing tests on a simulated lunar surface. However, these tests could have been in support of the unmanned Luna missions.

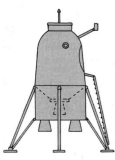

Destination Moon

If the Soviets had decided to fly men to the lunar surface using a lunar module separate from the main command craft, the mission profile might have looked something like this. Two launches are required for the mission: first, an unmanned SL-15 booster is launched to place the unmanned lunar landing complex in Earth orbit, and this is followed by the launch of a manned Heavy Zond spacecraft atop a Proton SL-13 booster. The Heavy Zond and the lunar landing complex would dock in Earth orbit (**1**) and then be boosted out towards the Moon (**2**). The final stage of the SL-15 would then be discarded (**3**).

A lunar braking module slows down the complex and completes entry into lunar orbit (**4**). Its work done, the module is separated (**5**) and the cosmonauts begin preparations for the manned landing. Once everything has been checked, cosmonauts transfer to the lunar lander and it separates from the Heavy Zond. The lander descends to the Moon (**6**) while the Heavy Zond continues to orbit (**7**).

At the conclusion of the lunar surface exploration, the lunar lander ascent stage ignites to bring the cosmonauts back into lunar orbit (**8**) where they dock with the Heavy Zond (**9**). With crew and scientific results transferred to the Zond, the lander is discarded (**10**) and the Heavy Zond engines are ignited for the return to Earth.

The return journey takes about three days (**11**) and as the final approach to Earth begins the Zond instrument module is separated and the initial entry into the atmosphere takes place (**12**). After a "skip" re-entry, the cosmonauts are recovered inside the Soviet Union (**13**).

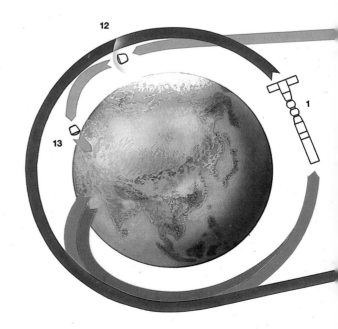

module until just prior to atmospheric re-entry.

These two reconstructions are typical of a number of other published and un-published ideas, but they do seem to rely more than a little on the approach taken by the American Apollo system. Of course, rendezvous was implied for the Soviet mission because of the historical emphasis placed on rendezvous techniques: witness the joint flights of Vostok 3-Vostok 4, Vostok 5-Vostok 6, Soyuz 1 to dock with (the unlaunched) Soyuz 2, Soyuz 3 rendezvous with Soyuz 2, Soyuz 4-Soyuz 5 docking and the Soyuz 6-Soyuz 7-Soyuz 8 group flight. However, a major requirement for the lunar module approach was an internal crew transfer system, and there is no evidence that any of the Soyuz craft used such a system until Salyut missions began in 1971. If an internal transfer system had been intended for the lunar mission, then surely it would have been built into Soyuz from the beginning?

The Direct Launch Approach

The first writer seriously to consider a totally different approach and publish his results was John Parfitt (1986, 1987). His reconstruction of the Soviet lunar mission grew out of dissatisfaction with the earlier analyses and as a result of new analyses of the Proton and giant Type G booster fami-lies. Parfitt's interpretation basically requires a single launch of a manned SL-15 booster, which would place a manned craft on the Moon and return that craft to Earth.

Parfitt – in conjunction with Alan Bond – estimated a mass of about 8,500kg for a fully fuelled three module Soyuz craft. From the data which had been derived from analysis and from Soviet sources, it was realised that the propellant load and dry mass of the third stage of the Proton SL-13 booster (designated P-3 stage) was sufficient to perform an LOI, lunar descent and ascent to lunar orbit with the fully fuelled Soyuz on top of the stage: the Soyuz itself would have the propellant for TEI. Working backwards again, the dry mass and propellant load of the Proton

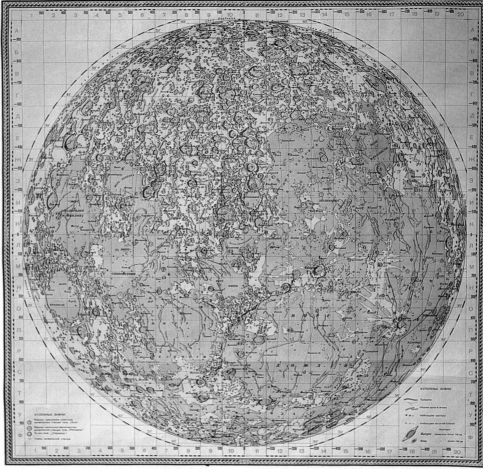

Above: *A Soviet map of the lunar surface based upon Earth-based observations. Curiously, the Soviets did not undertake a series of lunar mapping missions as did the Americans before Apollo.*

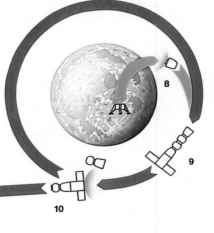

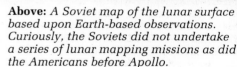

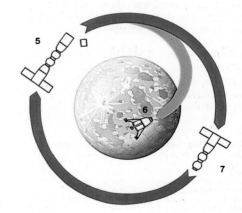

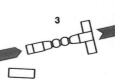

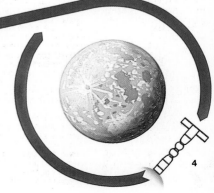

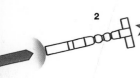

An Alternative Lunar Module

It now seems probable that the Soviets had no designs for separate lunar landing and main command craft. A fully-fuelled Soyuz spacecraft of the type which flew in 1966-1970 atop a modified Proton third stage could have performed the task. The cosmonauts would have used the orbital module as an airlock for their exits to the lunar surface. A major problem would seem to be the height of the orbital module above the lunar surface; if a cosmonaut had fallen from the elevated hatch he could have been seriously injured.

The Parfitt-Bond Model

This diagram illustrates the profile for a Soviet manned lunar mission involving a Soyuz spacecraft descending to the lunar surface. The mission begins with the launch of a multistage SL-15 booster, carrying the cosmonauts inside a fully fuelled Soyuz spacecraft. The lunar mission complex

comprising the second and third stages of the Proton booster and the Soyuz spacecraft are placed into an initial Earth orbit (1). The Proton second stage burns to initiate the launch towards the Moon (2) and is then separated (3). The lunar lander (Soyuz plus Proton third stage) coasts towards the Moon (4) and after a three day journey performs

lunar orbit injection (5). After further spacecraft checks and trajectory trimming, the lander brakes itself out of lunar orbit and soft-lands on the Moon (6).

While on the Moon, cosmonauts transfer to the lunar surface for their exploration, setting up experimental packages at the landing site. With their work complete, the Proton

stage ignites (7) to place the spacecraft back into lunar orbit, and it is then discarded. Further research can be conducted in lunar orbit, before the Soyuz instrument module engine ignites to fire the spacecraft out of lunar orbit, back towards the Earth (8).

As the Earth approaches, the orbital module is separated (9) and the

spacecraft oriented for any final course corrections. The instrument module is then separated (10) and a double-skip re-entry into the atmosphere is completed (11) with landing inside the Soviet Union (12).

The spacecraft performances do allow for some variations to this profile to be contemplated. Two examples are that the

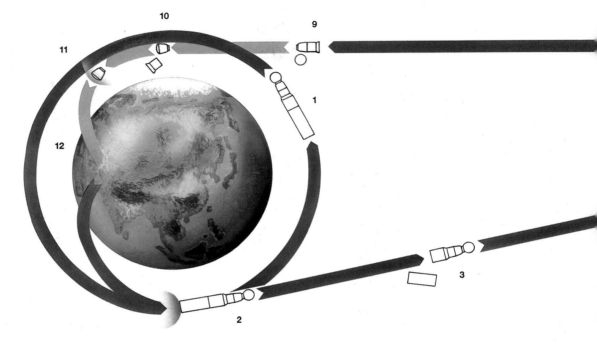

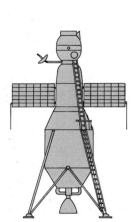

booster's second stage (the P-2) was just right to perform TLI for the fully fuelled P-3 stage with its Soyuz payload.

The giant SL-15 booster probably used three stages to reach Earth orbit and a drawing of this booster with its payload which appeared in the American *Aviation Week & Space Technology* magazine suggested that the payload size was the same as the stacked P-2 stage, P-3 stage and Soyuz. Calculations by Parfitt and Bond showed that the SL-15 could place this combination in Earth orbit.

The mission profile for the manned lunar landing follows directly from the above calculations. The three stage SL-15 would place the lunar mission complex into a low Earth orbit with cosmonauts on board (three could possibly have flown). When the final SL-15 stage separated, the P-2 stage would ignite for TLI and be discarded. The P-3 stage would then perform any mid-course corrections using either its main rocket engine system or its system of vernier engines, followed by LOI. A large burn would take the P-3 stage and Soyuz out of lunar orbit and the P-3 verniers would be used for the final targeting of the lunar landing. After the period of lunar exploration, the P-3 would again fire launching the spacecraft from the Moon, being eventually cast off after orbital injection. The Soyuz would then continue in lunar orbit until its mission was over. The orbital module would either be separated in lunar orbit or retained until just prior to atmospheric entry: the Soyuz propulsion system would perform TEI.

The Parfitt-Bond approach is so different

from all previous analyses, that it has yet to gain wide acceptance. However, it does have advantages over the other hypotheses. The single launch of a manned SL-15 booster minimises the number of launches, of course. The P-2 and P-3 stages would already have been proven on the Proton booster flights before being flown atop the SL-15. No complicated rendezvous and dockings would be necessary, which would also explain why the original Soyuz docking system gave the strong impression of being added on as a hasty afterthought to the basic design.

Finally, this approach would answer some otherwise unresolved questions about the Soyuz orbital module design. While the cosmonauts were flat on their backs during the launch and landing of a Soyuz, the orbital module work areas required them to be standing along the longitudinal axis of Soyuz. A simpler approach would have been to maintain the orbital and descent module work axes parallel; why are they different? The explanation would seem to be that during a lunar landing mission, the cosmonauts could stand in the orbital module, looking down on the lunar surface during the final approaches.

The other puzzle relating to the orbital module design is the placement of the entry hatch under the side of the module. This means that at launch the crew have to climb up into the orbital module and then drop into the descent module: more logically, one would surely place the hatch on the top surface of the module? However, the placement makes sense when one

considers that it would have been used by the cosmonaut(s) to descend to the lunar surface from the orbital module when on the Moon. One or two cosmonauts could remain in the pressurised descent module while the orbital module was de-pressurised for the lunar surface exploration.

Cosmonauts for the Mission

In 1981 the Soviet cosmonaut V.I. Sevastyanov gave details of a group of cosmonauts which had assembled in 1967 to begin training for a lunar mission. He was originally the training manager for the following men:

Military commanders:	V. F. Bykosvky, G. T. Dobovolsky, P. R. Popovich. P. I. Klimuk, A. A. Leonov
Military engineer:	Y. P. Artyukhin
Civilian engineers:	G. M. Grechko, O. G. Makarov, N. N. Rukavishnikov

It is unclear whether Sevastyanov himself was actually training for a mission. In any case, he left this post in late 1967 and was replaced by P.I. Belyayev.

Formed in the middle of 1967, the cosmonauts completed recovery training in the Black Sea (the Zond lunar missions being the only ones which had completed intentional splash-downs, and this suggests that any manned Soyuz lunar craft might also land in this way) and training in helicopters (in preparation for the lunar landings?).

By the summer of 1969, however, some of the cosmonauts in this team were being readied for missions within the Soyuz Earth orbital programme, and presumably

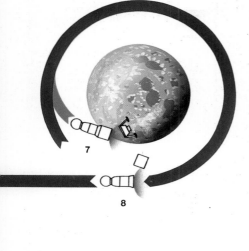

the training group had been fully disbanded by the end of that year as more men began to train for the Soyuz-Salyut missions due to begin in 1971.

Conclusions

The Zond missions successfully tested Soviet ability to fly the Earth-Moon-Earth route during 1968, and after the Soyuz 1 accident in 1967 the Soviet manned space programme seemed to have recovered by the end of 1968. But, as history recounts, there were no manned lunar missions launched by the Soviets. All the elements of the mission were test flown successfully

in Earth orbit or beyond, with one exception: the SL-15 booster was never flown successfully, and because of this single failure the Soviet manned lunar programme was finally cancelled in 1972-1973.

Soyuz had already been re-directed towards space station operations, while the Proton P-2 and P-3 stages were performing reasonably reliably in the early 1970s. The last Zond mission had flown in 1970 and during 1970-1971 the final series of tests of the fully fuelled Soyuz propulsion system were successfully conducted. Possibly everything relied upon the SL-15 launch in November 1972 – and that failed.

All of the elements of the original manned lunar landing programme are flight proven apart from the giant booster. One may legitimately wonder whether they will be re-assembled for a lunar attempt when the new giant Soviet launch vehicle Energiya becomes operational.

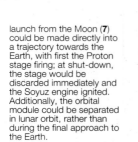

launch from the Moon (7) could be made directly into a trajectory towards the Earth, with first the Proton stage firing; at shut-down, the stage would be discarded immediately and the Soyuz engine ignited. Additionally, the orbital module could be separated in lunar orbit, rather than during the final approach to the Earth.

FURTHER READING

Anon, "Soviet Space Capability: Big Surprises Coming?" in *Space Markets*, Autumn 1986, pp176-177, 178 (early report of the work of Parfitt and Bond, noted below).

Clark, Phillip S., "Topics Connected With The Soviet Manned Lunar Programme" in *JBIS*, vol 40 (1987), pp 235-240.

Feoktistov, K.P., *Cosmic Apparatus*, Military Publishers, Moscow, 1983.

Grahn, Sven & Oslender, Dieter, "Cosmos 146 And Cosmos 154" in *Spaceflight*, vol 22 (1980), pp121-123.

Johnson, Nicholas L., "Apollo and Zond – Race Around the Moon" in *Spaceflight*, vol 20 (1978), pp403-412.

Novikov, N., "How They Fuel Space Ships" in *Conquest of Space in the USSR 1974*, pp106-117.

Oberg, James E. "Russia Meant To Win The "Moon Race" in *Spaceflight*, vol 17 (1975), pp163-171, 200.

Oberg, James E., "The Hidden History Of The Soyuz Project" in *Spaceflight*, vol 17 (1975), pp282-289.

Oberg, James E., "Zond Moonflight Controversy" in *Spaceflight*, vol 18 (1976), p75.

Parfitt, John A. & Bond, A., "The Soviet Manned Lunar Landing Programme" in *JBIS*, vol 40 (1987), pp231-234.

Vick, Charles P., "Russia's Moon Plan?" in *Spaceflight*, vol 21 (1979), p430.

Vick, Charles P., "The Soviet G-1-e Manned Lunar Landing Programme Booster" in *JBIS*, vol 38 (1985), pp11-18.

Vladimirov, B.P., "Rocket Launching Facilities at Baykonur Described" in *Zemlya I Vselennaya* 1978, pp64-71.

Woods, David R., "A Review Of The Soviet Lunar Exploration Programme" in *Spaceflight*, vol 18 (1976), pp273-290.

Woods, David R., "Lunar Mission Cosmos Satellites" in *Spaceflight*, vol 19 (1977), pp383-388.

Details of the Zond Tass statements were taken from:

Petrov, G.I. (ed), *Conquest Of Outer Space In The USSR*, 1967-1970, Nauka Publishers, Moscow, 1971.

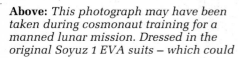

Above: *This photograph may have been taken during cosmonaut training for a manned lunar mission. Dressed in the original Soyuz 1 EVA suits – which could* have been used as the basis of a lunar suit – the Soyuz 2 cosmonauts, Yeliseyev and Khrunov, are seen posing next to a model of the Moon.

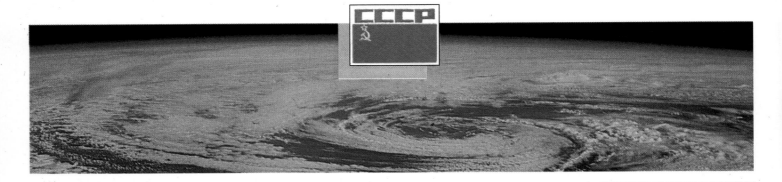

Chapter 5: **The First Soyuz Flights**

A break of twenty months occurred between the last Voskhod manned flight and the first launch in the Soyuz programme. During this period, the Americans flew the ten manned flights of their Gemini programme, which involved spacewalks, dockings, in-flight manoeuvres and a long-duration (14 days) mission. It is now known that a long duration Voskhod 3 was planned for 1965-1966, but it was cancelled.

Although the Soyuz spacecraft had been on the drawing board since 1962, it was not until November 1966 that the first orbital flight took place (in unmanned mode). The manned debut came in April 1967 and this resulted in the death of the single crewman. There followed a break of nearly eighteen months, during which unmanned testing continued in orbit, before manned missions were resumed.

The 1968 and 1969 mission all involved multiple spacecraft launches, while the final mission to be considered in this chapter was a solo mission in the summer of 1970. With the exception of four manned Soyuz flights (Soyuz 13, 16, 19 and 22) all the subsequent flights were connected with the Salyut/Soyuz space station programme.

Unmanned Tests of Soyuz

As the break in manned Soviet missions after Voskhod 2 lengthened, it became almost a sport among Soviet spaceflight analysts to scrutinise launches in the unmanned Cosmos programme for signs of the possible testing of a new manned spacecraft. The manned Vostok and Voskhod missions had all been flown in orbits inclined at 65 degrees to the equator; the lunar and interplanetary missions had flown at the same inclination. In the latter months of 1965, lunar and then Venus probes were launched at 51.8 degrees to the equator, thus implying an increased lift capacity of the Vostok-derived SL-6 booster. (This is because the closer that a launch is to being due east, the more payload can be placed in orbit; flights at 65° use a more northerly launch direction and so have a reduced payload capacity.) In February-March 1966 the biological Cosmos 110 mission also flew at 51.8 degrees, suggesting that manned missions might be launched at the lower inclination also.

The first mission which suggested the testing of a new manned spacecraft was Cosmos 133, launched in November 1966

Above: *An unexplained picture of Komarov undertaking EVA training, although his Soyuz 1 mission did not call for him to perform EVA. This is one of the mysteries which still surround Soyuz 1.*

into a 181-233km (as announced) orbit: two days later the satellite was de-orbited in a recovery attempt. Cosmos 140 followed just over two months later with an announced orbit of 170-241km. Once more, the spacecraft was returned after two days in orbit. The success of these missions will be judged in connection with Soyuz 1.

Although the new spacecraft was a manoeuvrable vehicle, American data (contained in the Two-Line Orbital Elements) do not suggest any in-orbit manoeuvres.

The First Spaceflight Fatality

Following the tests of Cosmos 133 and Cosmos 140, together with Cosmos 146 and Cosmos 154 launched within the Zond programme (see Chapter 4), Western observers concluded that a new manned

Soviet space mission was imminent. It was claimed that the mission would be the most spectacular yet, and if it were to be successful, the Soviet Union would have regained the initiative in the "race" for the Moon following the Apollo 1 fire which had killed three US astronauts in January 1967.

As the planned launch date drew near, the rumours became more specific: two manned spacecraft would be launched on a spectacular space mission involving veteran cosmonauts Komarov and Bykovsky, among others. A docking and a crew transfer were thought to be planned.

It was in this climate that Soyuz 1 was launched in the early hours of 23 April 1967 with a single cosmonaut on board: Vladimir M. Komarov, who was on his second space mission. The Soviet launch announcement gave the orbital altitudes as 201-224km, and the flight objectives as being: to test a new piloted spaceship; to test the systems and units of the spaceship's design in orbital flight; to perform extensive scientific and physical-technical experiments and investigations; to continue further medico-biological experiments and investigations of the effect of various factors of space flight on the human organism.

Four and a half hours after the launch, a further bulletin was issued, stating that all was well with the mission, with a similar announcement coming two hours later. It was said that between 10.30 and 18.20 (GMT) Soyuz 1 would be outside the area of radio visibility from the Soviet Union, and Komarov would be resting in this period. The final in-flight report came soon after Soyuz 1 had completed its thirteenth orbit.

The Soviets indicated that Komarov was not wearing a spacesuit inside Soyuz 1, although unusually there were no television transmissions released from the spacecraft. No details of the Soyuz design were released, and the overall impression gained by observers was that the Soviets were planning a second Soyuz launch about a day after Soyuz 1 was orbited, and that the two spacecraft would dock.

There was a break of more than twelve hours before any more news was released, and when it came the report stunned the world. Not only had there been no Soyuz 2 launch, but Komarov had died during the Soyuz 1 re-entry. The Soviet announcement made no allusion to any in-flight problems or to a planned second launch.

Above: *Left to right, Gagarin, Khrunov, Komarov, Yeliseyev and Bykovsky training for Soyuz 1/Soyuz 2 with Khrunov and Yeliseyev due for an EVA.*

Below: *From the same series of training shots as the above picture, Gagarin with Khrunov in his EVA suit. Gagarin was the Soyuz 1 back-up commander.*

Left: *Komarov arrives at Tyuratam for the Soyuz 1 launch. Also there are Yeliseyev and Khrunov, as well as Nikolayev.*

After reviewing what had been previously announced, it continued.

"On 24 April when the programme of tests was completed, the suggestion was made to him [Komarov] that he should terminate the flight and land. After performing all the operations connected with the transition to landing conditions, the spaceship safely went through the very difficult and responsible braking stage in the dense layers of the atmosphere and completed deceleration from satellite speed to a low velocity.

However, when the main canopy of the parachute opened at a height of 7 kilometres, according to preliminary data, as a result of tangling of the parachute's cords the spaceship fell at a high velocity, this being the cause of the death of Colonel Vladimir Komarov."

Ever since the end of the mission, Western observers have been trying to decide what went wrong and what had been planned for the mission.

First of all, is there any evidence that a

second Soyuz was due to be launched? Strangely enough, there is plenty of Soviet photographic evidence. A number of photographs have been released over the years, showing Komarov training with Bykovsky and two (then) rookie cosmonauts, Yeliseyev and Khrunov. The latter two were dressed in spacesuits, and they flew the Soyuz 4/Soyuz 5 docking mission in 1969, when they completed a transfer from Soyuz 5 to Soyuz 4. Since they were training with Komarov in February-March 1967 (as some group photographs have been accurately dated from Soviet sources), they must then have been training for the un-launched Soyuz 2 mission.

Some further photographic evidence comes in the form of pictures of Komarov arriving at Tyuratam for the Soyuz 1 launch. As usual for cosmonauts arriving for a launch, Komarov is garlanded with flowers . . . as are Bykovsky, Khrunov and Yeliseyev. Therefore, we may reasonably suppose that the latter three had also arrived for a space launch.

There is one other obscure hint that a second manned Soyuz was ready for

launch. All previous (and subsequent) first flights have been identified by the programme name only, rather than including the spacecraft number (eg, Vostok and Voshkod rather than Vostok 1 and Voshkod 1). However, the first Soyuz was identified at launch as Soyuz 1, suggesting that Soyuz 2 would appear very soon.

If we accept this hypothesis, and drawing upon the evidence of elapsed times between launch and docking in other Soyuz manned dockings, it is possible to draw up a flight plan for Soyuz 1 with Soyuz 2 based upon a docking on the first Soyuz 2 orbit (see table).

There are persistent reports that Soyuz 1 was in difficulty before the landing was attempted. According to one story – which originates from a Soviet source – after orbital injection the two "wings" of solar cells failed to deploy, thus depriving the craft of its power supply. This would explain why there were no television pictures from Soyuz 1 and no real in-flight activity detailed by the Soviets. Within perhaps a couple of hours of the Soyuz 1 launch, the Soyuz 2 mission had been perforce scrapped, and all possible effort was put into arranging a safe return for Soyuz 1. Recovery seems to have been attempted on the sixteenth, seventeenth and eighteenth orbits, with the first two re-entry burns being called off: possibly Komarov could not stabilise the Soyuz for the de-orbit manoeuvre? After orbit 18, Soyuz would not have been able to come down anywhere near the main Soyuz landing zone, so that orbit was the final chance to recover Soyuz until the next day (if Komarov survived that long).

The recovery attempt of Soyuz 1 was delayed so long that it became one of the few returns to be tried on a southbound pass over the landing site. Because of its problems, Soyuz 1 was probably unstable as it began re-entry, and the spacecraft's spinning may have fouled the parachute lines, resulting in the descent module crashing into the ground, killing Komarov.

The re-entry phase would certainly have been a worrying time anyway for the Soviets, according to a report by Viktor Yeviskov (*Re-Entry Technology And The Soviet Space Program*). To quote his claim:

"Before a manned space launch with a new type of re-entry vehicle, or when heat shield materials were being replaced, several space flight tests were carried out. The first Soyuz re-entry vehicle underwent four tests prior to

ORIGINAL SOYUZ 1/SOYUZ 2 FLIGHT PLAN

DATE (1967)	TIME (GMT)	EVENT
23 Apr	00.35	Soyuz 1 launch
24 Apr	00.10	Soyuz 2 launch
	01.00	Docking, followed by crew transfer
	05.30	Undocking
25 Apr	00.01	Soyuz 1 recovery
	23.36	Soyuz 2 recovery

Notes: These data are based upon a first orbit rendezvous between Soyuz 1 and Soyuz 2, as accomplished on the subsequent unmanned Cosmos dockings, and restricting the duration of the Soyuz missions to a maximum of two days like the Cosmos 133 and Cosmos 140 missions. The Soyuz 1 launch time was announced and the Soyuz 2 launch time is based upon the subsequent missions which involved rendezvous/docking manoeuvres. The docking and undocking times are estimated, and should be correct to within ten minutes or so. The landing times should be within five minutes of the planned values, and depend upon the orbital manoeuvres planned for the mission.

Vladimir Komarov's flight on board the Soyuz 1. None of the tests was successful. During the first test flight the heat shield burned during descent. This was due to a defect in the stopper in the frontal shield, where the module had been mounted on the lathe for machining. The module was thoroughly damaged. The three other failures were due to a breakdown in the temperature control system, malfunctioning of the automatic controls of the altitude [attitude? – author] control jets and burning of the parachute lines (the parachute of the Soyuz re-entry vehicle was ejected by pyrotechnical cartridge). In those cases, the heat shield served well."

The four missions to which Yevsikov refers must be the two Soyuz tests – Cosmos 133 and Cosmos 140 – and the two Zond-related tests – Cosmos 146 and Cosmos 154 – which used Soyuz modules. Clearly, Soyuz was not ready to carry men, and it is surprising that the test programme was not slowed down as each unmanned test threw up new problems. However, as the long gap in manned missions extended, it seems probable that there were political pressures to fly a manned mission as soon as possible. Sergei Korolyov had died in January 1966, and the designers who followed him had not the authority over the politicians that Korolyov had commanded. Indeed, this is Yevsikov's view,

"Some launches were made almost exclusively for propaganda purposes. An example, timed to celebrate international solidarity, was the ill-fated flight of Vladimir Komarov in Soyuz 1 on 1 May 1967 [author's note: of course, this date is incorrect – Soyuz 1 was launched on 23 April]. The management of the Design Bureau knew that the vehicle had not been completely debugged: more time was needed to make it operational. But the Communist Party ordered the launch despite the fact that four preliminary launches had revealed faults in the co-ordination, thermal control and parachute systems. It was rumoured that Vasily Mishin, the deputy chief designer who headed the enterprise after Korolyov's death in 1966, had objected to the launch."

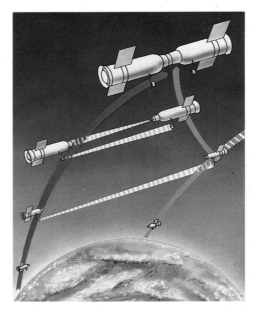

Cosmos Docking Sequence
This diagram is derived from a Soviet drawing of the Cosmos 186 docking with Cosmos 188 in 1967. The Soviets did not want to reveal the design of the Soyuz spacecraft (which had serious problems on its maiden manned flight, leading to Komarov's death), and therefore the true Soyuz design was airbrushed out of these pictures. Apparently the spacecraft were simple cylinders with a forward hemispherical (return ?) module. It was not until after the launch of Soyuz 3 a year later that the correct design of Soyuz was revealed, and undoctored pictures of the Cosmos dockings were released with the docking of Soyuz 4 and Soyuz 5.

The loss of Soyuz 1 stopped the Soviet manned space programme in its tracks in the same way that the Apollo 1 fire stopped the American programme. Curiously enough, both programmes began unmanned testing again within a month in the last quarter of 1967 and manned flights resumed in the same month in 1968.

Further Cosmos Tests

With the exception of the Zond-related propulsion test under the Cosmos 159 in May 1967, there was a break of six months in the Soviet manned space programme after the failure of Soyuz 1 before any further tests took place. On 27 October 1967, Cosmos 186 was launched; the announcement was given in the standard formula for the Cosmos satellites, the only anomaly being the inclusion of a launch time in the announcement. The orbital altitude was said to be 209-235km, suggestive

of an unmanned Soyuz mission. Some orbital manoeuvres took place during the first three days of the flight – the first time that a Soyuz had manoeuvred in orbit.

The launch of Cosmos 188 was announced on 30 October and the launch announcement stated that it had docked with Cosmos 186 at 09.20 – the first time that two unmanned satellites had docked in orbit (previous dockings had involved unmanned Agena targets with manned Gemini craft in the American programme). After Cosmos 188 had been placed in orbit – altitude 200-276km, as announced - Cosmos 186, as the active satellite of the pair, began a search for Cosmos 188, the docking actually occurring some 68 minutes after Cosmos 188 entered orbit.

The two satellites remained docked until 12.50, and after separation a phasing manoeuvre allowed the satellites to continue independent missions. Cosmos 186 was recovered on 31 October and Cosmos 188 remained in orbit until 2 November.

Television photographs of the docking and undocking were released, giving the Western public its first (limited) views of the Soyuz craft. These showed Soyuz to be generally cylindrical with a pair of solar cell "wings". Drawings of the satellites about to dock were released, and again these depicted Soyuz as cylindrical with a hemispherical front module. Later the same pictures were re-issued after the Soyuz 4/Soyuz 5 docking, and it was realised that the 1967 versions had been retouched to hide the true three-module Soyuz design. The correct design of Soyuz was not revealed until Soyuz 3 in October 1968.

A further unmanned docking experiment took place in April 1968. On 14 April Cosmos 212 was placed in an orbit with the announced altitudes 210-239km. Again, unusually, the launch announcement included a launch time. The following day Cosmos 213 was launched into a 205-291km (as announced) orbit, and at 10.21 the two spacecraft docked. Once more, it was the first satellite of the pair – this time, Cosmos 212 – which was the active satellite in the rendezvous and docking. The docking on this occasion took place only 47 minutes after the launch of the second satellite. The craft remained docked until 14.11 when the spacecraft separated to undertake independent missions. Both

SOYUZ MISSIONS, NOVEMBER 1966–AUGUST 1968									
LAUNCH DATE AND TIME	**RECOVERY DATE AND TIME**	**SATELLITE**	**CREW**	**MASS** (kg)	**ORBITAL EPOCH**	**INCL** (deg)	**PERIOD** (min)	**ALTITUDE** (km)	
1966 28 Nov 11.00*	30 Nov 10.21*	Cosmos 133	–	6,300 ?	**1966** 29.00 Nov	51.83	88.43	174-220	
1967 7 Feb 03.20*	9 Feb 02.52*	Cosmos 140	–	6,300 ?	**1967** 7.44 Feb	51.68	88.55	170-235	
23 Apr 00.35	24 Apr 03.23	Soyuz 1	V. M. Komarov	6,450	23.39 Apr	51.64	88.54	194-211	
27 Oct 09.30	31 Oct 08.20	Cosmos 186	–	6,400 ?	27.45 Oct 29.47 Oct 30.21 Oct 30.95 Oct	51.67 51.71 51.67 51.64	88.33 88.72 88.70 88.97	175-209 171-251 180-240 192-255	
30 Oct 08.12*	2 Nov 09.10*	Cosmos 188	–	6,400 ?	30.40 Oct 30.95 Oct	51.65 51.64	88.79 88.97	180-249 192-255	
1968 14 Apr 10.00	19 Apr 08.10*	Cosmos 212	–	6,400 ?	**1968** 15.15 Apr	51.64	88.32	178-205	
15 Apr 09.34	20 Apr 10.11*	Cosmos 213	–	6,400 ?	15.45 Apr 19.04 Apr	51.66 51.65	88.91 89.16	187-254 190-276	
28 Aug 09.59*	1 Sep 09.03*	Cosmos 238	–	6,450 ?	28.53 Aug 30.19 Aug	51.70 51.65	88.55 89.10	196-210 207-253	

Notes: The launch and recovery times are as announced by the Soviet Union, other than the cases marked '*': these figures are estimated, using the Two-Line Orbital Elements, and should be correct to plus/minus 2 minutes. Masses are estimated for the Cosmos flights, and should be correct to within 100kg. The orbital data are derived from the Two-Line Orbital Elements, rather than being those announced by the Soviet Union.

satellites returned to Earth after five days.

The way seemed clear for the resumption of manned Soyuz missions, but one more unmanned freight was first completed. Cosmos 238 was launched on 28 August 1968, this time with only the standard Cosmos launch announcement (altitude 199-219km). The satellite manoeuvred in orbit and was recovered after four days matching the Soyuz 3 duration which would follow. This mission might have been a simple final test before manned flights resumed, or it might have possibly been an intended unmanned Soyuz 2, which, when the manned Soyuz 3 mission was possibly delayed, resulted in the use of the Cosmos "cover" identity for the already-launched satellite.

Soyuz 2 and 3

The second manned phase of the Soyuz programme began in October 1968. On 25 October (three days after the return of the first manned Apollo mission) Western tracking stations picked up the launch of a new Soviet satellite. No Soviet announcement was immediately made, however. The following day, the launch of Soyuz 3 carrying Georgii T. Beregovoi was revealed, and the initial orbital altitude was announced as 205-225km. The launch announcement said the Soyuz 3 had carried out a rendezvous with the unmanned Soyuz 2 satellite. The two craft were brought to within 200m of each other, with Soyuz 3 being the active partner.

News of the launch of Soyuz 2 on the previous day was only revealed in the Soyuz 3 launch announcement, and one may speculate that if Soyuz 3 had been delayed, the 25 October launch would have become Cosmos 250 instead. The initial orbital altitude of Soyuz 2 was 185-224km, as announced.

On 27 October at 7.56 Soyuz 3 completed its sixteenth orbit. During the day a further rendezvous between Soyuz 2 and

Right: *The first view of the Soyuz launch vehicle came with the Soyuz 3 mission. Confirming western analyses, the Vostok booster was used, with a new upper stage.*

Below: *A television picture of Beregovoi aboard Soyuz 3 during his four-day mission. He provided a television tour around his spacecraft for viewers.*

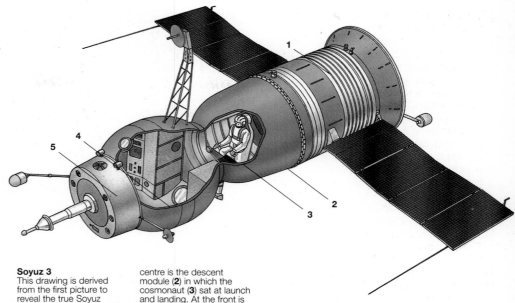

Soyuz 3
This drawing is derived from the first picture to reveal the true Soyuz design, depicting Soyuz 3. On the right is the instrument module (**1**) with the main propulsion system and solar panels for electrical power. In the centre is the descent module (**2**) in which the cosmonaut (**3**) sat at launch and landing. At the front is the orbital module (**4**) and the original Soyuz docking system (**5**) which was not designed to allow internal crew transfers between Soyuz spacecraft.

Soyuz 3 was completed. Following the rendezvous, the orbital altitudes were announced as: Soyuz 2, 181 – 231km; Soyuz 3, 179 – 252km.

The following day Soyuz 2 returned to Earth. At 07.25 the retro-rocket was ignited and by 7.51 the descent craft had entered the dense layers of the atmosphere and was back on the ground. The spacecraft had completed a ballistic re-entry, rather than a more controlled re-entry required to achieve lower deceleration overloads. Soyuz 3 continued to orbit alone.

During 28 October, Beregovoi undertook a series of Earth observations, noting three regions of forest fires and a thunder storm building up in the equatorial regions. A further orbital manoeuvre was completed, resulting in the announced parameters 199-244km. The next day Soviet observatories noted an increase in solar activity, but the resulting radiation increases were well within the safety limits set for space missions.

Regular television transmissions were returned to Earth, the Soviets possibly having learned a few lessons from the Apollo 7 mission a few weeks earlier. With the exception of Soyuz 1, live television pictures had been returned to Earth (although not publicly released "live"). With Soyuz 3 Beregovoi took the opportunity of giving viewers a conducted tour around the interior of Soyuz. They were shown both the orbital and descent modules (of course, the rear instrument module could not be seen), and Beregovoi is claimed to have said that although he did not need one, a spacesuit was carried onboard Soyuz. On 30 October, Soyuz 3 returned to Earth. After a retro-fire lasting for 145 seconds, the spacecraft landed near the town of Karaganda.

With Soyuz 3, the Soviet manned programme regained its confidence, and its success may have encouraged the Soviets to consider a manned flight around the Moon in December 1968, (see chapter 4). It is not known whether Soyuz 3 was intended to dock with Soyuz 2 or whether the simple rendezvous manoeuvres that they undertook were planned. If the former were to be true, then Soyuz 3 fell short of one of its flight objectives, but overall it represented a successful return to manned space missions after a break of eighteen months.

Docking in Orbit

Only 10 weeks after Soyuz 3, a new manned space mission was undertaken by the Soviet Union – the shortest gap between non-related manned space missions to that time. With the delays in the manned lunar programme proper, the testing of Soyuz in Earth orbit was evidently being given a high priority.

During the Soyuz 3 telecasts, it was clear that there were three seats in the Soyuz descent module, and it was therefore initially surprising to learn that Soyuz 4 only contained a single crew member. Soviet reports said that Soyuz 4 was "launched into orbit" at 10.39 Moscow Time, 07.39 GMT on 14 January 1969. However, the reception of radio transmissions by the Kettering Group and an analysis of the

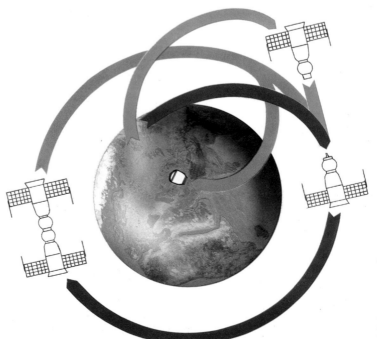

Soyuz 4 and 5 Docking Procedures
The illustration (left) is based upon a Soviet drawing showing the docking of Soyuz 4 with Soyuz 5. A day after Soyuz 4 was launched, Soyuz 5 was placed into orbit and a further day later the spacecraft began a mutual search and rendezvous operation. During this sequence, both spacecraft were manoeuvring, but with Soyuz 4 acting as the "active" spacecraft performing the major orbital changes to approach the "passive" Soyuz 5. This mission in January 1969 successfully completing the goals (in a revised form) which had been scheduled nearly two years earlier for accomplishment by Soyuz 1 with the un-launched Soyuz 2 manned mission. The Soyuz 1 mission was probably scheduled to dock after a day in orbit, rather than two days, like Soyuz 4 with Soyuz 5. The unmanned Cosmos docking missions certainly completed more rapid dockings than this.

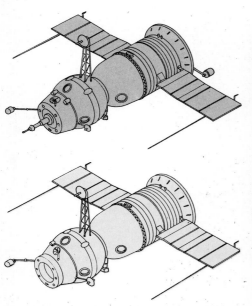

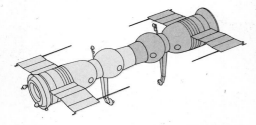

Soyuz 4 and Soyuz 5
Above are drawings of Soyuz 4 (with the docking probe) and Soyuz 5 (with the docking drogue). The spacecraft had launch masses of 6,626kg and 6,585kg respectively, but when docked the combined mass of the complex was 12,924kg, indicating that some 287kg of propellant had been used in the docking

manoeuvres. In the docked configuration (below) the only way to perform a transfer from one spacecraft to another was via a spacewalk. Although a cumbersome way of performing a transfer, it allowed the lunar surface spacesuit to be tested in orbit for the first time (although it was modified from the original Soyuz 1/Soyuz 2 spacesuits).

Left: *The launch of Soyuz 4 and Soyuz 5 in January 1969 marked the first winter launch in the Soviet manned space programme, suggesting that the flights had to be urgently completed.*

Two-Line Orbital Elements show that this is actually the orbital injection time: the actual launch time was 07.29, plus or minus one minute. Similarly, when Soyuz 5 was launched the following day the orbital injection time of 10.14 Moscow Time was announced (07.14 GMT); the launch time is calculated as 07.05 GMT.

Soyuz 4 was launched with Vladimir A. Shatalov on board, and the initial orbital altitude was announced as 173-225km. On the fifth orbit, a course correction was completed, resulting in a 207-237km orbit (as announced). This was in preparation for the launch of Soyuz 5.

Soyuz 5 was launched the following day, with a crew of Boris V. Volynov (commander), Aleksei S. Yeliseyev (flight engineer) and Yevgeni V. Khrunov (research engineer). Although not known at this time, Yeliseyev and Khrunov had previously trained for Soyuz 2 in 1967. The initial orbit was announced as 200-230km. After the launch of Soyuz 5, it was said that Shatalov and Volynov had confirmed that they had begun their programme of joint research. On its fifth orbit, Soyuz 5 com-

Above: *Soyuz 5 in the assembly shop (MIK) at the Tyuratam launch site. Clever photography gives the impression that, compared with a man, the spacecraft is much larger than it actually is.*

Below: *The first Soviet photograph of a manned spacecraft in orbit. This shows Soyuz 4 manoeuvring close to Soyuz 5 (from where the picture was taken) prior to the final docking manoeuvre.*

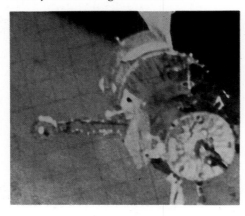

pleted a course correction, and the new orbit was announced as 211-253km.

On the two unmanned dockings and the Soyuz 2/Soyuz 3 mission, the initial rendezvous had taken place during the first orbit of the second craft. However, on this and future missions docking manoeuvres would be conducted at a more leisurely pace. On its 32nd orbit, Soyuz 4 manoeuvred again, into a new 201-253km orbit and on 16 January at 08.20 Soyuz 4, acting as the active spacecraft, docked with the passive Soyuz 5. The automatic rendezvous had begun at 07.37 and the distance between the craft was reduced to 100m. A "soft" docking was achieved first, followed by a "hard" docking as electrical connections were completed between the two spacecraft. In an exaggerated claim the Soyuz 4/Soyuz 5 complex was announced as the world's "First Experimental Space Station", and the orbital altitudes were announced as 209-250km. The mass of the docked spacecraft complex was 12,924kg.

Once the docking had been accomplished, the next step in the mission was being prepared. Khrunov and Yeliseyev, with the help of Volynov, donned space-suits in preparation for the first transfer from one spacecraft to another. Volynov retreated back into the Soyuz 5 descent module and shut the connecting hatch to the orbital module. During the 35th orbit of Soyuz 4 while over South America, Khrunov climbed outside Soyuz 5 and crawled over to and entered the Soyuz 4 orbital module, which Shatalov had de-pressurised ready for the transfer. As the complex flew over the Soviet Union Yeliseyev then climbed outside Soyuz 5 and transferred to Soyuz 4. The whole transfer operation took about an hour.

Once Khrunov and Yeliseyev were inside Soyuz 4, they re-pressurised the orbital module, took off their spacesuits and transferred to the descent module, joining Shatalov. Immediately, preparations for the undocking of the two Soyuz craft were begun. Undocking came at 12.55 and the spacecraft continued in independent flight.

Soyuz 4, now with three cosmonauts, returned to Earth on 17 January, after a flight of three days. Later that day, Soyuz 5 completed a manoeuvre, changing the orbital altitude to 201-229km (as announced), and on 18 January it also returned to Earth.

Some two years later than the Soyuz flight schedule had intended, the first docking of manned Soyuz craft and the partial transfer of a crew from one craft to another had been accomplished. Although the mission had been carried out much as Western analysts had anticipated, in retrospect some anomalies emerged which are discussed below. After the joint Soyuz 4/Soyuz 5 flight, further missions were expected to follow quickly, but no more launches took place until October 1969. It is now believed that the summer months were given over to the preparation of equipment for a manned lunar mission rehearsal as recounted in the previous chapter.

Mission Anomalies

Two aspects of the Soyuz 4/Soyuz 5 mission deserve closer attention: the Soyuz 5 spacesuit designs and the Soyuz docking system. Photographs of the crew transfer from Soyuz 5 to Soyuz 4 showed that the cosmonauts were not wearing large back-packs, but had their air supplies strapped to their legs. In itself, this would be strange,

but it also shows a design change from the original Soyuz 1/Soyuz 2 mission, planned two years earlier. Photographs released after the Soyuz 4/Soyuz 5 mission showed Yeliseyev and Kubasov in training during 1967 (with Komarov and Bykovsky) wearing spacesuits which incorporated back-packs with an air supply; nothing was strapped to the spacesuit legs. Therefore, why the change in design?

There is one possibility, which is admittedly speculative. The original Soyuz 1/Soyuz 2 design may have featured an orbital module entrance/exit hatch of a larger diameter, which would allow a cosmonaut with a bulky back-pack to climb outside, and for some (unknown) reason the diameter of the hatch may have later been reduced. As a result of this, there would not have been sufficient clearance for the back-pack as the cosmonaut climbed outside Soyuz, resulting in the repositioning of the back-pack contents on the cosmonaut's legs.

The second anomaly concerns the docking system. If the Soyuz spacecraft had always been intended for a docking and crew transfer, then an appropriate docking and internal transfer system would have been incorporated in the orbital module design from the start. As it is, the docking system used on Soyuz 4/Soyuz 5 was designed almost as if to prevent an internal

Above: *Few pictures of the crew transfer from Soyuz 5 to Soyuz 4 have been released. This television picture shows one of the cosmonauts performing the first Soviet spacewalk for four years.*

transfer. This anomaly has repercussions for any concepts for a manned lunar landing profile, and has therefore been noted in chapter 4.

Three Soyuz In Orbit

In the early 1970s it was speculated that Soyuz 6 had been slated for a solo mission in April 1969, but had been delayed until after the summer activity connected with the manned lunar effort. It was then combined with an already scheduled docking mission between Soyuz 7 and Soyuz 8. Whether this story is correct or not, the joint missions of Soyuz 6, Soyuz 7 and Soyuz 8 have raised many questions in the minds of Western observers.

On 11 October 1969 Soyuz 6 was launched with a two-man crew Georgi S. Shonin (commander) and Valerii N. Kubasov (flight engineer). The programme for the mission was announced as:

1 Overall checking and testing of the systems on-board and the improved design of the Soyuz spacecraft and rocket complex.

2 Further perfecting of the systems of manual control, orientation and stabilisation of the spacecraft during complex flight regimes and testing of autonomous navigation devices.

3 Taking a large number of scientific observations, photography of geolog-

Above: *Kubasov and Yeliseyev in the Soyuz assembly building with an orbital module behind them. This possibly dates from the Soyuz 4/Soyuz 5 period when Kubasov was Yeliseyev's back-up.*

Crew Transfer
Based upon NASA artwork, this diagram shows the crew transfer from the Soyuz 5 orbital module (centre left) to the Soyuz 4 module (centre right). Only one cosmonaut was completely outside the spacecraft at any one time. While Volynov remained inside the pressurised Soyuz 5 descent module (left) and Shatalov in the Soyuz 4 module (right), the orbital modules were de-pressurised and Khrunov transferred to Soyuz 4. Once he had transferred, Yeliseyev followed him across the docked spacecraft. The diagram clearly shows that there was no tunnel between the craft.

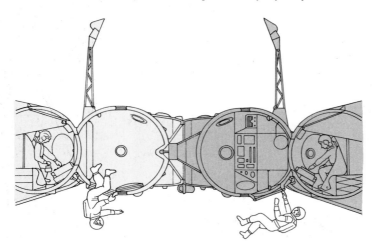

ical and geographical features on the Earth and investigation of the atmosphere for the purposes of national economy.

4 Scientific investigations of the physical characteristics of near-Earth space.

5 Conducting medico-biological investigations of the effects of spaceflight on the human organism.

6 Experimenting with methods of welding metals in high vacuum and in the state of weightlessness.

The initial orbital altitude was announced as 186-223km; on the fourth orbit a course correction was completed, but the orbital parameters were not announced.

The following day, 12 October, saw the launch of Soyuz 7 with three cosmonauts: Anatoli V. Filipchenko (commander), Vladislav N. Volkov (flight engineer) and Viktor V. Gorbatko (research engineer). A shorter list of programme objectives was announced for Soyuz 7:

1 Manoeuvring in orbit.

2 Navigational investigations jointly with Soyuz 6 in group flight.

3 Observation of heavenly bodies and the Earth's horizon, determination of actual brightness of the stars, measurements of illumination by the Sun and other scientific experiments.

While Soyuz 7 was on its second orbit, the orbits of it and Soyuz 6 were announced: Soyuz 6, 194−230km; Soyuz 7 207-226km.

The third Soyuz launch in three days came on 13 October, when Soyuz 8 was orbited with two cosmonauts on board. They were Shatalov and Yeliseyev, being re-cycled quickly after the Soyuz 4/Soyuz 5 mission. Once more, a list of objectives was announced for the flight:

1 Simultaneous complex scientific observations (with Soyuz 6 and Soyuz 7) in near-Earth space.

2 Final touches to the complicated system of control for the group flight of three spacecraft simultaneously.

3 Joint orbital manoeuvring to solve a number of problems connected with manned space flights.

The initial orbit of Soyuz 8 was announced as 205-223km. Shatalov was announced as the overall commander for the group flight.

Meanwhile, activity on Soyuz 6 and Soyuz 7 was continuing. On its 32nd orbit, Soyuz 6 completed an orbital manoeuvre, while the crew of Soyuz 7 were engaged in a series of experiments in Earth observations. Although they were to conduct joint work, the three craft had their own specific areas of scientific work to undertake: Soyuz 6, medico-biological investigations; Soyuz 7, photography of the Earth and sky in differing spectral bands; Soyuz 8, investigations of the polarisation of sunlight reflected by the atmosphere.

On 15 October – when a Soyuz 7/Soyuz 8 docking had been anticipated – all three spacecraft manoeuvred, approaching one another at differing distances. Soyuz 7 and Soyuz 8 seem to have made a close approach, and spent time photographing each other. The average orbits for the three

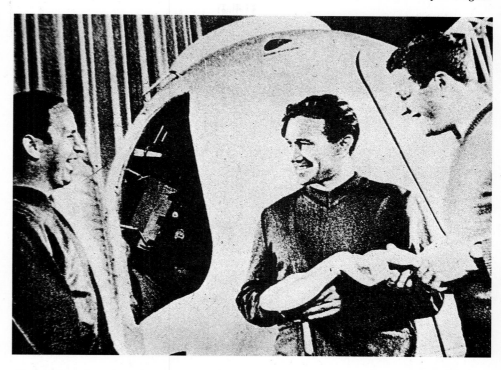

Above: *The three Soyuz 7 cosmonauts training for their mission with the "Volga" simulator. Soyuz 7 has never been confirmed as a docking mission, but Volga was used to simulate dockings.*

Below: *Kubasov on the ground, training with the Vulkan smelting furnace. On Soyuz 6, Vulkan was used for the first welding experiments in orbit just before the Soyuz 6 recovery.*

spacecraft were announced as 51.7 deg, 200-225km.

16 October was the final day in orbit for Soyuz 6. During the spacecraft's 77th orbit the orbital module was depressurised and a series of welding experiments took place, using the Vulkan equipment (mass, about 50kg), operated by Kubasov in the descent module. After the orbital module had been re-pressurised, several samples of welded material were taken into the descent module for return to Earth. The return of Soyuz 6 was uneventful. During the mission Soyuz 6 had spent 4h 24m co-orbiting with Soyuz 7 and Soyuz 8.

The flights of Soyuz 7 and Soyuz 8 continued for a further period; two manoeuvres were completed by Soyuz 8. The recovery of Soyuz 7 came on 17 October, after it had spent 35h 20m – co-orbiting with Soyuz 8 but – surprisingly – not having completed a docking. Finally, Soyuz 8 returned to Earth on 18 October.

It was said at the time of launch that Soyuz 6 had no docking system, but no such claims were made for Soyuz 7 and Soyuz 8, and in fact there is some photographic evidence which suggests that Soyuz 7 and Soyuz 8 were intended to dock in orbit. If they carried docking equipment, then a docking mission would be a logical inference. The actual reason for a triple Soyuz mission is unclear, unless we were simply witnessing two independent missions (a delayed solo Soyuz 6 and joint Soyuz 7/Soyuz 8) flown together.

It was at the time of this mission that a Soviet spokesman announced that the Soviet Union no longer had plans for manned lunar missions, and would be concentrating on manned orbital stations in Earth orbit. Therefore, the triple mission may be seen as a measure which filled a

Right: *The American Apollo 11 astronauts, Aldrin and Armstrong, with Sevastyanov and Nikolayev who flew on Soyuz 9. Armstrong was in the Soviet Union when Soyuz 9 was launched.*

gap in the flight programme, while missions were being re-directed. Certainly, the next flight was to be directly connected with a space station programme.

A Long Solo Mission

The flight of Soyuz 9 was to mark the end of the first major phase of the Soyuz programme. With the exception of a small programme in the mid-1970s, all future manned missions would be launched in the orbital station programme. Soyuz 9 was launched on 1 June 1970, while Neil Armstrong – the first man to set foot on the Moon – was visiting the Soviet Union. The crew comprised A.G. Nikolayev (who had flown on Vostok 3) and Vitali Sevastyanov. The main goal of the mission was to investigate the effects of a long period of weightlessness on the human body. At that time, the longest manned mission had been the American Gemini 7, which had clocked up 14 days in orbit. Soyuz 9 was to exceed that record by a substantial margin of over three and a half days or 25 per cent.

The initial Soyuz 9 orbit was announced as 207-220km, but on the fourteenth orbit a manoeuvre was completed, raising the orbit to 213-267km: a further three orbits later a second manoeuvre resulted in a 247-266km orbit – high enough to prevent orbital decay without the need for regular small manoeuvres.

Although the cosmonauts were flying a primarily biological mission, they also undertook research into observing and photographing the Earth and the sky. In order to test the accuracy of new equipment designed for orbital determination, on 5 June, between the 49th and 52nd circuits, ground stations measured the orbit as accurately as they could. According to these measurements the specification for the 50th circuit was:

Inclination	51.722°
Orbital Period	89,398 min
Perigee	241.638km
Apogee	261.064km

SOYUZ MISSIONS, OCTOBER 1968–JUNE 1970								
LAUNCH DATE AND TIME	**RECOVERY DATE AND TIME**	**SATELLITE**	**CREW**	**MASS** (kg)	**ORBITAL EPOCH**	**INCL** (deg)	**PERIOD** (min)	**ALTITUDE** (km)
1968 25 Oct 09.00*	28 Oct 07.51*	Soyuz 2	–	6,450 ?	27.33 Oct 27.82 Oct	51.67 51.66	88.23 88.63	177-196 184-230
26 Oct 08.34	30 Oct 07.25	Soyuz 3	G. T. Beregovoi	6,575	26.78 Oct 28.38 Oct 28.68 Oct	51.65 51.66 51.68	88.42 88.55 88.86	186-206 173-232 205-231
1969 14 Jan 07.29*	17 Jan 06.53	Soyuz 4	V. A. Shatalov	6,626	14.55 Jan 15.17 Jan	51.73 51.73	88.26 88.77	161-215 206-222
15 Jan 07.05*	18 Jan 08.00	Soyuz 5	B. V. Volynov Y. V. Khrunov A. S. Yeliseyev	6,585	15.53 Jan 15.72 Jan 16.77 Jan	51.67 51.69 51.68	88.60 88.94 88.88	189-222 211-234 201-237
11 Oct 11.10	16 Oct 09.52	Soyuz 6	G. S. Shonin V. N. Kubasov	6,577	11.52 Oct 11.95 Oct 13.06 Oct 14.54 Oct 15.71 Oct	51.70 51.69 51.68 51.69 51.65	88.42 88.74 88.80 88.64 89.07	169-224 192-232 201-230 203-212 215-242
12 Oct 10.45	17 Oct 09.26	Soyuz 7	A. V. Filipchenko V. V. Gorbatko V. N. Volkov	6,570	12.81 Oct 14.54 Oct 16.88 Oct	51.64 51.66 51.65	88.79 88.84 88.94	202-227 211-223 201-244
13 Oct 10.19	18 Oct 09.10	Soyuz 8	V. A. Shatalov A. S. Yeliseyev	6,646	15.95 Oct 16.75 Oct 17.37 Oct	51.63 51.63 51.65	88.79 88.70 88.90	201-227 208-212 204-235
1970 1 Jun 19.00	19 Jun 11.59	Soyuz 9	A. G. Nikolayev V. I. Sevastyanov	6,590	1.97 Jun 2.15 Jun 3.52 Jun 14.37 Jun 16.40 Jun	51.64 51.65 51.68 51.68 51.68	88.53 89.14 89.54 89.06 88.78	176-228 208-256 244-259 222-234 210-219

Notes: The launch times are as announced by the Soviet Union, other than for Soyuz 2, Soyuz 4 and Soyuz 5. Soyuz 2 was unmanned: for Soyuz 4 and Soyuz 5 the orbital injection times (07.39 and 07.14) were announced rather than launch times. All of the landing times are as announced by the Soviet Union. The orbital data are derived from the Two-Line Orbital Elements, rather than being those announced by the Soviet Union.

Above: *Nikolayev displays the chess board carried on Soyuz 9. The Soyuz crew played games of chess with mission controllers during their flight of 17.7 days, the longest flight to that time.*

Below: *The night launch of Soyuz 9 – the first night launch of a manned space mission. This was dictated by the mission duration and the requirement of landing in the afternoon.*

Experimental work in orbit continued throughout the flight, until, on 15 June, preparations began for the return to Earth. On orbit 208 a manoeuvre lowered the orbit to 215.1-231.4km. Recovery came on 19 June, with the descent module touching down after a flight lasting for 424h 58m 55s.

This flight clearly gave the Soviet Union useful information in preparation for the up-and-coming space station programme, although the results would be reduced in clarity because the Soyuz was spinning during its mission, and therefore the cosmonauts were being subjected to a form of artificial gravity. In fact, the constant spinning of the spacecraft resulted in the crew being in a very bad physical shape when they returned to Earth.

So ended the first period of the manned Soyuz programme, which at the time seemed a series of rather confused missions, whose overall objective it was difficult to determine. The role of Soyuz within the still-born manned lunar programme was not widely known in the West, and the Soyuz missions seemed to proceed at a painfully slow pace. In retrospect, it would seem that up to the middle of 1969 the Soyuz flights were simply the preliminary testing of the Soviet manned lunar craft, while the Soviets waited for their lunar booster to be ready for flight. With the loss of the first SL-15 booster in July 1969, Soyuz was partially re-directed towards an Earth orbital programme of basic research, although the lunar option was kept open until 1970 or 1971 – depending upon the flights of the SL-15 (which failed again). By the time Soyuz 9 flew in the summer of 1970, Soyuz had been committed to the role of ferry for a manned orbital station.

FURTHER READING

The press releases for the period October 1967 to December 1970 are contained in *Conquest of Outer Space in The USSR, 1967-1970* which was referenced at the end of chapter 4.

The following are useful sources for information connected with Soyuz 1.

Clark, P.S., "Soyuz Enters The Third Decade" in *Space*, Sep-Oct 1987, pp60-61, 63-64.

Gatland, K.W., "The Soviet Space Programme After Soyuz 1" in *Spaceflight*, vol 9 (1967), pp294-298.

Mandrovsky, B., *Soyuz 1 – Facts and Speculations: An Evaluation*, Aerospace Technology Division, Library of Congress, 16 May 1967.

Oberg, J.E., "Soyuz 1 Ten Years After: New Conclusions" in *Spaceflight*, vol 19 (1977), pp183-188.

Yevsiksov, V., *Re-Entry Technology and the Soviet Space Program (Some Personal Observations)*, Delphic Associates, December 1982, pp4, 59-60.

The following booklet is devoted to Soyuz 4/Soyuz 5, but includes some photographs from the original Soyuz 1/Soyuz 2 training sessions (uncredited to the earlier mission, of course!):

Transfer in Orbit, Novosti Press Agency, 1969.

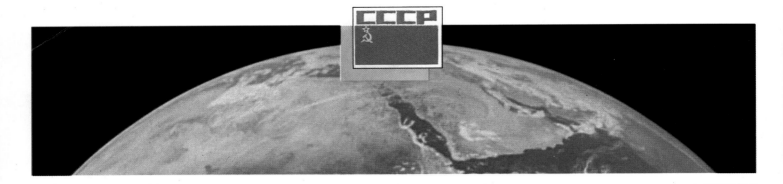

Chapter 6: **The First Space Station**

During the 1960s Soviet statements about the future of the manned space programme placed an almost equal emphasis on the establishment of a manned laboratory in Earth orbit and the project to fly men to the Moon. However, compared with the manned lunar programme, the space station programme seems to have had a low priority. One must wonder whether many of the Soviet statements were similar to plans by NASA for advanced manned missions during the late 1960s which were never any more than paper studies.

There are some reports that in the mid-1960s the Korolyov design bureau housed a full-scale mock-up of a large multi-storey space station, possibly similar in concept to the American Skylab which evolved from the Apollo Applications Programme. This station could have been a contender as a giant Type-G payload to be flown to Earth orbit, but the failure of the booster's flight programme meant that this laboratory never flew – if indeed it was actually a project funded for flights.

In 1969 the Soviet Union lost the "race" to place a man on the Moon. On 24 October 1969 Mstislav V. Keldysh – the President of the Soviet Academy Of Sciences and the "Chief Theoretician" of the Soviet space programme – admitted that "We [the Soviet Union] no longer have a timetable for manned Moon trips". He also revealed that the Soviet Union would be directing its space programme towards the establishment of a manned station in orbit around the Earth.

The go-ahead for the programme which would result in the launch of Salyut in April 1971 was given in late 1969 and construction of the station began in 1970. Since eighteen months is a very short interval between the initiation of design of a space station and launch, it seems possible that Salyut actually formed part of the manned lunar orbital programme, and design changes for Earth orbital operations were instigated in late 1969.

Salyut 1 Design

The essential design of Salyut 1 has been retained through the later evolutions: Salyut 4, Salyut 6, Salyut 7 and the Mir space station core. The engineers involved in the design were from the Korolyov bureau, and some of them subsequently flew missions to Salyuts. Bearing this in

Above: *The Salyut 1 orbital station in the assembly shop. Looking underneath the station, the hole on the right exposed the main experiments to space. On the left a pair of solar cells are folded, with the docking port clearly shown.*

mind, it is not surprising that some equipment from Soyuz was incorporated into Salyut 1.

The Soviet descriptions of Salyut 1 have to be read carefully, because the numerical information given often refers to the combined Salyut-Soyuz complex. The length of the two-spacecraft complex is normally given as 23m, although some Soviet sources suggest that it was 21.4m (it is possible that the latter figure excluded protruding antennae which are included in the former figure). Since Soyuz was about 7m long – excluding the protruding docking probe – the length of Salyut 1 would seem to be about 14.5m, excluding antennae. The mass of the complex was 25.6 tonnes, with Salyut comprising 18.9 tonnes and Soyuz 6.7 tonnes. Again, one Soviet source gives the Salyut 1 mass as 18.5 tonnes, but this might refer to the mass when Soyuz 11 docked, after propellant had been consumed by the station required to maintain its orbital altitude.

The basic configuration of Salyut 1 was that of four cylinders of differing diameters and lengths. Three were pressurised and the fourth which housed the propulsion system was open to the raw space environment. These four elements will be considered in detail next.

The Transfer Compartment

The Soyuz ferry craft was designed to dock at the front end of Salyut, on the transfer compartment. This cylinder was 2m in diameter and 3m long. At the front of the transfer compartment was situated the drogue half of the docking system; for the first time, the Soviets were flying a docking system which allowed the cosmonauts to transfer between two spacecraft without having to perform spacewalks.

At the opposite end of the transfer compartment was a hatch which led to the main work compartments. Some photographs suggest that a hatch cover opened into the work compartment, and since models of Salyut 1 showed a hatch similar to one used for spacewalks on Salyuts 6 and 7 on the side of the transfer compartment, it is possible that there was an internal hatch to separate the transfer and work compartments. If that were the case the transfer compartment could be depressurised and spacewalks carried out from it. One Soviet drawing even exists which shows a Salyut 1 design station without a Soyuz attached but with a cosmonaut performing a spacewalk. Whether this represents a hard plan or is simply a flight of the artist's imagination is unknown.

The transfer compartment was used to house some of the scientific equipment. These included the pressurised part of the Orion-1 telescope, cameras and biological experiments (the Salyut 1 experiments are detailed later). On the exterior of the transfer compartment were the external part of Orion-1 (in a hemispherical depression embedded in the transfer compartment); two solar panels; antennae for the rendez-vous radio system; optical lights for the station's orientation during the manual docking of a Soyuz ferry with Salyut; an external television camera; panels associated with the heat regulation system units; ion sensors for the Salyut orientation system; panels with sensors for the detection of micrometeorites.

The Work Compartments

The Salyut work compartments were split into three sections. After leaving the transfer compartment, the cosmonauts would enter a larger cylinder 2.9m in diameter and 3.8m long. This attached to a conical frustrum 1.2m long, leading to the largest compartment of the station: a cylinder 4.1m long and 4.15m in diameter.

Situated along the length of the work compartments were equipment compartments. The instrument and cable networks installed on the compartment frames were covered by removable panels. To assist the orientation of the cosmonauts, each surface of the work compartment area was painted a different colour. The front and rear were light grey, one side was apple green, the other was light yellow and the floor was dark grey.

In the smaller of the two work compartments was a table for eating, attached to which was a (replaceable) tank of water; the primary water storage was in tanks on the starboard and port sides. A pair of refrigerator systems were installed in the large diameter work area, one on the port side and the other on the starboard side.

The cosmonauts spent their spare time in the small diameter work area: here were housed a tape recorder and pre-recorded cassettes, a sketch pad and a small library. Sleeping bags were located port and starboard in the large work compartment, but if they so wished the cosmonauts could sleep in the orbital module of the Soyuz ferry craft. The sanitary and hygienic unit was aft, in the large work compartment: it was separated from the rest of the compartment, and had its own ventilators. The surface of the sanitation area was covered in washable material.

The size and shape of the large work compartment was very reminiscent of the Proton satellites, launched in 1965-1968. Exact dimensions are not available from Soviet sources, but the core of Proton 4 was about 4m in diameter and 3m long. One could speculate that the Proton 4 cylinder was modified for Salyut. The conical frustrum connecting the two main sections of the work compartments housed the equipment used by the cosmonauts for physical exercises and medical research.

The whole station was well-lit inside, with special lights being carried to assist during television transmissions to the ground control. The gyroscopes for the navigation system were installed in the forward part of the work compartment.

The Salyut Work Stations

In order for the cosmonauts to control Salyut systems and operate the various scientific experiments seven work stations were provided.

Number 1 station was the central element in this part of Salyut's design. The control of the on-board systems and – in part – the scientific equipment was concentrated here. The station was located in the lower part of the smaller work compartment, and was equipped with two chairs, panels for the automatic orientation and navigation of the laboratory and optical viewers for manual orientation.

Number 2 station was designated the "astropost" and was also located in the smaller work compartment. It was designed for working with the manual astro-orientation and astro-navigation systems. It also included a control handle for the orientation of Salyut, a means of holding the cosmonaut in his work position, and a viewing port.

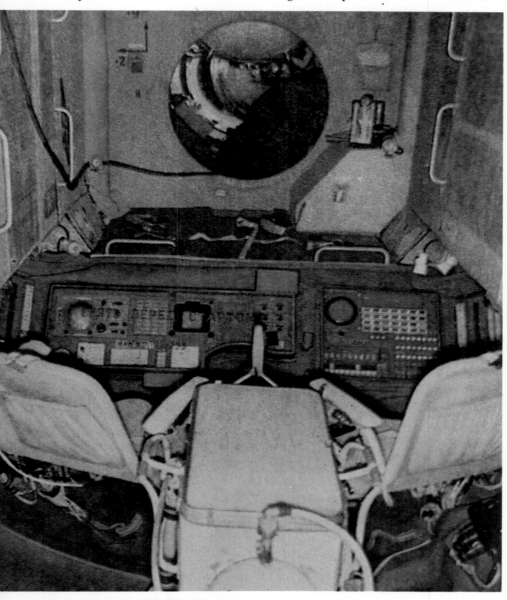

Above: *The view inside the first Salyut work compartment, looking towards the main control panel (basically an unmodified Soyuz panel) and through the hatch into the transfer compartment which leads to the Soyuz ferry.*

Left: *The right hand side of the Salyut control panel shown above. Although Salyut 1 housed three cosmonauts, there were seats for only two men at the control panel. The globe shows the position of Salyut over the Earth.*

Above: *A "still" taken from a Soviet film, showing the launch of Salyut 1 in the early hours of the morning, local time. No true photographs of the launch have been released by the Soviets.*

Left: *The four stage variant of the Proton launch vehicle. The upper fourth stage (slightly smaller diameter) was not carried on the three stage variant used to launch the Salyut stations.*

The Salyut 1 Orbital Station, with a Soyuz Ferry Docked
1 The familiar shape of a Soyuz ferry, being used for the first time without a toroidal propellant tank and with an internal transfer system.
2 One of four Salyut solar panels: two at the front and two at the rear.
3 Rendezvous antennas.
4 EVA/access hatch (on the hidden side of Salyut).
5 Orion stellar telescope.
6 Atmospheric regeneration system
7 Forward work compartment.

8 Movie camera.
9 Photographic camera.
10 Biological research equipment.
11 Food refrigeration unit.
12 Rear work compartment.
13 Attitude control engines.
14 Modified Soyuz propulsion system (designated KTDU-66).
15 Propellant tanks for KTDU-66.
16 Micrometeorite panel.
17 Not shown on Soviet diagrams, but in-flight photographs and films

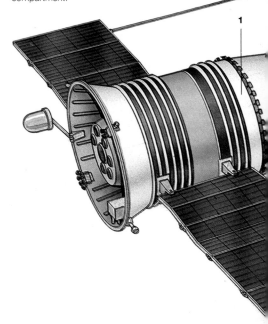

Number 3 station was designed to control scientific apparatus, and it was located in the central part of the large work compartment. It contained a control panel and viewing port.

Number 4 station was used for controlling not only the scientific experiments but also medical equipment, which was located in the frustrum connecting the two work compartments. The station comprised scientific experiments, a viewing port, a chair and the main medical research equipment.

Number 5 station was the position from which the Orion-1 telescope was controlled, and was located in the transfer compartment. It housed control panels, a viewing port and a sight and arm system for guiding the telescope.

Number 6 station was a second "astropost", virtually identical with Station 2 except that it included a chair and a shutter

was provided to convert it into a "warm room" (Soviet terminology). It was located on the port side of the small work compartment.

Number 7 station – the final one – was a control for the scientific apparatus studying "space about the Earth" - presumably the general low Earth orbit environment. It contained control panels for equipment, a viewing port and a means of anchoring a cosmonaut while he was working at the Station. It was located on the starboard side of Salyut's small work compartment, symmetrically with Station 6.

The Propulsion System

The fourth compartment of Salyut 1 was the only one that was inaccessible to the cosmonauts. Located at the rear of the station, it was a cylinder with a diameter of 2.2m and a length of 2.17m. It housed the Salyut propulsion system, which was a modification of the Soyuz KTDU-35. The Salyut engine - designated KTDU-66 – had been modified to include larger propellant tanks for the UDMH and nitric acid propellants which permitted a total burn time of 1,000 seconds, implying a propellant load of 1,490kg. The engine was recessed slightly within the rear cylinder, and sited within the recess were two sets of four small vernier engines for attitude control of the station. Further sets of verniers were mounted outside, on the rim of the cylinder.

The rear compartment had a second set of Soyuz solar cells mounted for the power supply. The Soviets normally quote a total solar cell area of 42m^2, but this includes the set of cells on the Soyuz ferry; the solar cells on Salyut 1 alone had an area of 28m^2.

Salyut 1 Experiments

Salyut 1 carried more than 1,300 separate instruments into orbit, the scientific payload being in excess of 1,200kg. Details of a number of specific pieces of scientific equipment have been released, and these are summarised below.

The Orion-1 telescope was the main astrophysical experiment carried on Salyut. It comprised two mirrors, one 280mm in diameter and the other 50mm in diameter. The instrument was designed to obtain spectrograms of stars within the range 2,000-3,000 Angstroem units (one Angstroem unit is one ten-millionth of a millimetre, and is the unit in which stellar spectra are measured). The mirrors were covered with aluminum, which would

allow the cosmonauts to re-surface them if they were damaged by the impact of micro-meteoritic material. Together with a spectroscope, the Orion-1 was able to obtain a resolution of 5 Angstroem units at an operating range of about 2,600 Angstroem units. The tracking system allowed the telescope to retain its orientation to within one second of an arc (1/3,600 of a degree). In order to operate Orion, one cosmonaut had to control the orientation of Salyut, while another guided Orion. The cosmonauts had to work the system quickly because on each orbit there was only a 30-35 minute period, during which the observations could be taken - while Salyut was in the Earth's shadow. A mechanical arm and airlock system allowed the cosmonauts to replace film cassettes in the telescope system.

Anna-3 was the gamma radiation telescope carried by Salyut, and it was able to detect radiation with energies in excess of 100Mev (mega-electron volts) with a positional resolution of one degree. This capability represented virtually a new area of astronomy, because gamma radiation is normally absorbed by the atmosphere, and therefore cannot be studied from ground level. Anna-3 consisted of Cherenkov counters for measuring gamma radiation, and was 600mm by 400mm in size with a mass of 45kg. The Soviets announced that the programme for Anna-3 was: to perform detailed studies of the telescope's basic fitness; to discover the possibilities of investigating gamma quanta with different orientations of the Salyut station; and to

determine the "background noise" in gamma ray astronomy – that is, the level of charged particles coming from outside and inside the station.

An FEK-7 photoemulsion camera was also carried on Salyut 1, the purpose of which was to study the multiple-charged particles of primary cosmic rays. Specifically, the camera was designed to search for the Dirac monopole, anti-nuclei and trans-uranium nuclei – the discovery of which elsewhere in the Universe would have had important implications for particle physics. Similar cameras had been flown on Cosmos 213, Zonds 5, 7 and 8 and Soyuz 5, but these had operated for far shorter periods than did the one on Salyut.

A number of sensors were mounted on the outside of Salyut to detect micrometeorites; they were situated on the propulsion system and the larger of the two work compartments. Although the chances of a significant micrometeorite hitting Salyut were very slim, as missions became longer and spacecraft got bigger, so the chances of a hit would increase.

Salyut carried several cameras which could operate in the optical or multi-spectral modes, and these were used for geological, meteorological, oceanographical and astronomical studies. Some joint work was planned between Salyut and the polar-orbiting Meteor-1 weather satellites.

Since Salyut 1 was planned to be the base for relatively long space missions (by 1971 standards, more than two weeks was a long mission), great emphasis was placed on medical experiments. Salyut carried a

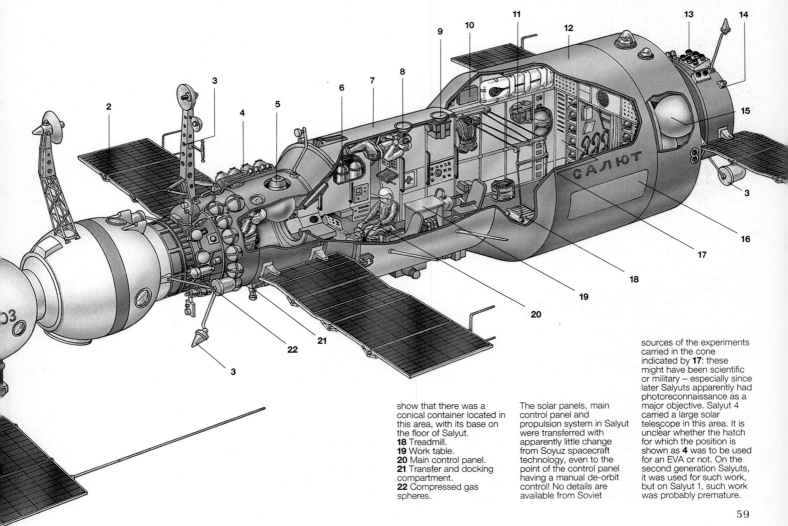

show that there was a conical container located in this area, with its base on the floor of Salyut.
18 Treadmill.
19 Work table.
20 Main control panel.
21 Transfer and docking compartment.
22 Compressed gas spheres.

The solar panels, main control panel and propulsion system in Salyut were transferred with apparently little change from Soyuz spacecraft technology, even to the point of the control panel having a manual de-orbit control! No details are available from Soviet

sources of the experiments carried in the cone indicated by **17**: these might have been scientific or military – especially since later Salyuts apparently had photoreconnaissance as a major objective. Salyut 4 carried a large solar telescope in this area. It is unclear whether the hatch for which the position is shown as **4** was to be used for an EVA or not. On the second generation Salyuts, it was used for such work, but on Salyut 1, such work was probably premature.

number of items which were to be used to keep the visiting cosmonauts physically fit, because in its weightless environment the men's leg and heart muscles were expected to waste away. "Penguin" suits were carried which forced the cosmonauts wearing them to use their muscles to prevent the suit from folding up into a foetal position which was its natural tendency. A treadmill was installed in the frustrum area, connecting the two work compartments, again to provide muscular exercise. The cosmonauts also used a Chibis unit, into which they placed their legs; the pressure within the unit was reduced, and this forced blood to circulate in the lower body, something which was not expected to happen properly in zero gravity conditions. Tests were performed regularly by the cosmonauts on their blood, exhaled air and vision.

Non-human biological experiments were also carried out on Salyut 1. The Soyuz 11 crew took fertilized frog eggs into orbit, and their development in orbit was monitored. The growth of small plants and insects and the genetic changes to the fruit fly (*drosophila*) were studied on Salyut. A small hydroponic farm called Oazis-1 was used to study the growth of high-order plants in orbit.

Other more minor experiments involved the determination of charged particle fluxes with a Cherenkov scintillation telescope and the study of high-frequency secondary-electron resonance and the ionospheric structure using an "Era" system.

Modifications to Soyuz

The Soviets state that as Salyut 1 was being constructed in 1970, the Soyuz spacecraft was being modified (from what intended mission is not stated!) to act as a ferry to orbital stations. There were only two flights of the original Salyut ferry, Soyuz 10 and Soyuz 11 in 1971. Outwardly, the Soyuz appeared to be the same as the vehicle that had flown previously, but a close look showed some modifications. The mass had increased slightly from a maximum of 6,646kg for the Soyuz 1-9

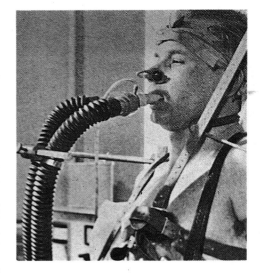

Above: *Training for the Salyut mission, a cosmonaut has his exhaled breath analysed while he exercises on a Salyut-type treadmill.*

series to 6,790kg for Soyuz 11 (the mass of Soyuz 10 was not announced). The toroidal propellant tank had been deleted, resulting in the propellant capacity being reduced from about 755kg to 500kg. The same basic KTDU-35 engine system was retained for missions.

The most important change in the Soyuz design was the installation of an internal transfer system with the docking probe unit. This allowed the Soyuz crew to transfer to Salyut without having to resort to a spacewalk between the craft, as would have been dictated by the original Soyuz design.

Salyut 1 was configured for operations with three men, and therefore one can assume that the standard Soyuz crew size would be three men, rather than one, two or three as had flown earlier Soyuz missions. There were no unmanned tests of the new Soyuz before manned operations began in April 1971.

Although few details are available, it

would seem that a number of Salyut mock-ups and mass models were assembled in support of ground testing prior to orbital operations beginning. No photographs of the Salyut 1 cosmonauts inside a Salyut trainer have been released, although such a trainer was surely used. Later Salyut cosmonauts were photographed with full-scale Salyut trainers, even though it is believed that some crews did not train in the Salyut with which they are shown.

Preparations for Flight

The cosmonauts trained for the Salyut docking operations using a system called Volga. When seen by American journalists in 1973, there were two complete Volga trainers. The cosmonauts would sit in mock-ups of Soyuz static descent craft with a full set of Soyuz controls. Outside were moving models of Soyuz and Salyut on rails, and these would respond to the cosmonaut control instructions: televisions pictures of the Soyuz model's view of the docking were supplied to the Soyuz descent craft periscope system to give as realistic a view as possible of docking with a Salyut. The Volga trainers were last seen in pre-flight training photographs of the cosmonauts launched to Salyut 3 in 1974.

Although still photographs have not been published by the Soviet Union, tests of a Salyut mock-up were shown in the Salyut 1 film "A Steep Road Into Space". The sequences showed the mockup surrounded by netting and illustrated how the payload shroud covering the small work compartment and the transfer compartment separated during a launch. Other photographs showed Salyut being assembled at Tyuratam using the same horizontal system used for Soyuz and even Vostok preparation. Interestingly, these photographs all show the side of Salyut which did not have the supposed EVA hatch. Spheres containing oxygen were mounted on the exterior of Salyut, suggesting the possibility of EVA work and the requirement to replenish air lost during a depressurisation.

In April 1971 the Soviet Union was celebrating the tenth anniversary of Yuri Gagarin's flight, and on 19 April Salyut was launched into orbit. A brief glimpse of the upper part of the three-stage Proton SL-13 launch vehicle was shown in "A Steep Road Into Space", but the lower part of the vehicle was not clearly shown. Launch came at about 01.39, and the Soviet Union announced an orbit of 200-222km. Salyut remained in a low orbit until the launch of Soyuz 10 nearly four days later.

The Cosmonauts

A group of nine cosmonauts was initially picked to train for missions to Salyut 1, with two visits being planned. It is possible that after the failure of Soyuz 10, a third launch would have followed if Soyuz 11 had been successful.

Nine cosmonauts trained for the Salyut 1 missions, although initially only eight were named and the ninth man was only shown in photographs. The originally-known men were G.T. Dobrovolsky, V.N. Kubasov, A.A. Leonov, V.I. Patsayev, N.N. Rukavishnikov, V.A. Shatalov, V.N. Vol-

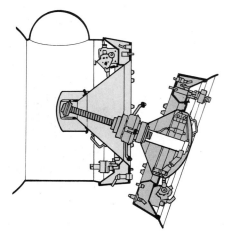

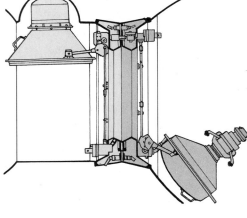

Crew Transfer System
First discussed with the Americans during ASTP talks in 1970, the Salyut/Soyuz internal transfer system was not flown for the first time until 1971. When the initial docking is made, there is quite a large degree of freedom for the probe on entry into the drogue of the Salyut docking system (left). After the initial docking contact, the spacecraft are aligned for a "soft" docking and then the Soyuz probe retracts to give a "hard" docking, with air-tight seals and electrical connections being made. The cosmonauts make a remote check of the atmosphere in Salyut, to ensure that the life support systems are working properly. This done, the Soyuz hatch is swung back into the Soyuz orbital module, and the Salyut hatch is swung forward with the cone slotting into a recess in the Salyut wall (right). The way is now clear for the cosmonauts to transfer from Soyuz to Salyut and begin operations in the station. This system is still in use on the Mir orbital complex.

Above: *The cosmonauts involved in the Salyut 1 mission. From left to right they are Volkov, Leonov (hidden), Rukavishnikov (on globe), Kubasov, Shatalov, Kolodin and Dobrovolsky.*

kov and A.S. Yeliseyev. In 1987 the name of the ninth man, Petr Ivanovich Kolodin was confirmed by the Soviets; he was a military engineer.

The crew assignments for Salyut 1 have been complicated, but the original groupings seem to have been:

Soyuz 10	Shatalov	Yeliseyev	Rukavishnikov
Soyuz 11	Leonov	Kubasov	Kolodin
Back-ups	Dobrovolsky	Volkov	Patsayev

However, Kubasov was found to have a lung problem and was grounded a week prior to the Soyuz 11 launch. Originally the Soviets contemplated simply replacing him with Volkov (his backup), but four days before the planned launch it was decided to replace the entire crew with the back-up team. The Shatalov crew - who had flown Soyuz 10 – were re-cycled in the training system as the new back-up team for Soyuz 11.

Kolodin was none too pleased about being passed over with Leonov and spent the days prior to the Soyuz 11 launch arguing with the mission managers, trying to get the decision reversed, but to no avail.

When Leonov and Kubasov were chosen in 1973 for the Apollo-Soyuz mission two years later, photographs of the cosmonauts in training were released, and among these were pictures of their Salyut 1 training (although they were not identified as such by the Soviets). Most photographs showed them singly training on the same equip-

Below: *Released at the time of ASTP, this picture shows Kubasov and Leonov in a 3-man Soyuz simulator, probably training for Salyut 1. The third man is not identified but might be Patsayev.*

ment and in similar situations as the flown Soyuz 11 crew: only one photograph showed them in a three-manned Soyuz trainer, and that showed the forehead of a third man, the identity of whom was the subject of much speculation among observers of the Soviet space programme. This man may have been Patsayev, although why he would have been training with Leonov and Kubasov is unclear. Pictures of Kolodin are available with the other Salyut 1 cosmonauts next to a terrestrial globe; one can speculate that his behaviour in arguing about the rescheduling of the Soyuz 11 crews meant that he was not assigned to any further space missions.

Soyuz 10: A Failed Mission

On 22 April at 23.54 Soyuz 10 was launched with Shatalov, Yeliseyev and Rukavishnikov. The first two men were on their third space missions, while Rukavishnikov – designated the Salyut research engineer – was on his first. Soyuz 10 was in a 208-246km orbit, which was higher than planned. At 10.55 on 23 April Shatalov manoeuvred the spacecraft, ready for the docking manoeuvres. Approximately a day after launch, the automatic rendezvous and docking system began to operate, with both Soyuz and the Salyut undertaking small manoeuvres and attitude changes. The automatic system brought the spacecraft to within 180m of one another, at which time Shatalov took over control of his craft. At 01.47 on 24 April Soyuz 10 docked with Salyut 1, and the Soviets were ready to begin manned operations with the world's first orbital space station.

However, this was not to be. Although the actual docking seems to have been completed without any problems, the crew were unable to enter Salyut. The reasons for this have never been clear. One rumour from an unknown source (as are many rumours connected with the Soviet space programme) suggested that the crew were unable to equalise the pressure in the

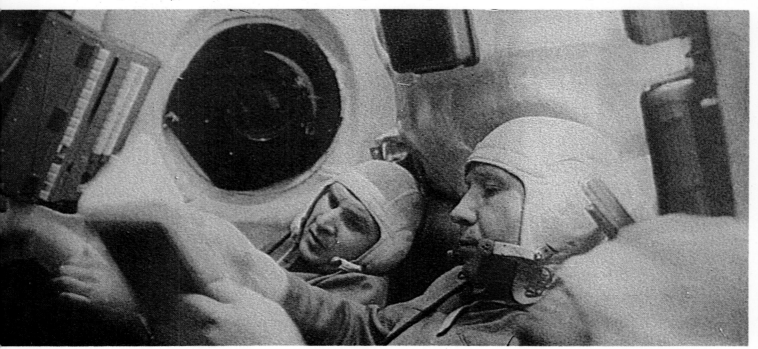

Soyuz Landing Windows

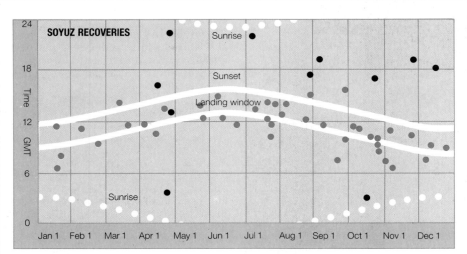

Soyuz landing windows will be an essential concept in the descriptions of missions to Salyut orbital stations, and they are described here.

The diagram above shows the landing times of all of the Soyuz and Soyuz-T missions to the end of 1986 plotted by landing date. Although there is some scatter with this data, a number of missions are believed to have been terminated earlier than intended and therefore they distort the overall picture. The blue dots represent the missions which terminated (or are believed to have terminated) earlier than required by the missions' flight plans, while the red dots show the remainder: the missions which flew for their planned durations and were recovered as planned.

The diagram also shows the times of sunrise and sunset at the normal Kazakhstan landing site by the sinusoidal curves, and for convenience parallel with the sunset line another line represents the time: sunset minus three hours. It is clear from the configuration of the red dots that a successful mission is ideally recovered before sunset and not usually significantly earlier than three hours before sunset. The majority of missions returned to Earth as the satellite was passing northbound over the landing site, but an occasional recovery has been made on a southbound trajectory (Soyuz 1, Soyuz-T 15).

From Soviet documentation, more exact limits on the landing times can be imposed. The Soviets say that their landings are governed by two conditions which ideally will be simultaneously satisfied: one is that the landing is in daylight and the other is that in-orbit retro-fire must take place in sunlight (for manual orientation of the spacecraft). The latter condition means that the lower limit on the landing time is not strictly speaking three hours before sunset, but the approximation presented here to try and define landing opportunities will not be greatly in error.

Up to now, Salyut has not been considered in connection with the landing opportunities, and the second diagram illustrates the result of adding the Salyut restrictions. Because the Earth is rotating underneath the Salyut orbit, a launch to Salyut can take place only when the orbital plane of Salyut is passing over the launch site at Tyuratam. Similarly, the recovery

Above: This graph shows how Soyuz landing times relate to the time of sunset. A successful mission is recovered in the 3 hours or so before sunset at the landing site. Unsuccessful missions do not conform.

Below: The sloping lines indicate launch times. Landing opportunities come as the launch line passes within the zone defined by sunset and sunset minus 3 hours. The graph allows mission durations to be estimated.

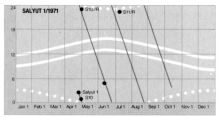

from a Salyut visit can only take place at a specific time each day as the orbit of Salyut passes over the landing zone: this is either the orbit of a launch opportunity or the following orbit.

This model is complicated because the orbital plane of Salyut rotates about the Earth (technically, it "precesses") and as a result launch and landing times come earlier as time goes on. On an average, the orbital plane of a Salyut rotates about the Earth after nearly 60 days, and therefore the launch times repeat after approximately two months. The sloping straight lines show for each date the nominal launch time to Salyut 1 during April-October 1971. A "landing window" covers the period that the landing times imposed by the Salyut orbital plane coincide with orbital passes over the landing zone between sunset and three hours before sunset.

In 1971 two landing windows were about 15-22 May and 14-22 July. The former would seem to be the planned landing period for Soyuz 10 (giving a planned mission duration of 22-29 days) and the latter was probably to be used for Soyuz 11 (planned duration 38-46 days). It therefore seems probable that both of these flights were terminated earlier than planned.

In the later Salyut and Mir chapters similar diagrams will be shown for each Salyut mission (or major residency for Salyuts 6 and 7 and the Mir complex), and these figures can provide the rationale for working out the planning of manned space launches.

Soyuz orbital module and the Salyut transfer compartment.

According to Soviet accounts, Soyuz 10 remained docked with Salyut 1 until 07.17 and then they separated from the station, their long mission abandoned. However, Part 2 of the Library Of Congress report "Soviet Space Programs, 1976-1980" indicates in Table 15 that on 24 April the docking and undocking events were:

First docking	01.47 GMT
First undocking	04.18 GMT
Second docking	05.47 GMT
Second undocking	07.17 GMT

There is no other unclassified source of this information. Soyuz 10 completed a fly-around of Salyut 1 before preparing for a premature recovery. This was to be the first night landing in the Soviet space programme involving cosmonauts, and Soyuz 10 touched down at 23.40 on 24 April.

From a consideration of the Salyut 1 landing windows, it seems clear that Soyuz 10 was slated for a mission lasting 22-29 days. If it had been achieved it would have been a record duration for a manned space mission. Bearing in mind the actual launch and the projected landing opportunities for Soyuz 11, the mission duration was pos-

Below: *An aerial view of the Soyuz 11 launch vehicle on the launch pad, prior to its flight on 6 June 1971. This was the last use of the shroud tower design that was first carried by Soyuz 1.*

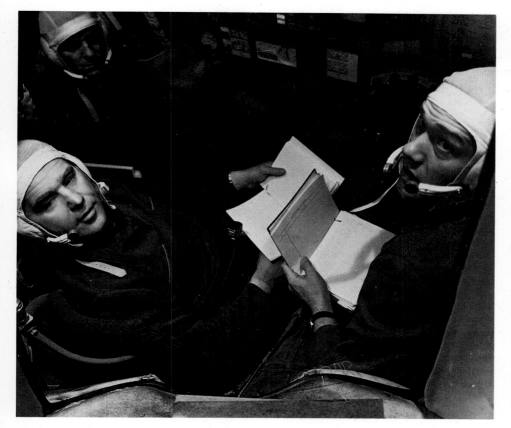

low would mean that Salyut would decay from orbit within days and be destroyed on re-entry into the atmosphere. Two days later Salyut had manoeuvred itself into a high, more circular orbit with an altitude of 253-276km.

Soyuz 11: Men on Board

The orbit of Salyut 1 slowly reduced in altitude due to the action of the Earth's atmosphere, and at the beginning of June it had dropped to nearly 200km circular. The Soviet Union was preparing for a new attempt to man Salyut, and in advance of the Soyuz launch Salyut was manoeuvred during the period 1 June to 6 June into a higher 210-235km orbit.

At 04.55 on 6 June, Soyuz 11 was launched into orbit with its crew: Dobrovolskys, Volkov and Patsayev (the latter having been selected for cosmonaut training specifically for this mission). Soyuz 11 entered orbit around the Earth at 05.04. At 10.50 the spacecraft corrected its orbit, resulting in an announced altitude of 185-217km.

On 7 June at 07.24 the rendezvous operations between Soyuz 11 and Salyut 1 began and the cosmonauts were able to see the orbital station through the periscope extending below the Soyuz descent module. At a distance of 150m the two spacecraft were aligned ready for the docking operations, and at a distance of 100m and closing at 0.9m/s Dobrovolsky took control of Soyuz. Initially the station seemed to drift to the right, but Dobrovolsky compensated for this; at a distance of

Above: *The crew of Soyuz 10 in the flight simulator. Left to right, they are Rukavishnikov, Shatalov and Yeliseyev, the third joint flight for the Shatalov-Yeliseyev team.*

sibly going to be closer to the upper Soyuz 10 limit.

After the docking between Salyut 1 and Soyuz 10, the complex circled the Earth in a 184-240km orbit. However, an orbit this

EXPERIMENTAL WORK COMPLETED ON SALYUT 1

DATE (1971)	ACTIVITIES AND NOTES	DATE (1971)	ACTIVITIES AND NOTES
7 June	1 Soyuz 11 docks with Salyut and manned operations begin. 2 Orbit of complex: 212-249km. 3 Communications maintained using the *Academician Sergei Korolyov* tracking ship and a Molniya-1 satellite.	17 June	1 Studies of high-energy electron resonance and high energy plasma. 2 Physical exercises and medical tests.
		18 June	1 Operations with Orion-1, with stella spectra being taken.
8 June	1 Checking and preparation of the scientific instruments. 2 Orbital correction at 08.02: orbit – 239-265km.	19 June	1 Spectrographic measurements of the horizon. 2 Checks of the Salyut gyroscope system. 3 Cardiovascular studies. 4 Investigations into the sensitivity of the eyes to different colours. 5 Measurements of the density of bone structures.
9 June	1 Medical and biological experiments. 2 Soyuz 11 powered down. 3 First use of Penguin suits. 4 Check composition of Salyut atmosphere. 5 Orbital correction at 07.06: orbit – 259-282km.		
		20 June	1 Rest day, but observations completed of Earth's land and water surfaces.
10 June	1 Most of the day given over to medical and biological experiments. 2 Tests on the density of body tissues. 3 Blood samples taken for subsequent analysis on Earth.	21 June	1 Continued work with Orion-1. 2 Studies of gamma radiation using Anna-3. 3 Measurements of the electron and general cosmic radiation background levels.
11 June	1 Spectrographic measurements of the Soviet Union to study the spectrum characteristics of natural formations and water surfaces. 2 First use of Anna-3 gamma radiation observatory.	22 June	1 Atmospheric studies using a manual spectrograph. 2 Biological observations of higher plant growth in the zero gravity environment. 3 Meteorological observations.
12 June	1 Medical and biological experiments. 2 Measurements of radiation with a view to the safety levels for the mission. 3 Studies of the cardiovascular system. 4 Tests of respiration, gas metabolism and energy expenditure. 5 Photographs of cloud formation.	23 June	1 Tests of the visual orientation system. 2 Observations of the Earth's surface. 3 Medical observations and experiments.
		24 June	1 Previous manned duration record exceeded (held by the Soyuz 9 crew). 2 Photography of Earth. 3 Astronomical photography. 4 Exercises and medical experiments.
13 June	1 Observations of cloud cover. 2 Geological observartions. 3 Measurements of primary cosmic rays. 4 Studies of the Oazis-1 hydroponic farm.	25 June	1 Continued experiments with Era. 2 Observations of Earth for meteorological research. 3 Physical exercises. 4 Preparations for the return to Earth.
14 June	1 Orbit – 255-277km. 2 Tests performed with the aim of developing an automatic navigation system. 3 Multi-spectral photography of geological formations. 4 Observations of cloud cover. 5 Joint meteorological observations with a Meteor-1 satellite. 6 Tests of the cosmonaut's eye sensitivity.	26 June	1 Observations of the intensity of charged particles and the spectra of cosmic ray nuclei. 2 Micrometeorite measurements. 3 Exercises and medical experiments. 4 Change water tanks of the cooling and drying unit of the sanitation system.
15 June	1 Multi-spectral photographs of the Caspian Sea area for use in the fields of agriculture, land improvement, geodesy and cartography: this was co-ordinated with aeroplanes over-flying the same areas. 2 Meteorological observations in conjunction with a Meteor-1 satellite. 3 Radiation measurements for biological studies. 4 Measurements of protons, neutrons and gamma radiation. 5 Cardiovascular measurements.	27 June	1 Check the operation of Soyuz 11. 2 Physical exercises and medical monitoring.
		28 June	1 Medical tests and exercises. 2 Checks of Salyut 1 and Soyuz 11 systems.
16 June	1 Experiments dealing with manual and automatic control of Salyut. 2 Checks of the accuracy of Salyut's ion orientation system. 3 Investigations into the composition of the upper atmosphere, measured by radio-frequency mass spectroscopy.	29 June	1 Medical checks and exercises. 2 Moth-balling of scientific equipment. 3 Soyuz 11 undocks from Salyut for crew recovery.

Notes: It is impossible to list all of the work undertaken during three weeks on a space station in a condensed form, and this listing – based upon Soviet announcements and the diary notes of the cosmonauts – represents only a summary of the major activities each day.

60m he reduced the closing velocity to 0.3m/s. At 07.49 15 sec a soft docking between Soyuz 11 and Salyut was achieved, and a hard docking was completed at 07.55 30 sec. The orbit of the combination was announced as 212-249km.

Later on the same day manned operations of Salyut 1 began. Patsayev became the first man to enter an orbital laboratory, followed by Volkov. The first few days of the mission were given over to activating Salyut's systems for use by the cosmonauts. On 8 June at 08.02 a manoeuvre was performed using the Salyut propulsion system, and the Soviets said that the resulting orbit was 239-265km. A further manoeuvre began at 07.06 the following day, and following the manoeuvre which lasted for 73 seconds the Soviets gave the orbit as 259-282km.

The accompanying table provides a summary of the activities undertaken by the cosmonauts during their stay on Salyut 1 and is derived not only from press announcements, but also the cosmonaut diaries which were published in Moscow (1973) in the book *Salyut Na Orbite*. Officially, 17 June was a quiet day, but the Kettering Group monitored the activation of Soyuz 11 during the day as if it were being prepared for a return to Earth. The Soviets merely said that some "minor correction work" had been undertaken that day. There have been rumours that a fire of some sort broke out on Salyut, and possibly Soyuz 11 was being readied for an emergency return?

The cosmonauts began their preparations for their scheduled return to Earth on 25 June, with a programme of physical

Above: *Television picture of Dobrovolsky and Volkov working aboard the Salyut 1 orbital station where they spent more than three weeks. Most probably a flight of about six weeks was planned.*

exercises that were more strenuous and of a longer duration than those previously undertaken. Checks were made of Soyuz 11, and the flight logs and experimental results were carefully loaded into the Soyuz descent craft. By 18.15 on 29 June the three cosmonauts had transferred back into Soyuz 11, and Volkov shut the connecting hatches for the last time. All the spacecraft seals were checked and three minutes after the command was given, at 18.28, Soyuz 11 undocked from Salyut, thus ending the first manned occupancy of an orbital station. At 22.35 retro-fire was completed, and the Soviet Union prepared to greet the returning heroes from orbit.

After retro-fire the Soyuz split into its three component parts and the descent craft continued its re-entry automatically. By 23.02 the parachute was open and at 23.17 the spacecraft touched down, just before sunrise. The recovery team were on hand to help the cosmonauts from the spacecraft, since it was expected that they would be weakened after more than three weeks in orbit. However, on opening the spacecraft hatch, they found the cosmonauts dead in their seats.

It seems that as the orbital module had separated from the descent module, a valve was jerked open and the precious life-sustaining atmosphere inside leaked from the descent craft to the vacuum of space. The Soyuz was not large enough to seat three cosmonauts in pressure suits, and so they had no protection as their air leaked out. There was a hand-operated pump which could be used to try to retain pressure inside the descent module, but this took too long to operate. When the bodies of the three cosmonauts were lain in state after the spacecraft recovery, it was seen that Patsayev had a bruise on his head, and it has been suggested by Geoffrey Perry that this contusion was caused as he tried to operate the pump in a desperate attempt to save their lives. However, the cosmonauts died with the valve having been less than half closed.

In 1971 the American Apollo programme was successfully flying after the drama of the Apollo 13 accident in April 1970. The Soviets had hoped to hail the successful Soyuz 11 mission to Salyut as their answer to Apollo. But, it was not to be. It would be more than two years before another manned mission was undertaken by the Soviet Union – although these delays were in part a result of failures in the Salyut programme itself, rather than being wholly due to the re-design work that had to be carried out on Soyuz.

Conclusion

If Soyuz 11 had been successful, it was reported that a Soyuz 12 visit would have been made to Salyut 1 in September 1971, possibly commanded by Filipchenko. However, in the event all crews were grounded. On 30 June Salyut was placed in a higher orbit, with a further manoeuvre to prevent natural decay taking place a month later. Two manoeuvres were completed in August, resulting in an orbit reaching out to more than 300km, while in the latter days of September the orbit was lowered.

Below: *A second view of the Soyuz 11 booster on the launch pad. During 1969-1971 Soviet coverage of their space programme was improving, with better photographs of missions being released.*

Above: *Patsayev, Dobrovolsky and Volkov pose inside the Soyuz simulator during their training for the Soyuz 11 mission. Originally they were the back-up crew, but Kubasov's illness resulted in their flying the mission.*

Below: *The ashes of Dobrovolsky, Volkov and Patsayev carried in Red Square, ready for immurement in the Kremlin Wall, with the remains of other Soviet heroes, including Komarov and Gagarin.*

Finally, on 11 October the propulsion system of Salyut 1 was activated for the last time and "the Salyut station ceased to exist above the Pacific Ocean". Rather than allow Salyuts to decay out of control, the Soviets preferred to de-orbit them so that any debris which survived re-entry would land in the Pacific Ocean, out of harm's way. This practice would also be used for the unmanned progress cargo ferries which began to support Salyut missions in 1978.

Salyut 1 itself seems to have performed well during its six month mission, although the loss of the Soyuz 11 crew stopped the programme in its tracks. While the American space station programme, scheduled for operations in 1973, would involve the launch of a single laboratory called Skylab, with only one backup vehicle available for a much later flight if the first failed, the Soviets planned a whole series of missions within the Salyut programme. Indeed, Salyut was destined to be the major Soviet space programme for the next fifteen years.

FURTHER READING

The following book is the Soviet account of the Salyut 1 programme, and includes Tass announcements of the Soyuz missions as well as extracts from the logs prepared by the Soyuz 11 cosmonauts while on board Salyut 1.

Patsayev, V.A. (editor), *Salyut Na Orbite*, Mashinostroyeniye Press, Moscow, 1973.

The same book is available as a NASA Technical Translation (ref NASA TT F-15, 450, March 1974) under the title *Salyut Space Station in Orbit*.

For information relating to the boxed feature on Soyuz landing windows, see:

Clark, Phillip S., "Soyuz Missions to Salyut Stations" in *Spaceflight*, vol 21 (1979), pp259-263.

Christy, Robert D., "Safety Practices For Soyuz Recoveries" in *Spaceflight*, vol 23 (1981), pp321-322.

LAUNCHES IN THE SALYUT 1 PROGRAMME

LAUNCH DATE & TIME	RE-ENTRY DATE & TIME	SPACECRAFT	MASS (kg)	CREW
1971 19 Apr 01.39*	(1971 11 Oct)	Salyut 1	18,900	—
22 Apr 23.54	24 Apr 23.40	Soyuz 10	6,800 ?	V. A. Shatalov A. S. Yeliseyev N. N. Rukavishnikov
6 Jun 04.55	29 Jun 23.17	Soyuz 11	6,790	G. T. Dobrovolsky V. N. Volkov V. I. Patsayev

Notes: The launch and landing times are known from official Soviet data in the cases of the two Soyuz missions: in the case of Soyuz 11 the landing time was not announced, but both the launch time and mission duration are known. For Salyut 1 the re-entry date is that for the station being de-orbited and destroyed. The launch time for Salyut 1 (marked *) is estimated, and is derived from the Two-Line Orbital Elements. The masses of Salyut 1 and Soyuz 11 are known from Soviet sources, that for Soyuz 10 being assumed to be similar to the Soyuz 11 mass.

MANOEUVRES OF THE SALYUT 1 ORBITAL STATION

PRE MANOEUVRE ORBIT		POST MANOEUVRE ORBIT	
Epoch (1971)	Altitude (km)	Epoch (1971)	Altitude (km)
Initial orbit		19.43 Apr	203-216
20.11 Apr	201-211		
Soyuz 10: docked and undocked 24 April			
Soyuz 10: redocked and undocked 24 April			
		24.10 Apr	184-240
25.82 Apr	186-220	26.26 Apr	253-276
1.95 Jun	198-204	3.60 Jun	187-221
3.97 Jun	188-215	4.53 Jun	208-217
5.02 Jun	208-212	6.13 Jun	210-235
Soyuz 11: docked 7 June			
8.16 Jun	206-227	8.41 Jun	225-262
8.41 Jun	225-262	9.90 Jun	257-264
Soyuz 11: undocked 29 June			
30.50 Jun	222-237	30.99 Jun	238-284
28.10 Jul	200-237	28.66 Jul	226-292
17.21 Aug	201p-251	18-20 Aug	248-299
18.20 Aug	248-299	19.26 Aug	285-314
24.19 Sep	265-300	25.94 Sep	221-266
10.69 Oct	177-182	11 Oct	De-orbited

Notes: This table lists all of the orbital manoeuvres completed by the Salyut 1 complex – either on its own or with Soyuz 11 attached. The data is based upon the Two-Line Orbital Elements, and for each manoeuvre the last orbital data before the change are shown, together with the first set of data after the change. The Soviet Union has not detailed every manoeuvre, and therefore this table represents a complete record. The last set of data was issued for 10.69 Oct, with Salyut being de-orbited at an unspecified time the following day. For convenience, the dockings and undockings of ferry craft are indicated.
This format is the standard for listing orbital data for all of the future Salyut missions.

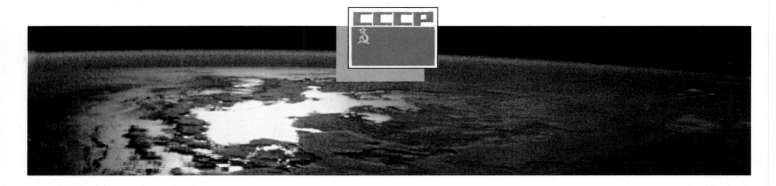

Chapter 7: The Military Salyuts

After the Soyuz 11 disaster there was a break of twelve months before a Soyuz was orbited again – and then it was an unmanned test in the Cosmos programme. During this lull in orbital missions a new design of Salyut was brought on-line and modifications of the Soyuz were made to make it more specifically a simple space station ferry.

During 1974 and 1975 two varieties of Salyut stations were flown, and these are generally identified as "military" and "civilian" programmes, although there was certainly some overlap in the experimental work undertaken by the crews. These labels will be used here for convenience. This chapter deals with the military Salyuts. It describes events that were happening at the same time as developments in the civil Salyut programme described in the next chapter, so forward references will be essential at some points.

There were three launches in the military Salyut programme. There was also a launch failure in 1972 which could have been either a military or civilian station: the failure will be discussed in the next chapter since it was probably a civilian Salyut mission.

The Military Salyut Design

In discussing the design of the military Salyut, one is placed at a great disadvantage. While there are detailed cutaways and drawings available for the civilian Salyut stations (Salyuts 1, 4, 6 and 7), the Soviet Union has not released a single sketch which hints at the correct design of Salyuts 2, 3 or 5. While the civilian stations were designed by scientists who were within the Korolyov design bureau, the military Salyuts were produced by the bureau headed by Chelomei, the designer who had produced the Proton booster. The same bureau is thought to have been responsible for the designs of Cosmos satellites which have flown within the unmanned military programme.

The civilian Salyuts have been described as being four stepped cylinders in shape, but the only description which we have for Salyut 3 says that it consisted of two cylinders and that with its solar cell "wings" extended it looked like a bird in flight. After Salyut 1, all of the civilian Salyuts had three solar cells, and one cannot describe a cylindrical craft with three wings as resembling a flying bird; in saying this, though, it is possible that too much is

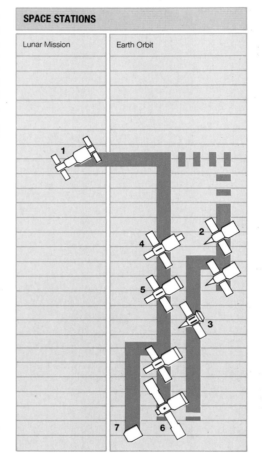

SPACE STATIONS

Lunar Mission	Earth Orbit

Salyut Variants
1 Hybrid Salyut 1 (1971).
2 Military Salyut 3 (1974).
3 Heavy Cosmos module (Cosmos 929, 1977).
4 Civilian Salyut 4 (1974).
5 Second generation civilian Salyut 6 (1977).
6 Third generation Mir orbital station core (1986).
7 Kvant science module for Mir complex (1987).

being read into a single Soviet comment which may have been more poetic than literal in intention. A few television photographs have been released of the crews inside Salyut 3 and Salyut 5, and these show a (purposely ?) darkened interior which gives the impression of being cramped.

The military Salyuts incorporated a re-entry module which returned to Earth after the manned missions had been completed; no such re-entry craft was carried on a civilian Salyut. As with the complete Salyut station, no description has been given of the re-entry craft and no recovery photographs have been released.

Some comments about the military Salyut design have appeared in the West-

ern aerospace press, and these have suggested that the Soyuz docked at the rear of the stations, and that the solar wings were on the rear module. The latter is very difficult to imagine because that would mean that they were mounted on a 4.15m diameter section (assuming that military Salyuts used a rear works compartment of the same dimensions as Salyut 3), outside the payload shroud, and would therefore need special protection to guard them against the rigours of launch and the ascent to orbit.

However, the rear docking of the Soyuz seems logical if the position of the re-entry vehicle is considered. Extrapolating from experiences with the later Salyuts 6 and 7 which allowed craft to dock at either the front or rear of the station, clearly the Salyut itself could perform no orbital manoeuvres while the rear port was occupied because of the danger of damaging the docked spacecraft when the propulsion system was operating. Similarly, the recoverable capsule of the military Salyut could not be located at the end which included the propulsion system for the same reason; it had to be placed at the end where the transfer compartment would normally be located. Therefore, the Soyuz had to dock at the rear of the military Salyuts – unless one suggests that an asymmetrical radial docking port was used.

It would seem logical that the military Salyuts used a relative of the Salyut 6 propulsion system for orbital manoeuvres. Rather than having the Soyuz KTDU-35 modifications, two nozzles would be mounted close to the edge of the rear Salyut bulk-head. This would ensure that the docking hatch for the Soyuz would not be in the "line of fire" when Salyut performed orbital manoeuvres.

Bearing in mind what we know about the design of the Heavy Cosmos module (see Chapter 10) and its descent craft, the accompanying diagram of a possible military Salyut design has been prepared. The conical descent craft used on the Heavy Cosmos is replaced by the Salyut transfer compartment at the front end. At the rear is the docking port and the manoeuvring system; in order to restrict the length of Salyut to about 13.5m, the large diameter work compartment has had to be reduced in length to about 2.5m. The small work compartment – attached to which are two solar panels – would contain the scientific equipment (whether that be a reconnais-

Above: *The spacecraft models which form part of the Volga docking simulator. The models dock under cosmonaut control, with realistic pictures being relayed to the cosmonauts.*

Above: *Soyuz launch vehicle in the MIK assembly building at Tyuratam. A new shroud tower was used for the ferry missions, starting in 1973.*

Left: *A statue at Zvezdny Gorodok – "Star Town" – where the cosmonaut training centre is located outside Moscow. A modern town, this is home and workplace for many of the cosmonauts.*

sance camera system or whatever). As a result the free area in Salyut for the cosmonauts is very limited; it would basically consist of the shortened large work compartment.

This design incorporates the various features we believe to be accurate: the rear docking of Soyuz (reported by *Aviation Week & Space Technology*), the appearance of a flying bird (two solar panels only), while the body of Salyut itself is made up of two cylinders when the descent craft is ignored. The solar panels, however, are not placed on the rear of the station, as claimed by *Aviation Week & Space Technology*. Unless the Soviet Union decides at some time to release more information about the military Salyut, its design cannot be definitively stated by Western observers.

Salyut 2 Failure

In late June 1972, a test flight of a modified Soyuz within the Cosmos programme took place. Named Cosmos 496, the mis-

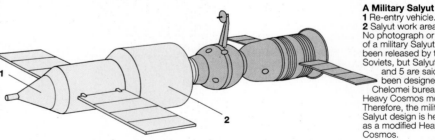

A Military Salyut
1 Re-entry vehicle.
2 Salyut work areas.
No photograph or diagram of a military Salyut has been released by the Soviets, but Salyuts 2, 3 and 5 are said to have been designed by the Chelomei bureau, like the Heavy Cosmos modules. Therefore, the military Salyut design is here shown as a modified Heavy Cosmos.

sion lasted for six days. After this, the Soviet Union felt confident enough to begin manned missions again. There followed a reported Salyut launch failure within a month; this is discussed in the next chapter.

The next Salyut to reach orbit was launched on 3 April 1973, nearly two years after Salyut 1 had been orbited. The spacecraft was placed in an announced orbit of 215-260km. The launch announcement said that Salyut 2 was of an improved design, and that it would be testing new on-board systems and equipment: it would be used for conducting scientific and technical research and experiments during its space flight. On 11 April the Soviets said that orbital corrections had been com-

LAUNCHES IN THE MILITARY SALYUT PROGRAMME				
LAUNCH DATE & TIME	**RE-ENTRY DATE & TIME**	**SPACECRAFT**	**MASS** (kg)	**CREW**
1973 3 Apr 09.00*	(1973 28 May)	Salyut 2	18,900 ?	–
1974 24 Jun 22.38*	(1975 24 Jan)	Salyut 3	18,900 ?	–
3 Jul 18.51	19 Jul 12.21	Soyuz 14	6,800 ?	P. R. Popovich / Y. P. Artyukhin
26 Aug 19.58	28 Aug 20.10	Soyuz 15	6,760	G. V. Sarafanov / L. S. Demin
1976 22 June 18.04*	(1977 8 Aug)	Salyut 5	18,900 ?	–
6 Jul 12.09	24 Aug 18.33	Soyuz 21	6,800 ?	B. V. Volynov / V. M. Zholobov
14 Oct 17.40	16 Oct 17.46	Soyuz 23	6,760	V. D. Zudov / V. I. Rozdestvensky
1977 7 Feb 16.12	25 Feb 09.38	Soyuz 24	6,800 ?	V. V. Gorbatko / Y. N. Glazkov

Notes: This table includes both the orbital station launches in the military Salyut programme and also the associated Soyuz ferry missions. Launch times marked '*' are estimated, and for orbital missions they are calculated using the Two-Line Orbital Elements. Dates in parentheses are the Salyut re-entry dates; other dates and times in the same column are actual recoveries. Salyut 2 failed before a crew could be launched.

pleted on 4 April and 8 April, and these resulted in a new orbit of 261-296km. Ground control was continuing to monitor the mission.

Naturally, Western observers were expecting a manned launch to Salyut 2 within two weeks of the launch, but none came. On 18 April the Soviet Union denied that manned visits were planned and finally on 28 April it was announced that Salyut 2 had completed its mission. It was claimed that this had been a test of an automatic Salyut variant of a new type; this announcement was received with scepticism in the West.

Observers who were following the progress of Salyut in orbit had realised that this station was different from Salyut 1

SALYUT 2/1973

(graph: vertical axis marked 0, 6, 12, 18, 24; horizontal axis marked Jan 1, Feb 1, Mar 1, Apr 1, May 1, Jun 1, Jul 1, Aug 1, Sep 1, Oct 1, Nov 1, Dec 1)

● Salyut 2

Below: *The Soyuz 14 crew, Pavel Popovich (left) and Yuri Artyukhin.*

Artyukhin was a military engineer; was their mission a military one?

MANOEUVRES OF THE SALYUT 2 ORBITAL STATION

PRE MANOEUVRE ORBIT		POST MANOEUVRE ORBIT	
Epoch (1973)	Altitude (km)	Epoch (1973)	Altitude (km)
Initial orbit		3.68 Apr	209-252
4,42 Apr	204-245	4.60 Apr	240-249
4.60 Apr	240-249	5.66 Apr	238-260
5.66 Apr	238-260	6.53 Apr	232-262
6.53 Apr	232-262	7.03 Apr	236-259
8.14 Apr	237-256	8.58 Apr	258-279
27.79 May	158-160	28 May	Orbital decay

Notes: Salyut 2 was never manned because of a rumoured partial disintegration during an orbital manoeuvre. The spacecraft actually decayed from orbit, rather than being de-orbited in a controlled manner.

because it was transmitting on frequencies more akin to an unmanned reconnaissance satellite than to a manned spacecraft. Furthermore, Salyut had been having serious difficulties in orbit. Certainly, as the Soviets claimed, Salyut had completed a series of orbital manoeuvres which concluded on 8 April (see table). It is claimed that a manoeuvre was made on 14 April, and this resulted in the partial disintegration of the station. However, the reality of this is uncertain.

The catalogues of objects in orbit associated with the launch list a total of 26 pieces; of course, two of these - designated A and B – were Salyut 2 and the third stage of the Proton SL-13 booster. It is impossible to say when each object was first tracked, but the following limits can be set:

Objects A-V	between 3-5 April
Objects W-Y	between 5-6 April
Objects Z-AB	between 12-19 April

(The objects are catalogued as A, B, C . . . X, Y, Z but excluding I and O, then continuing AA, AB, AC . . . AZ, BA, BB . . . BZ, etc).

Clearly, the majority of Salyut fragments were catalogued within two days of the launch, and they might have been the result of the ejection of instrument covers once the station had reached orbit, or alternatively the rocket body may have partially disintegrated. The public record does not say. The objects catalogued later may indeed have appeared after 5 April, but they may also have been small objects which were not immediately tracked in orbit. The latter is not unknown as other missions bear witness. For instance, one can cite the discovery of a fragment desig-

nated 1958-Beta-3 from Vanguard 1 (launched in February 1958) in September 1965.

Certainly it would appear that until perhaps 18 April the Soviet Union was planning a manned mission to Salyut 2; if there had been a partial disintegration of Salyut on 14 April, with three new (large ?) pieces separating, this incident may have made the Soviets abandon the station.

Although there were no launches to Salyut 2, the launch time allows the landing opportunities for manned missions to be estimated. Most probably, Salyut was

designed to operate for six months or so (as did Salyut 1, Salyut 3 and Salyut 4), in which case the appropriate landing windows are 24 May-29 May, 20 July-29 July, 21 September-1 October. These dates are about two weeks before the landing windows which would have applied if Cosmos 557 had successfully operated as a Salyut after its launch in May 1973 (see next chapter).

Manned missions to Salyut 3, Salyut 4 and Salyut 5 were launched between 8 and 15 days after the Salyut reached orbit, and therefore a manned launch to Salyut 2

Above: *Interior shots from the military Salyuts are rare and normally not of good quality. Here, Popovich is shown inside Salyut 3 during a television broadcast; little detail of Salyut can be discerned.*

Below: *The Soyuz 14 cosmonauts waving farewell as they begin their lift ride to the top of their launch vehicle.*

of its true mission by the Soviet Union, and it would be fourteen months before the next named Salyut was placed in orbit.

Salyut 3 and Soyuz 14

In September 1973 the first manned test of the new Soyuz ferry was completed, apparently clearing it for use with a new Salyut. Soyuz 12 remained in orbit for just two days. This was followed in December by a "solo" Soyuz 13 mission which was apparently the first manned flight of the variant which would be used for the Apollo-Soyuz mission in 1975. A final unmanned check of Soyuz was completed in May 1974 under the Cosmos 656 identity, and manned space station missions were set to resume.

By this time the American manned space programme had virtually completed operations with Apollo spacecraft. The final mission to Skylab had finished in February 1974, and apart from the Apollo-Soyuz mission in 1975 there would be no more manned American flights until the first flight of the *Columbia* Space Shuttle in April 1981. Therefore, although the days of trying to compete with the Americans in space had been over politically for some years, Soviet space planners knew that the arena of manned spaceflight would be theirs alone for the rest of the decade.

Late at night on 24 June 1974 a three stage Proton SL-13 was launched from Tyuratam and soon Salyut 3 was in orbit; the announced altitudes were 219-270km. During the next week a series of manoeuvres was completed, resulting in Salyut achieving a 268-272km orbit on 2 July (see table). This should be compared with the Salyut 2 orbit after the 8 April 1973 manoeuvre — 258-279km. Salyut 3 was clearly a follow-on to Salyut 2, and its use of military transmission frequencies confirmed this.

With Salyut 3 in its operating orbit, Soyuz 14 was launched on 3 July at 18.51. Two cosmonauts were carried: Col Pavel Popovich and Lt-Col Yuri Artyukhin. The presence of a military engineer (Artyukhin) hinted at the probably military nature of the mission. Soyuz 14 manoeuvred in orbit and slowly approached Salyut 3. On 4 July at 20.35 Soyuz 14 successfully docked with Salyut 3. The approach had been guided automatically down to a distance of 100m; after that the docking was completed manually. Artyukhin, followed by Popovich, entered Salyut 3 at 01.30 on 5 July.

During the early part of the mission solar activity increased, and the resulting rise in radiation striking the station outside the atmosphere's protection raised the question of the cosmonauts' safety. It was decided, however, that the increased radiation levels were within safe limits and the flight could continue.

The table overleaf provides a summary of the work undertaken by the Soyuz 14 cosmonauts as gleaned from Soviet announcements. Clearly, this work-load is far below that of which the cosmonauts were capable, and therefore one can only assume that the majority of their time was taken up with tasks connected with the more clandestine side of the mission. A 10m focal length camera is said to have been carried on Salyut, and this would probably have allowed the production of images with a ground resolution of 30cm. There have been claims in the Western aerospace press that objects were laid out at Tyuratam launch centre for Salyut 3 to photograph in order that the high resolution camera system could be calibrated.

A number of specific scientific experiments have been described, most of which were connected with monitoring of the cosmonauts'' health and physical condi-

MANOEUVRES OF THE SALYUT 3 ORBITAL STATION			
PRE MANOEUVRE ORBIT		**POST MANOEUVRE ORBIT**	
Epoch (1974)	Altitude (km)	Epoch (1974)	Altitude (km)
Initial orbit		25.06 Jun	211-244
25.24 Jun	208-240	25.74 Jun	213-253
27.04 Jun	213-252	27.16 Jun	251-268
28.10 Jun	250-266	28.16 Jun	265-271
1.40 Jul	266-267	2.15 Jul	268-272
Soyuz 14: docked 4 July			
		5.14 Jul	267-271
16.56 Jul	265-269	18.18 Jul	265-273
Soyuz 14: undocked 19 July			
		20.30 Jul	264-272
15.23 Aug	261-266	16.66 Aug	258-262
22.14 Aug	258-261	24.26 Aug	258-286
Soyuz 15: docking failure 27 August			
		28.13 Aug	257-286
21.07 Sep	251-274		
Salyut 3: capsule recovery 23 September			
		23.63 Sep	249-280
22.61 Oct	235-259	24.10 Oct	261-285
24.10 Oct	261-285	24.23 Oct	255-294
(1975) 22.39 Jan	218-239	(1975) 24 Jan	De-orbited

Notes: The first available Salyut 3 orbit after each docking and recovery is shown, whether or not a Salyut manoeuvre was completed.

could have been intended during the period 11 April to 18 April. This suggests that the first mission would have lasted for about a month, since a thirty day flight would have been launched within the period 24 April — 29 April, somewhat later than subsequent Salyut missions would suggest.

Whatever had been intended for Salyut 2, it slowly decayed from orbit, re-entering the atmosphere on 28 May. Although a second Salyut had been planned for 1973, it failed after launch and was named Cosmos 557. No acknowledgement was made

Launch and Landing Windows
This is the launch and landing graph for Salyut 3 during 1974. Soyuz 14 clearly came down during a nominal landing window, while equally clearly Soyuz 15 did not. Looking at the landing opportunities available in September, would seem to suggest that a flight of three weeks was probably intended. Within days of the predicted Soyuz 15 landing, a capsule from Salyut 3 was returned to Earth (23 September), ending the scheduled operations with the station.

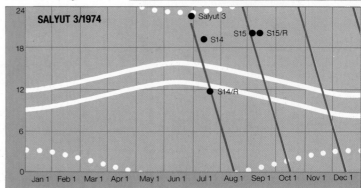

SALYUT 3/1974

Salyut 3 · S14 · S15 ·· S15/R · S14/R

Jan 1 · Feb 1 · Mar 1 · Apr 1 · May 1 · Jun 1 · Jul 1 · Aug 1 · Sep 1 · Oct 1 · Nov 1 · Dec 1

tion. Polinom-2M was used to study the performance of the heart and circulatory system in orbit, and it produced electrocardiograms and seismocardiograms for study back on the ground. Levkoi-3 allowed the cosmonauts to measure intracranial pressure and the ability of blood vessels to carry blood in zero gravity. Blood composition was monitored using Amak-3; blood samples were taken from the cosmonauts for examination back on Earth. The capacity of the lungs and the inhalation/exhalation rates were measured using a portable Rezeda-5 instrument. The working of the vestibular system in weightlessness was investigated using an Impulse instrument. The suits used during their exercise periods were designated NRK-1 (Atlet) and TNK-2 (Penguin).

A number of instruments were tested on Salyut 3 with a view to introducing improved operational systems on later missions. The only named one seems to have been Priboy, which tested the purification of water after it had been condensed for Salyut's atmospheric moisture.

After a stay of two weeks on Salyut, the Soyuz 14 crew prepared to return to Earth. They undocked from Salyut 3 on 19 July at 09.03; retro-fire came at 11.35, and Soyuz 14 touched down safely at 12.21. After all the set-backs of 1971-1973, the first successful space station mission was over.

Soyuz 15: Failure to Dock

After the success of the Soyuz 14 visit to Salyut 3, the Soviets were planning a slightly longer visit to the station. On 16 August it was announced that Salyut 3 was continuing in orbit and that the station was still in operating condition; the orbital altitude was given as 260-278km. Although not stated, a small orbital adjustment had just been made to Salyut. A further manoeuvre was made without comment on 22-24 August (see table).

The orbital adjustments of Salyut alerted observers to the probability that another Soyuz launch was imminent. On 26 August at 19.58 Soyuz 15 was launched, carrying two rookie cosmonauts: the commander was Lt-Col G. Sarafanov and the flight engineer was Col-Engineer L. Demin.

Once more, an "all-military" crew was being launched to Salyut. The purpose of the mission was to continue the research and experiments initiated during the Soyuz 14 mission to Salyut 3.

A Tass announcement was made on 27 August, stating that twelve orbits of the Earth had been completed, and that a course correction had resulted in Soyuz 15 entering a 254-275km orbit. At this point all seemed to be going well with the mission. The next announcement came on 28 August at 07.00. It read:

"A Tass special correspondent reported from the flight control centre that the second working day of the cosmonauts Sarafanov and Demin on board Soyuz 15 had ended at 08.00 Moscow Time on August 28, by which time the spaceship had made 22 orbits around the Earth. Under the second day's programme Sarafanov and Demin carried out experiments to perfect the technique of piloting the ship in different flight situations. During manoeuvring, Soyuz 15 many times approached the Salyut 3 station. The cosmonauts checked all of the ship's systems, made observations on the stages of approach to the station and inspected the station when approaching it. The crew, who reported that they felt fine, were concluding the flight and preparing the spacecraft for the return to Earth."

The cosmonauts remained in orbit until late on 28 August; at 19.25 they instigated retro-fire and Soyuz 15 landed at 20.10.

The Soviets claimed in retrospect that Soyuz 15 was launched with the prime objective of testing a new automatic rendezvous and docking system, and that the complete docking manoeuvre with salyut should have taken place automatically. However, the automatic system failed, and although the cosmonauts could have docked with Salyut manually they returned to Earth. Somehow, this does not quite ring true.

Most probably, the automatic search and rendezvous system worked successfully until the time came for manual control to be taken at a distance of about 100m from Salyut, and at this point some (unspec-

ified) part of the system repeatedly failed. Since Soyuz only carried a limited quantity of fuel and needed to have its batteries re-charged by Salyut if it were to remain in orbit for more than two days, the mission had to be curtailed when the docking manoeuvre failed.

Because the mission was recalled early, the crew did not return to Earth under the normal landing conditions; again, the Soviets claimed that a night-time landing had been planned for the mission. However, if Soyuz 15 had come down during the next available landing window (lasting between 12-20 September) a mission of 17-25 days duration is indicated. The available evidence of the relationship between earlier successful Soyuz durations and their launch times suggests that a flight lasting for about three weeks had been intended. It is possible that the crew were partially to blame for the failure. This deduction is supported by the fact that while Popovich was retired from active training, the commanders of the successful Soyuz 21 and Soyuz 24 military Salyut visits trained for later Salyut missions, while the commanders of the unsuccessful Soyuz 15 and the later Soyuz 23 – which also failed in a Salyut docking – were given no more missions to the end of 1986.

The End of Salyut 3

The early Salyuts had only limited operating lifetimes. Salyut 1 would have completed all of its planned manned activity within five months of its launch, and it remained in orbit for only six months. Possibly mission planners wondered whether they could squeeze another manned visit into their Salyut 3 schedule; this would have called for a Soyuz 16 launch (a designation actually given to the dress rehearsal mission for Apollo-Soyuz in December 1974) on about 29 October when the lighting conditions would again be right for a three week visit to Salyut. However, this would have meant operating Salyut in a manned mode for two months longer than originally planned, and the Soviets decided not to risk this. On 19 September it was announced that the Salyut 3 programme would soon be concluded. The original programme of research would be

OPERATIONS OF SALYUT 3 WITH SOYUZ 14				
DATE (1974)	ACTIVITIES AND NOTES	DATE (1974)	ACTIVITIES AND NOTES	
5 July	1 Soyuz 14 docks with Salyut 3. 2 On-board systems de-mothballed and activated. 3 Orbit – 265-276km.		3 Cultivation and observation of on-board biological specimens. 4 Check of manual control systems. 5 Further observations of polarisation properties of Earth's atmosphere.	
6 July	1 Continued activation of Salyut systems. 2 Medical examinations undertaken, including electrocardiograms.	13 July	1 Photographic session of geological and morphological formations. 2 Preliminary check of Soyuz 14 systems. 3 New orientation system checked.	
7 July	1 Rheograph used to measure the blood circulation in the brain and arteries. 2 Physical exercises. 3 First use of Penguin suits.	14 July	1 Photography of Soviet Central Asia for geological work. 2 Observations of glacier movements.	
8 July	1 Observations of the polarisation of sunlight which is reflected from the Earth's surface and atmosphere for Earth resources investigations.	15 July	1 Further Earth resources photography. 2 Studies of cloud formation. 3 Tested navigation system which used Earth's horizon as a reference point.	
9 July	1 Checks of Salyut's atmosphere. 2 Observations of Earth's horizon to determine atmospheric properties.	16 July	1 Spectroscopic photography of Earth. 2 Tests of the Priboy regenerative apparatus. 3 Joint meteorological observations with a Meteor-1 satellite.	
10 July	1 Pulmonary ventilation examined in conjunction with determining cosmonaut energy expenditure.	17 July	1 Major preparations for return to Earth. 2 Soyuz 14 systems checked. Possibly with the Soyuz performing a small orbital manoeuvre of the complex. 3 Research material transferred to Soyuz 14.	
11 July	1 Announced that the mission had reached its half-way point. 2 Spectrographs of the Earth's horizon. 3 Priboy system tested for regeneration of water from condensation of moisture from Salyut's atmosphere.			
12 July	1 Partial rest day, but some work undertaken. 2 Cardiovascular and vestibular examinations.	18 July	1 Salyut systems mothballed. 2 Soyuz 14 undocks.	
Notes: The above listing is a summary of the experimental and other work undertaken by the Soyuz 14 cosmonauts aboard Salyut 3. This is not an extensive programme, and therefore one must assume that for most of their time the cosmonauts were undertaking work which was not announced – presumably connected with the supposed military nature of the mission.				

completed on 23 September, and after that date a further series of experiments would be undertaken during an additional research programme.

Seven days later, the announcement came that the Salyut 3 programme had been completed on 23 September, and that on that date:

"*. . . . the recoverable module containing materials of research and experiments was separated from Salyut 3. The motors were switched on at the set time and the module began its descent to Earth. The motors were discarded before entry into the dense layers of the atmosphere and the parachute system was activated at an altitude of 8.4km. The recoverable apparatus landed in the predetermined area of the Soviet Union.*"

Calculations suggest that the recovery would have been at about 09.36 (to within two minutes), just after the normal Soyuz landing window had closed – a window which should have been used by Soyuz 15.

A brief summary of the work undertaken by Salyut 3 was published on 7 October, noting that the orbit was 249-293km and that the extended programme of work was being successfully carried out. A further statement was made on 28 October, and this indicated that an orbital manoeuvre had been completed on 27 October at 23.14, resulting in a new orbit with altitudes 268-299km. The data in the table listing Salyut 3 manoeuvres suggest that a further orbital manoeuvre may have been completed later that day as well.

Five months after its launch, another progress report was given for Salyut 3; on 29 November it was said that as part of the extended period of work, studies were being undertaken into the aerodynamic characteristics of the station and that the life support systems were also being monitored. The orbit was said to be 247-293km. On 25 December it was announced that the termination of the Salyut 3 flight was approaching as the extended programme of research drew to a close. The orbit was quoted as 235-270km. A further progress report about the flight came on 24 January 1975 when it was announced that earlier that day Salyut 3 had been de-orbited and had ceased to exist over the Pacific Ocean.

However, the Soviets were not without a space station in orbit. On 26 December 1974 the civilian Salyut 4 had been launched and even as Salyut 3 was burning up, the first crew were on board the new station.

Salyut 5 in Orbit

During 1975 the energies of the Soviet manned space programme were divided between operations with the civilian Salyut 4 and the Apollo-Soyuz joint mission with the Americans. The main Salyut 4 programme ended in February 1976 when the unmanned Soyuz 20 returned to Earth after remaining in orbit for three months - most of the time actually docked with Salyut 4. Since the second visit to Salyut 4 had lasted for two months, it was expected by outside observers that a three month manned mission would follow the Soyuz 20 flight.

Above: *The launch of Soyuz 21, carrying the first two-man crew to Salyut 5. Although the mission initially went well,* *the cosmonauts returned earlier than planned, because of problems with Salyut's life support system.*

It has been noted that launches in the military and civilian Salyut programmes seemed to alternate: Salyut 2 (military), Cosmos 557 (civilian), Salyut 3 (military), Salyut 4 (civilian). Therefore, it was expected that when Salyut 5 was launched it would be part of the military Salyut series. The launch of the new Salyut actually came on 22 June 1976, almost exactly two years after the Salyut 3 launch. The initial orbit was announced as 219-260km but a series of manoeuvres was undertaken in preparation for the first manned visit (see table). Significantly, Salyut 5 was positioned in the standard "military Salyut" orbit of about 260-270km altitude, some 70km below the operating altitude of the civilian missions. The transmission frequencies were the same as those used on Salyut 2, Salyut 3 and the unmanned reconnaissance satellites, totally different from those used for the civilian missions, and further evidence that this was a military flight.

An announcement on 29 June noted that an orbital manoeuvre had been completed, resulting in a 220-275km orbit; by the time of the launch of the first crew, the orbit would be more circular.

Soyuz 21: The First Visit

Two weeks after the launch of Salyut 5, its first crew was launched. On 6 July at 12.09 Soyuz 21 lifted off from Tyuratam with a two-man crew: commander Col Boris Volynov and flight engineer Lt-Col Vitaly Zholobov: they had been the back-up crew for both Salyut 3 manned missions. The mid-day launch of Soyuz 21 implied that this would not simply be a short visit to Salyut 5, although consideration of the landing windows shows that this would not be the rumoured three

MANOEUVRES OF THE SALYUT 5 ORBITAL STATION			
PRE MANOEUVRE ORBIT		POST MANOEUVRE ORBIT	
Epoch (1976)	Altitude (km)	Epoch (1976)	Altitude (km)
Initial orbit		22.99 Jun	208-233
22.99 Jun	208-233	24.04 Jun	213-245
26.39 Jun	212-243	26.89 Jun	215-257
3.32 Jul	213-254		
Soyuz 21: docked 7 July			
		8.11 Jul	264-274
Soyuz 21: undocked 24 August			
		24.99 Aug	259-269
9.24 Oct	250-256	10.11 Oct	254-272
Soyuz 23: docking failure 15 October			
		16.40 Oct	255-268
(1977) 10.77 Jan	226-237	(1977) 14.35 Jan	231-259
17.46 Jan	230-257	20.38 Jan	257-260
4.43 Feb	254-254	5.92 Feb	254-260
Soyuz 24: docked 8 February			
		8.91 Feb	253-259
22.08 Feb	249-251	22.71 Feb	249-253
23.64 Feb	248-253		
Soyuz 24: undocked 25 February			
Salyut 5: capsule recovery 26 February			
		26.62 Feb	248-256
3.40 Mar	246-255	5.45 Mar	253-274
14.39 Apr	240-257	15.69 Apr	254-269
15.69 Apr	254-269	16.32 Apr	257-266
8.08 Aug	182-185	8 Aug	De-orbited

Launch and Landing Windows
Following the launch of Salyut 5 in June 1976, two manned visits were launched to the station, although only the first crew boarded the station. The crew of Soyuz 21 spent nearly 50 days in orbit, but landing window considerations suggest that a mission closer to nine weeks was intended. In October, Soyuz 23 was launched amid rumours that a 90-day mission was planned. As if repeating the experience of Salyut 3, Soyuz 23 failed to dock with Salyut.

Launch and Landing Windows
Following the failure of Soyuz 23 it was thought that Salyut 5 might be abandoned, but this was not to be. On 7 February 1977 Soyuz 24 was launched towards the station – this being the longest interval between a space station launch and the first manned visit. A short 18 day visit was planned, repeating the planned Soyuz 23 duration and the day after the crew returned to Earth, the Salyut 5 capsule was recovered.

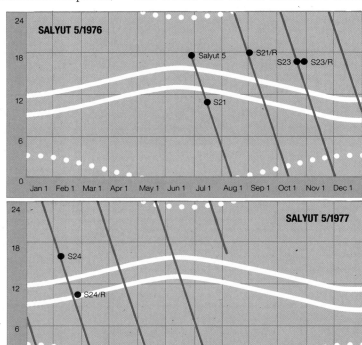

SUMMARY OF THE SALYUT 5 EXPERIMENTS			
DISCIPLINE/ NAME	DESCRIPTION	DISCIPLINE/ NAME	DESCRIPTION
Science-Technology		Technical Experiments	
Kristal	Study of the growth of monocrystals in zero gravity.	Astroizmerital	Testing and evaluation of a new navigation system.
Reaktsiya	Study of the behaviour of melted metals and alloys in space.	Priboi	Check of the humidity of the Salyut atmosphere.
Fizika/Sfera	Experiment to form perfect metallic spheres in a weightless condition.	Stroka	Operations with a ground-space telex system.
Fizika/Potok	Study of the capillary strength of heightened surface tension in fluids when in a zero gravity environment.	Medical Experiments	
		Chibis	Vacuum suit to cover the lower part of the body, subjecting it to a negative pressure, forcing blood into the legs.
Fizika/Diffuzia	Study of the mixing of substances in contact with each other due to the thermal movement of their particles.	Impulse	Measurements of the vestibular system in zero gravity, to investigate disorientation in zero gravity.
Biological Experiments		Levkoi	Measurement of cerebral blood pressure.
Aquarium	Study of the development of viviparous guppy fish in a weightless condition.	Tonus	Measurement of muscles, in order that any wasting could be quickly counteracted.
Terrarium	Study of turtles during a space flight compared with control turtles which remained on Earth.	Rezeda	Measurements of breathing capacity.
Kultivator	Using drosophila (fruit-flies), the study of the changes in the properties and structure of chromosomes caused by the space environment.	Amak	Taking of and analysis of blood samples.
		Polinom-2M	Multi-purpose equipment to monitor blood circulation, frequency of breathing, body temperature and heart functions.
Bioblock	Study of plant growth in orbit in three 'biofixators': biofixator 1 – inseminated eggs of the danio rerio aquarium fish; biofixator 2 – crepis plant seeds; biofixator 3 – mushroom spores.	Palm-2M	Measurements of cosmonaut reaction times.
Astronomical Observations		Un-named experiments	Calibration of taste buds in zero gravity. Observations of the voice timbre to determine the cosmonaut's emotional state. Determinatiion of the degree of eye muscle relaxation. Investigation of optical illusions.
ITS-5	Infra-red telescope to make observations of the Sun, Moon and planets.		
Earth Resources			
	Various un-named experiments and series of observations.		

Notes: Like the table which deals with Salyut 3/Soyuz 14, this list does not include all of the experiments performed on Salyut 5 because much of the cosmonauts' time was taken up with experiments connected with the military nature of the mission.

month space mission. The optimum landing opportunity occurred during 3-12 September which presupposed a mission duration of 59-68 days – the same order-of magnitude as the last manned mission. Soyuz 18 had spent 63 days in orbit during the Salyut 4 mission in 1975.

The initial orbit of Soyuz 21 was announced as 193-253km, and following a correction the altitude was raised to 254-280km. On 7 July Soyuz 21 slowly approached Salyut 5 and following a final approach which lasted 10 minutes, Soyuz 21 docked with Salyut at 13.40. The crew soon transferred into Salyut. Ahead of the Soyuz 21 crew was a full period of experimental work, and more information has been released about these experiments than was the case for Salyut 3. Of course, there were also duties to be undertake which related to the military nature of the mission, but these were not alluded to by the Soviet Union.

The relevant table provides a list of the experiments which have been announced as forming part of the Salyut 5 mission; most of the work was undertaken by the Soyuz 21 crew, but some work was continued by the Soyuz 24 cosmonauts who manned the station in February 1977. Of course, this list excludes the military observations which are believed to have been a significant part of the mission.

The Soviet Union provided regular progress reports on the Soyuz 21 mission, but compared with the wealth of information which had been provided about the Salyut 4 visits in 1975, the Soyuz 21 reports lacked detail and revealed little. As with the Soyuz 14 visit to Salyut 3, the Salyut 5 cosmonauts were involved in some televised reports of the mission, but no details of the Salyut interior could be discerned. Western observers were shocked when the impending termination of the mission was suddenly announced on 24 August. For earlier Soyuz missions to Salyuts, the

period between the announcement and actual recovery was quite lengthy:

    Soyuz 11, 4 days
    Soyuz 14, 7 days
    Soyuz 17, 2 days
    Soyuz 18, 12 days

In the case of Soyuz 21 the announcement that the mission would be ending came at 10.04, and within twelve hours the cosmonauts were back on Earth. Undocking from Salyut occurred at 15.12 and the landing was achieved at 18.33.

Above: *One of the rare television pictures of Volynov and Zholobov aboard Salyut 5 during their Soyuz 21 visit. The lack of Salyut 5 publicity contrasts with the detailed Salyut 4 coverage in 1975.*

There was much speculation in the West as to what had gone wrong in orbit. The "planned" 90 day orbital flight reports reared their heads again, but were not totally refuted at the time. There was speculation that the cosmonauts in the small living area of Salyut had been suffering from the effects of sensory deprivation, which was possible. Finally, a report appeared in *Aviation Week & Space Technology* declaring that the cosmonauts had completed an emergency evacuation into Soyuz 21 after the Salyut atmosphere had developed an acrid odour. No comment was made by the Soviet Union, but when the next cosmonauts boarded Salyut they were wearing breathing masks.

Soyuz 23: Another Failure

In the period between Soyuz 21 returning to Earth and the next launch to Salyut 5, the back-up spacecraft for the 1975 Apollo-Soyuz mission flew an Earth resources mission in September 1976 as Soyuz 22. However, the main thrust of the Soviet manned space programme remained Salyut. On 9-10 October Salyut 5 performed a manoeuvre to raise its orbit, indicating that a fresh manned visit was imminent.

The launch of Soyuz 23 came on 14 October at 17.40 with two rookie cosmonauts aboard: commander Lt-Col V. Zudov and flight engineer Lt-Col Engineer V. Rozdestvensky. The usual orbital manoeuvre was completed during the first ten hours of the flight, and the new orbit was announced as 243-275km. The launch announcement said that the Soyuz 23 mission was to continue the work undertaken by the Soyuz 21 crew on Salyut 5. How-

Below: *Zholobov (foreground) and Volynov in the Soyuz simulator, training for their Salyut 5 mission. On a light note, Zholobov was the first Soviet cosmonaut to wear a moustache at launch!*

ever, the new crew were not destined to board Salyut 5. To quote an announcement made by Tass on 16 October:

"The cosmonauts carried out the programme for the second day. At 21h 58 min Moscow Time [18.58 GMT] on October 15 the spaceship Soyuz 23 was put into the automatic regime for the approach to Salyut 5. Docking with the Salyut 5 station was cancelled because of an unplanned operation of the approach control system of the ship. The crew are completing the mission and preparing to return to Earth."

Once more, a Soviet attempt to re-man an orbital station had been frustrated. At 17.02 the Soyuz 23 retro-rocket fired and Soyuz 23 returned to Earth at 17.46 on 16 October. However, the landing did not run quite as planned. Although it returned to the same general landing area as other Soyuz craft, Soyuz 23 completed the first splashdown by cosmonauts, when it landed in Lake Tengiz. It was ironic that some of Rozdestvensky's pre-cosmonaut assignments had been as head of a naval diving team in the Black Sea! In fact, Lake Tengiz was covered by a layer of ice, which Soyuz 23 broke, but this meant that the rescue team were unable to get close to the spacecraft for many hours, and it cannot have been the happiest of experiences for the recovery team, nor indeed for the crew.

Once more there were rumours that this had been scheduled as a three month mission, but the landing windows suggest that this was unlikely. The landing opportunity

Below: *The manned mission control centre at Kaliningrad. The main display shows a recovery ground track with the rescue aeroplane positions.*

of 31 October – 7 November implies a duration of 17-24 days, while the next window of 23-29 December implies 70-76 days. It is possible that the December landing opportunity was preferred, although future events put this in doubt.

Soyuz 24: A Final Visit

After the Soyuz 23 failure, Salyut 5 continued to orbit the Earth unmanned. On 25 October 1976 the orbit was said to be 259-272km and all on-board systems were reported to be operating normally. The next progress report was released on 22 November when it was said that during the unmanned part of the mission Salyut had taken photographs of the Earth and experiments had been undertaken with the infrared telescope-spectrometer ITS-5; radiation from the Earth and Moon had been studied. The final announcement of the year came on 22 December, when the orbit was given as 232-263km; all the equipment was operating properly, but no hints were given of future plans. Two reports were made in early 1977: a mission review was published on 11 January, while on 21 January the orbit was said to be 256-275km, following manoeuvres on 14 and 18 January.

Despite the hint given by these manoeuvres, Western observers were surprised when Soyuz 24 was launched at 16.12 on 7 February. The crew were Col Viktor Gorbatko and Lt-Col Engineer Yuri Glazkov who was on his first mission. This was the final crew of purely military cosmonauts to be launched, and was the last visit to a military Salyut. The launch of Soyuz 24 was described as "routine", suggesting that no "space spectaculars" could be expected. These cosmonauts had been

the back-ups for the Soyuz 23 mission. After the standard course correction, the Soyuz 24 orbit was 218-281km. The usual approach to Salyut was completed in two parts; Soyuz initially approached to within 80m of Salyut automatically, and then the cosmonauts took over manual control for the docking which occurred on 8 February at 17.38.

In apparent confirmation of the reports suggesting that there were problems with the Salyut 5 environmental control system which caused the early return of Soyuz 21,

Above: *The Soyuz 23 cosmonauts – Zudov and Rozdestvensky – in their flight pressure suits prior to their launch to Salyut 5. They failed to dock.*

when the Soyuz 24 cosmonauts entered Salyut 5 for the first time they were wearing breathing apparatus and were carefully sampling the atmosphere in Salyut. The preliminary tests of the atmosphere must have proved that there was not a serious problem, and the breathing apparatus was quickly abandoned.

Most of the work undertaken by the Soyuz 24 crew was a continuation of the research begun by the Soyuz 21 crew. On 12 February the orbit of the complex was announced as 253-274km, and on 16 February the half-way point of the mission was announced. There was a major innovation on 21 February when the bulletin for that day said:

"A progress report on Salyut 5 . . . described how the cosmonauts, by using the special multi-functional system on board the space station, could carry out a complete or partial change of the atmosphere in the station. This operation was carried out during the routine communications session, when the cosmonauts said that the atmosphere inside the station was fine and breathing was easy."

There have been rumours that spacewalks had been planned for Salyut missions before 1977, but they were cancelled due to the work load imposed on the cosmonauts. While this experiment might have been connected with spacewalks, more probably it was a preliminary test anticipating such work on Salyut 6. The cancelled spacewalks were probably scheduled for the civilian Salyut 4.

Everything about the Soyuz 24 mission was described as "routine", and coverage of the mission was low-key, even for the military programme. Two manoeuvres were completed in readiness for the Soyuz recovery. The first, on 23 February, was a test of the Soyuz 24 propulsion system, and the second the following day was completed by Salyut 5. On 22 February the cosmonauts began to prepare Soyuz 24 for the return journey to Earth, transferring experimental results to the Soyuz descent craft. On 24 February the orbit of Salyut 5 was said to be 248-269km, and the loading of Soyuz was being continued. Undocking from Salyut came on 25 February at 06.21 and landing came at 09.38 – although the recovery announcement was so terse that these times were not even quoted.

As with Salyut 3, a recoverable capsule was carried on Salyut 5. On 2 March a bulletin announced:

"While reporting on their work on Salyut 5, the cosmonauts described an operation carried out on the space station on February 26, a day after their return to Earth. A returnable capsule containing material from the scientific experiments carried out during the flight was detached from Salyut 5 and returned to Earth. This capsule has, like the Soyuz descent module, its own braking motor and parachute system, and operates on command from Earth."

The landing opportunity for the capsule was 09.28 - virtually the same time of day as the Salyut 3 capsule had returned.

Unmanned Operations

There were no more manned visits planned to Salyut 5, but its work was not yet over. An announcement on 22 March gave the orbit as 250-273km following a manoeuvre on 5 March. Scientific research was still being undertaken. The first anniversary of the Salyut 5 launch was celebrated with an announcement reviewing the mission. The final act in the story came on 8 August 1977 when Salyut 5 was de-orbited, allowing any debris to come down over the Pacific Ocean. For the first time since the launch of Salyut 3 in 1974, no Salyut station was orbiting the Earth.

The end of the Salyut 5 mission also marked the end of the second generation Salyut programme (Salyut 1 having been the sole first generation Salyut flight). Overall, the programme had met with mixed fortunes. There had been three launches of the military Salyut variants, and one of the missions had been a total write-off (Salyut 2). Of the five manned visits, two had been terminated early because of docking failures and one – while not a failure – had been recalled earlier than planned.

In some quarters it was expected that experience with the military Salyuts might lead to an unmanned Salyut variant which could function as a large reconnaissance platform, like the American Big Bird satellites which had a high resolution, long-lived, capsule return capability. Certainly, it would be surprising if the technology developed during the Salyut programme were not to be applied elsewhere. In the middle of 1975 the first of the fourth generation photo-reconnaissance satellites was launched, and by the end of the 1970s the missions were slowly reaching operational status. The satellites were launched by the Soyuz SL-4 vehicle, but had long lives (two months) and were able to return capsules to Earth. Perhaps some of the technology had been incorporated from the military Salyuts?

A further possibility is that as orbital stations become more modular in design, one may expect that specialist military modules will be launched, to be occasionally tended by cosmonauts on the main space station complex.

FURTHER READING

There have been no books published which specifically deal with the military Salyut programme. During the years that the missions were flown *Spaceflight* carried a series of mission reports for the Salyut series prepared by Gordon Hooper which provide convenient summaries of the day-by-day operations on the Salyut 3 and Salyut 5 stations.

More general review articles dealing with the military Salyuts are:

"Soviet Launches Of More Military Salyuts Expected" in *Aviation Week & Space Technology*, 4 Dec 1978, p17.

Clark, Phillip S., "The Design Of Salyut Orbital Stations" in *Spaceflight*, vol 23 (1981), pp257-258.

Johnson, Nicholas, "The Military And Civilian Salyut Space Programmes" in *Spaceflight*, vol 21 (1979), pp364-370.

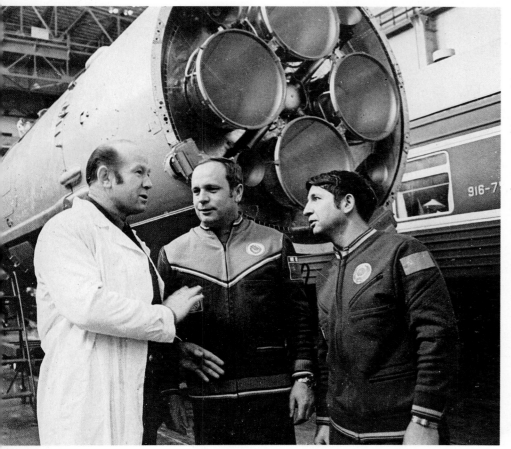

Left: *Leonov (left) advising Viktor Gorbatko and Yuri Glazkov about their forthcoming Soyuz 24 launch. They are in the assembly shop (MIK) at Tyuratam in front of the booster third stage.*

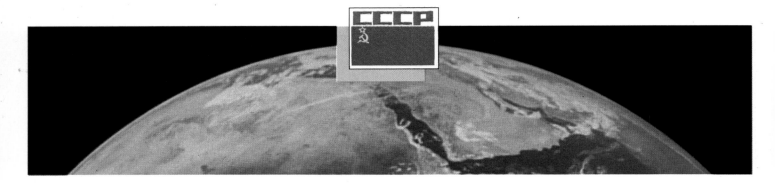

Chapter 8: The Civilian Salyuts

The civilian Salyut programme was less successful than the military one with which it alternated in that only one of the three stations launched actually hosted cosmonauts. Salyut 4 in 1975 hosted two pairs of cosmonauts, while a third pair failed to reach orbit.

Following the Soyuz 11 accident the spacecraft underwent some major design changes, while Salyut also underwent modification. Before the new ferry actually participated in a Salyut mission, four unmanned tests and a single manned mission were conducted to verify its safety. Manned flights had been set to resume in 1972, but the loss of a Salyut that summer followed by the abandonment of two Salyuts – one military, one civilian - meant that there was ample time for further testing of Soyuz before Soyuz 12 was launched on a manned test flight.

Details of the Salyut 4 orbital station have been released by the Soviets, and we may logically assume that the two earlier failed missions shared the same design. The basic structure of Salyut 1 was retained; externally the only major difference being the replacement of the two pairs of Soyuz solar panels on Salyut 1 with three steerable solar panels mounted on the exterior of the small work compartment of Salyut 4. Each panel was about 3m by 7m, giving a total area in excess of $62m^2$; this compares with $42m^2$ for the Salyut 1/Soyuz combination.

Internally the layout of Salyut 4 was virtually identical with that of Salyut 1, although instruments and equipment had been improved because of the longer missions planned. While interior photographs of Salyut 1 revealed a large conical instrument container in the work compartment

to which the Soviets did not generally allude (although the cone does seem to have been Station 3), the Salyut 4 instrument was readily identified as a solar telescope, OTS-1. The experiments carried on Salyut 4 are noted in the accompanying table, and discussed in connection with the Soyuz 17 and Soyuz 18 missions.

At the rear of Salyut 4 was the standard Soyuz propulsion system, KTDU-35. Because Salyut 4 would be operating at a higher altitude than did Salyut 1, its orbital decay would be diminished, requiring fewer manoeuvres to raise the orbit and thus less on-board propellant. It will be recalled that Salyut 1 carried twice the propellant load of a normal Soyuz KTDU-35 engine.

The Soyuz Ferry

During 1970 modifications had been made to the original manned Soyuz to convert it to a ferry craft for Salyut; these were noted in chapter 6. Following the loss of the Soyuz 11 crew, further changes were made. As a direct result of the Soyuz 11 accident, the spacecraft was converted into a two-manned vehicle, the place of the third cosmonaut being taken by an extra life-support system. The two cosmonauts would now wear pressure suits at launch, docking, undocking, and re-entry, and these would be connected to the extra life support system in Soyuz.

The second major change to Soyuz was possibly not a direct result of the Soyuz 11 loss. Its solar panels were deleted, and from launch until docking with Salyut it had to rely on chemical batteries located in the instrument module. Once the Soyuz had docked, these batteries could be recharged from the power generated from

the Salyut electrical system. Of course, any Soyuz which failed to dock with Salyut was constrained to return to Earth within about two days before the chemical batteries were exhausted. In such circumstances, once the docking attempt was abandoned, the crews powered down the Soyuz during the second day in orbit.

The mass of the new "ferry Soyuz" is normally quoted at 6,800kg by the Soviets, although some listings do include individual ferry Soyuz masses (normally, the combined Salyut/Soyuz masses are quoted). The propellant mass was 500kg, the same as for Soyuz 10 and Soyuz 11. A list of the test flights of the Soyuz ferry vehicle during 1972-1974 is given in tabular form.

The 1972 Missions

Following the Soyuz 11 loss a period of almost a year elapsed before there was another launch in the Salyut-Soyuz programme. In March 1972 the Agence France Presse reported that a new Salyut would be launched in May, and that two visits – each involving two cosmonauts – would be

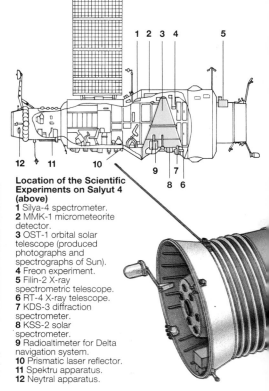

Location of the Scientific Experiments on Salyut 4 (above)
1 Silya-4 spectrometer.
2 MMK-1 micrometeorite detector.
3 OST-1 orbital solar telescope (produced photographs and spectrographs of Sun).
4 Freon experiment.
5 Filin-2 X-ray spectrometric telescope.
6 RT-4 X-ray telescope.
7 KDS-3 diffraction spectrometer.
8 KSS-2 solar spectrometer.
9 Radioaltimeter for Delta navigation system.
10 Prismatic laser reflector.
11 Spektru apparatus.
12 Neytral apparatus.

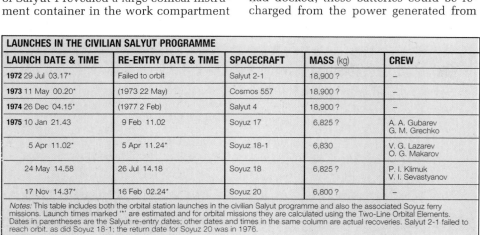

LAUNCHES IN THE CIVILIAN SALYUT PROGRAMME				
LAUNCH DATE & TIME	**RE-ENTRY DATE & TIME**	**SPACECRAFT**	**MASS** (kg)	**CREW**
1972 29 Jul 03.17*	Failed to orbit	Salyut 2-1	18,900 ?	–
1973 11 May 00.20*	(1973 22 May)	Cosmos 557	18,900 ?	–
1974 26 Dec 04.15*	(1977 2 Feb)	Salyut 4	18,900 ?	–
1975 10 Jan 21.43	9 Feb 11.02	Soyuz 17	6,825 ?	A. A. Gubarev G. M. Grechko
5 Apr 11.02*	5 Apr 11.24*	Soyuz 18-1	6,830	V. G. Lazarev O. G. Makarov
24 May 14.58	26 Jul 14.18	Soyuz 18	6,825 ?	P. I. Klimuk V. I. Sevastyanov
17 Nov 14.37*	16 Feb 02.24*	Soyuz 20	6,800 ?	

Notes: This table includes both the orbital station launches in the civilian Salyut programme and also the associated Soyuz ferry missions. Launch times marked '*' are estimated and for orbital missions they are calculated using the Two-Line Orbital Elements. Dates in parentheses are the Salyut re-entry dates; other dates and times in the same column are actual recoveries. Salyut 2-1 failed to reach orbit. as did Soyuz 18-1; the return date for Soyuz 20 was in 1976.

TEST LAUNCHES OF THE SOYUZ FERRY

LAUNCH DATE AND TIME	RECOVERY DATE AND TIME	SATELLITE	CREW	MASS (kg)	ORBITAL EPOCH	INCL (deg)	PERIOD (min)	ALTITUDE (km)
1972 26 Jun 14.53*	2 Jul 13.54*	Cosmos 496	–	6,675 ?	26.74 Jun 28.74 Jun	51.72 51.74	89.65 89.20	184-331 185-285
1973 15 Jun 06.00*	17 Jun 06.01*	Cosmos 573	–	6,675 ?	15.43 Jun 16.55 Jun	51.54 51.55	89.53 89.27	191-312 191-286
27 Sep 12.18	29 Sep 11.34	Soyuz 12	V. G. Lazarev O. G. Makarov	6,720	27.63 Sep 27.81 Sep	51.75 51.58	88.60 91.24	181-229 327-344
30 Nov 05.20*	29 Jan 05.29*	Cosmos 613	–	6,675 ?	30.34 Nov 1.08 Dec 3.49 Dec 5.47 Dec	51.61 51.60 51.60 51.62	89.04 89.11 89.13 91.06	187-268 188-273 194-270 253-400
1974 27 May 07.25*	29 May 07.50*	Cosmos 656	–	6,675 ?	27.43 May 27.74 May 28.49 May	51.82 51.66 51.60	89.59 90.12 90.19	186-323 193-367 194-374

Notes: This table lists all of the tests of the Soyuz ferry prior to its use for the first Salyut 3 visit. Because of its extended duration, it is possible that Cosmos 496 had solar panels, but the other flights did not. Launch times and landing times for the Cosmos flights are calculated from the Two-Line Orbital Elements and are marked * (Cosmos 613 was recovered in 1974); the Soyuz 12 times were officially announced. All the orbital data are derived from the Two-Lines.

made. The commanders for the missions would be rookie cosmonauts Vinogradov and Voronin and their flight engineers would be Sevastyanov and Kubasov respectively; Popovich was said to be back-up commander. Interestingly, this was one of the first reports that the new Soyuz would carry only two men.

The May launch date was not met, but on 26 June an unmanned Soyuz was launched. Named Cosmos 496, the apogee of the orbit in excess of 300km was unusual. Two days later, apogee was reduced by about 25km, and after a further four days the spacecraft was recovered. In retrospect, the 6-day mission of Cosmos 496 was strange in view of the limited lifetime of the batteries in the new ferry Soyuz. Possibly only the modified descent module was carried on this flight attached to a Soyuz 11 class instrument module with solar cells? Certainly an accurate painting of the civilian Salyut released in 1973 showed a Soyuz ferry with solar wings, although since the Soyuz also sports the old Soyuz 4 docking system, that part of the painting is not totally accurate. However, it does raise the question as to whether the new Soyuz ferry without solar panels was intended to be introduced in

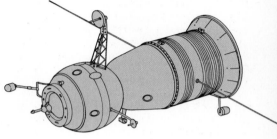

Salyut 4 Civilian Orbital Station
1 Soyuz manned transport craft.
2 EVA/access hatch (on side hidden in this illustration).
3 Rendezvous antenna.
4 One of three steerable solar panels.
5 Gas storage for life support system.
6 Food and storage lockers.
7 Attitude control jets.

8 KTDU-35 main propulsion system (identical with that on Soyuz).
9 Rendezvous transponder.
10 Propellant tanks for main propulsion system.
11 OST-1 apparatus.
12 Chibis lower body, negative pressure suit.
13 Treadmill.
14 Table (can be folded for storage).
15 Main control console (based upon that for Soyuz).

three larger steerable panels. A hatch cover was incorporated into the forward work compartment bulkhead, and this would have allowed the transfer compartment to act as an airlock during spacewalk activities. Although spacewalks were apparently planned (for the Soyuz 18 mission ?), they were cancelled – possibly as a result of the April launch abort and the delay in re-manning the station.

Soyuz Modifications
The modified Soyuz ferry, introduced after the Soyuz 11 accident, is seen above. The solar panels are deleted, with whip antennas in their former position. The batteries on Soyuz would be recharged from the Salyut power supply after docking with the station. The crew size was reduced to two men in pressure suits, the space for the third man going to extra life support systems.

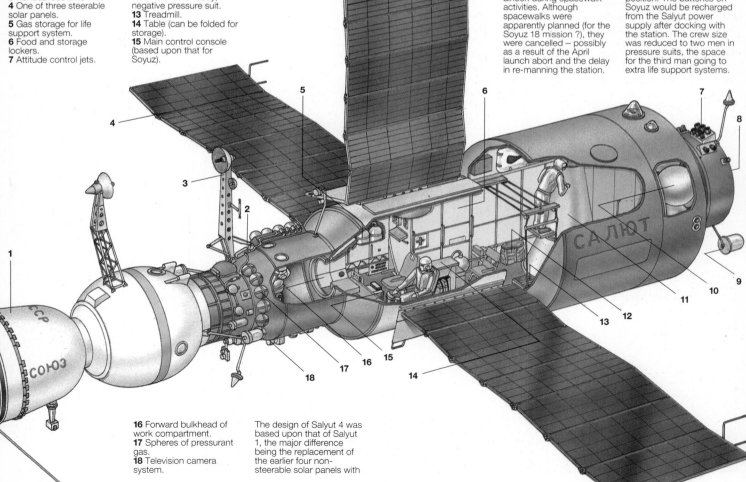

16 Forward bulkhead of work compartment.
17 Spheres of pressurant gas.
18 Television camera system.

The design of Salyut 4 was based upon that of Salyut 1, the major difference being the replacement of the earlier four non-steerable solar panels with

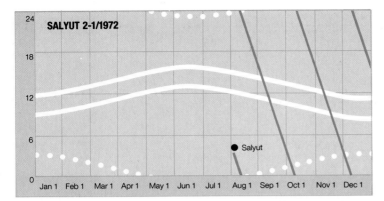

SALYUT 2-1/1972

24 — 18 — 12 — 6 — 0

● Salyut

Jan 1 Feb 1 Mar 1 Apr 1 May 1 Jun 1 Jul 1 Aug 1 Sep 1 Oct 1 Nov 1 Dec 1

1972 or in 1973 for Salyut missions.

In the early morning of 29 July the second Salyut orbital laboratory was launched, but it failed to reach orbit after a second stage engine failure in the Proton SL-13 booster. From the launch time, the landing windows can be calculated, and two missions can be "guestimated". The two landing windows for 1972 would have been September 4 – 13, and November 9 – 16. For Salyut 1 the planned durations were 4-6 weeks; assuming that two 30-day missions were planned for 1972, the launch opportunities would have been 5 – 14 August and 10 – 17 October. There is no way of proving that this is what the Soviets were planning, but it does show the opportunities which were available in 1972.

Possibly Vinogradov, Sevastyanov, Voronov and Kubasov were ready for flights to the July Salyut, had it attained orbit. Vinogradov and Voronov have not made spaceflights to the end of 1987, although they were known to be still in training during 1974-1975 because the Apollo-Soyuz astronauts met them in Moscow. If Popovich was involved in this programme he was soon switched to the military Salyut training group, while Kubasov and Sevastyanov probably trained for the next civilian Salyut attempts. Kubasov eventually flew on Apollo-Soyuz, while Sevastyanov flew to Salyut 4.

There are rumours of another Salyut being readied for launch in September-October, but booster problems prevented the launch itself. However, public sources do not confirm these reports.

Cosmos 557: A Failed Salyut

In 1973 the Americans were planning the launch of their first (and, so far, only) flown space station programme, the Skylab. Skylab 1, the orbital laboratory, was launched in May 1973, and this was followed by three-manned visits lasting for 28

Above: *Although dated 1977, cropped versions of this picture appeared in 1973 showing the new design of the Salyut orbital station. Curiously, the otherwise accurate painting shows the Soyuz still with solar panels.*

MANOEUVRES OF THE COSMOS 557 ORBITAL STATION			
PRE MANOEUVRE ORBIT		**POST MANOEUVRE ORBIT**	
Epoch (1973)	Altitude (km)	Epoch (1973)	Altitude (km)
Initial orbit		11.13 May	214-249
11.13 May	214-249	11.75 May	217-245
21.17 May	163-178	22 May	Orbital decay

Notes: Cosmos 557 was a Salyut which failed soon after orbital injection – if not during the launch phase itself. The manoeuvre shown above may or may not have taken place: it is possible that the first set of Two-Line Orbital Elements for the satellite is incorrect. The satellite decayed rather than being de-orbited towards a controlled re-entry.

COSMOS 557/1973

24 — 18 — 12 — 6 — 0

● Cosmos 557

Jan 1 Feb 1 Mar 1 Apr 1 May 1 Jun 1 Jul 1 Aug 1 Sep 1 Oct 1 Nov 1 Dec 1

days, 59 days and 84 days respectively. It would seem that the Soviet Union was planning to fly two Salyut stations in 1973 and to man them simultaneously in an attempt to steal some of Skylab's thunder. One of these was Salyut 2 which failed after some days in orbit, and the other was Cosmos 557, a Salyut which must have failed either during the launch or very soon after orbital injection (hence the designation Cosmos rather than Salyut). One of these missions would have been military and the other civilian in nature and had the Soviet plan actually been successful, their policy on the release of information concerning the two missions would have been interesting.

Salyut 2 was considered in the previous chapter. The second Salyut planned for the year was launched on 11 May, to be called Cosmos 557. As noted in the table of Cosmos 557 manoeuvres, there might have been a small orbital manoeuvre completed, but it is also possible that the Goddard Two-Line Orbital Elements are in error, and the initial orbit was erroneously tracked. Cosmos 557 decayed from orbit eleven days after it was launched. As with the July 1972 failure and Salyut 2, it is possible to project forward the orbit of Cosmos 557, and so to calculate the landing windows which could have been used. These were 6 June – 13 June; 6 August – 15 August; 14 October – 22 October.

The later Salyut 4 mission was scheduled to receive two visits, one of 30 days and another of about 60 days, duration. However, one cannot necessarily assume that missions of a similar duration were planned for the Cosmos 557 Salyut. Salyut 4 took about six days to attain its operational orbit, and the first manned visit began 12 days later. Therefore, we may reasonably assume that the first manned flight to Cosmos 557 was scheduled to begin about two-and-a-half weeks after the station's launch; that is on 28 May. That launch date would have allowed a visit lasting up to 16 days with a recovery on the last day of the landing window. This is less than the duration of the projected first visit to Salyut 2 suggested in the previous chapter, but the Cosmos 557 crew would have been launched just before the end of the first Salyut 2 mission, and would therefore have maintained a continued Soviet manned presence in space.

A second visit to Cosmos 557 may have been scheduled for the middle of August, allowing a recovery after two months during the October landing window. Of course, as with Salyut 2, these calculations do not claim to represent what the Soviets were actually planning; it is only possible to indicate the options that were open to the Soviet mission planners.

We may also speculate about the crews to be launched to Cosmos 557 if it had been flown successfully. Most probably one visit would have been completed by Lazarev and Makarov who later flew Soyuz 12 and who were launched to Salyut 4 two years later. When the Soviets announced the crews for the Apollo-Soyuz Test Project (ASTP) mission, they released photographs of the prime crew comprising Leonov and Kubasov in an advanced stage of training (although this was not stated). Since they had trained together for Salyut 1, they were probably also scheduled for a Salyut mission in 1973 before the failures caused their reassignment to ASTP.

Since the summer of 1973 might have seen the simultaneous operations of both a military and a civilian Salyut, it is interesting to summarise the differences between the two programmes. This is provided in the accompanying table.

Testing Soyuz

Following the failures of the military Salyut 2 and the civilian Salyut/Cosmos 557, the Soviet space planners continued to test the Soyuz ferry designed to carry cosmonauts to the orbital stations. It is often assumed that all the ferries to be used following the Soyuz 11 accident would have shed their solar panels, but this may not have been the case. Just after the Salyut 2 launch, a painting by Leonov and Sokolov was released which showed the new civilian Salyut design and the probe of an approaching Soyuz ferry; the section of the painting depicting the ferry itself was originally not shown. Some years later, the uncropped painting was published, and this showed a Soyuz 11-type ferry approaching the Salyut. Perhaps, therefore, the Soyuz ferries launched in 1972 and 1973 were still scheduled to carry solar cells? Alternatively, the painting may represent a view of the Salyut-Soyuz programme shown as it would have evolved if the Soyuz 11 accident and re-design had not taken place.

On 15 June an unmanned Soyuz was launched under the identity of Cosmos 573. The initial orbit was like that of Cosmos 496, launched nearly a year earlier, and the following day the apogee was lowered – again mirroring the events of the Cosmos 496 mission. However, Cosmos 573 was in orbit for only two days and was

COMPARISON OF MILITARY AND CIVILIAN SALYUT MISSIONS

	MILITARY PROGRAMME	CIVILIAN PROGRAMME
Operating altitude (km)	250-280	320-350
Transmission frequencies (MHz)	19.944 143.625	15.008 922.75
Crew composition	Military commander Military engineer	Military commander Design bureau civilian engineer
Capsule recovery	Capsule carried and recovered	No capsule carried
Experiments	Earth resources Biological/medical Technology Mainly small items of equipment required	Astronomical Biological/medical Technology Major scientific equipment required
Information revealed	No sketches or cut-away diagrams; few interior pictures	Full cut-away diagram released; detailed interior pictures

Notes: Although not claimed to be conclusive, this table is an attempt to show that there are sufficient differences between the presumed military and civilian Salyut programmes for them to have been different programmes in reality. All information is from Soviet sources, other than the transmission frequencies which have been monitored by the Kettering Group over the years.

Left: *Lazarev and Makarov who flew the Soyuz 12 ferry test, following the delays in the Salyut programme. Probably they were scheduled for a Cosmos 557 visit.*

Below: *The Salyut training mock-up, with the orbital module of the Soyuz trainer on the right hand side of the picture. This was a full-scale Salyut, in which the cosmonauts trained for their flights.*

then successfully recovered.

The Soviet Union realised that following the Cosmos 557 failure a long gap would ensue before a new Salyut would be ready for launch. Therefore, it was decided to fly two manned Soyuz test missions in order to re-establish a manned presence in space. The first was Soyuz 12, and the second was Soyuz 13 (considered in the next chapter). On 27 September, only a few days after the second American Skylab visit had been completed, Soyuz 12 was launched on a solo mission. The crew comprised two cosmonauts: V.G. Lazarev and O.G. Makarov. Their mission was basically to test the re-designed Soyuz ferry – without solar panels – with the minimum of scientific research being undertaken. Apart from testing the Soyuz systems, the only experimental work carried out by the Soyuz 12 crew was photography of the Earth. Makarov took Earth resources photographs using a hand-held multispectral camera, while Lazarev simultaneously took photographs of the same areas using a normal camera. Drawings of the spacecraft indicate that it carried a docking system; these raise the question of whether it was originally due to fly to Salyut 2 or Cosmos 557.

The initial orbit, 194-249km, was typical of a manned Soyuz. However, within seven hours Soyuz had manoeuvred to a higher circular orbit with announced parameters of 326-345km. This altitude was reminiscent of the apogees of the unmanned Cosmos 496 and Cosmos 573 missions and was close to the altitude scheduled for the civilian Salyut programme. During the first day in orbit, it was announced that the missions would last for two days – probably to forestall rumours of a flight being terminated earlier than scheduled when re-entry took place. Accordingly, Soyuz 12 was successfully recovered after a flight lasting for slightly less than 48 hours.

Two other unmanned Soyuz tests flew before Salyut operations were resumed. At the end of November, Cosmos 613 was launched and over a period of six days was manoeuvred to a 250-400km orbit. It was then powered down to simulate conditions when docked with a Salyut, and was later recovered after a flight lasting for 60

MANOEUVRES OF THE SALYUT 4 ORBITAL STATION			
PRE MANOEUVRE ORBIT		**POST MANOEUVRE ORBIT**	
Epoch (1974)	Altitude (km)	Epoch (1974)	Altitude (km)
Initial orbit		26.29 Dec	215-252
27.04 Dec	211-250	27.41 Dec	215-286
30.21 Dec	211-284	29.40 Dec	276-344
30.21 Dec	277-342	30.40 Dec	338-351
(1975)		(1975)	
Soyuz 17: docked 11 January			
		12.53 Jan	336-349
Soyuz 17: undocked 9 February			
		9.75 Feb	334-346
22.74 Mar	330-340	22.93 Mar	337-350
31.87 Mar	337-349	1.82 Apr	339-351
Soyuz 18-1: launch failure 5 April			
		5.82 Apr	338-350
12.48 May	332-348	14.06 May	348-355
22.38 May	347-354	22.83 May	343-351
Soyuz 18: docked 25 May			
		26.57 May	339-349
24.47 Jul	335-344	25.55 Jul	335-360
25.55 Jul	335-360	25.62 Jul	342-361
Soyuz 18: undocked 26 July			
		27.19 Jul	341-362
3.54 Nov	330-351	4.8 Nov	344-353
Soyuz 20: docked 19 November			
		20.24 Nov	342-351
(1976)		(1976)	
Soyuz 20: undocked 16 February			
		16.39 Feb	335-341
(1977) 2.22 Feb	186-187	(1977) 2 Feb	De-orbited

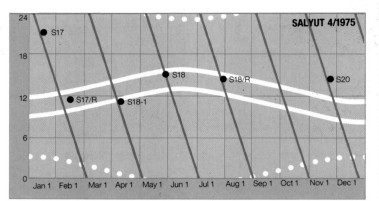

SALYUT 4/1975

Launch and Landing Windows
This is the launch and landing graph for Salyut 4 during 1975. The launch date of Soyuz 17 clearly indicated a four week flight, while that for the April failure (Soyuz 18-1) suggested that a flight of 8-9 weeks was planned. The replacement mission flew for nine weeks, the longest Soviet manned flight to that time. No further manned flights took place, but the unmanned Soyuz 20 flew for three months docked with Salyut, ready for longer manned flights in 1977.

SALYUT 4 EXPERIMENTS	
NAME	**DESCRIPTION**
Astronomical Experiments	
OST-1	Orbiting Solar Telescope, for photographs and spectrographs of the Sun
Filin-2	X-ray spectrometer
RT-4	X-ray telescope
ITS-K	Infra-red spectrometric telescope
SSP-2	Solar spectrometer
MMK-1	Micro-meteorite detector
Emissiya	Measurement of neutral particles in the atmosphere
Medical Experiments	
Chibis	Physical conditioning suit
Rezed-5	Pulmonary ventilation recorder
Polynom	Monitoring of body parameters
Amak-3	Blood analyser
Plotnost	Bone tissue density monitor
Tonus-2	Muscular microelectric stimulator
Levka-3	Blood vessel monitor
Biological Experiments	
Various experiments: Oazis, Bioterm-2M, Bioterm-3, Bioterm-4, KM, FKT	

Notes: This list identifies most of the main experiments carried out during the Salyut 4 mission; sometimes it is difficult from Soviet sources to identify specifically experiments which are actually discrete and those which are part of a larger experiment (eg, one of the instruments on the OTS-1 telescope experiment).

days. The success of this procedure meant that the new Soyuz ferry was now qualified to support a manned mission lasting for up to two months.

Following this mission, it was a surprise when another unmanned Soyuz was launched under the identity of Cosmos 656. Unlike the earlier unmanned tests, Cosmos 656 did not manoeuvre extensively, although some minor manoeuvres were completed during the first day. Cosmos 656 was recovered after two days in orbit. A month later, Salyut 3 was launched, and it was followed into space by the Soyuz 14 and Soyuz 15 spacecraft as detailed in the previous chapter. In December Soyuz 16 was launched as a manned rehearsal for the 1975 ASTP mission.

The Launch of Salyut 4

On 26 December 1974 – the day following the announcement that the Salyut 3 flight would soon be terminated – a Proton SL-13 booster was launched from Tyuratam bearing a new space station as its payload. The launch announcement came nearly 4 hours after the event, quoting the orbit as 219-270km. The announcement also stated that the purpose of the station was to test further the design and on-board systems and to conduct scientific and technical studies and experiments in orbit. The next announcement relating to Salyut 4 came on 6 January 1975 when it was revealed that following a series of orbital corrections the altitude was now 343-355km. In fact, the table of manoeuvres shows that these had actually been completed during the first four days in orbit, while the arrival of first crew for the station was awaited.

Salyut 4 was the first civilian Salyut to be successfully operated, and manned missions took place during its first eight months in orbit. Unlike the previous Salyut 3 and the subsequent Salyut 5, which were in the military programme, the emphasis with Salyut 4 was on a programme of scientific experiments, particularly in the astronomical field. The table provides a list of the experiments which have been announced for Salyut 4.

Soyuz 17: A Month in Orbit

On 10 January (11 January Moscow Time) the first visiting mission to Salyut 4 was launched. The two-man crew comprised Lt-Col A.A. Gubarev and G.M. Grechko (a civilian engineer), who had been paired as back-ups for the Soyuz 12

mission in 1973; they were probably involved with the planned Cosmos 557 visits during the same year. The initial Soyuz 17 orbit was not announced, but following a course correction the altitude was 293-354km. The docking of Soyuz with Salyut 4 was completed without any problems. The first phase of the docking was automatic, and bringing the Soyuz to a distance of 100m from Salyut. From this point on, the crew completed the manoeuvre manually, the docking coming at 01.25 on 12 January. After docking the cosmonauts checked the seals of the hatches before opening them and moving into Salyut.

Almost daily progress reports were released during the mission and the cosmonauts were clearly involved in a programme that concentrated on scientific and biological experiments. This was the first Soviet manned flight that was planned to last significantly longer than the Soyuz 9 mission in 1970 (17.7 days), and although the Soviets had access to the results of the longer American Skylab visits of 1973 and 1974, they clearly wished to obtain their own data relating to the human body's adaptation to zero gravity.

Commenting of Salyut 4 on 14 January, the former cosmonaut and Salyut designer Dr Konstantin Feoktistov said that the operating altitude of 350km would ensure that the propellant consumption would be reduced to half that required for the lower orbit Salyuts which had to compensate more frequently for the affects of orbital decay.

On 16 January the two Salyut cosmonauts switched on the solar telescope for the first time and began to test it. The previous day, it had been confirmed that a new teleprinter named "Stroka" was being used by the cosmonauts for communications. This allowed the ground controllers to send instructions to the cosmonauts which could be read and acted upon at a later time, rather than having continually to interrupt the crew during their work. The Filin X-ray telescope was used on 17 January to observe the Crab Nebula (the remains of a supernova – an exploding star – which had erupted in 1054).

It was revealed on 19 January that Salyut 4 was using ion sensors to govern the station's orientation. It was claimed that these provided the most efficient form of orientation and had the most rapid reaction time. The sensors detect the ion flow around the Earth and can orient the long axis of the station relative to the known direction of that flow. An innovation in what was an otherwise very regular programme of research came on 3 February, when the cosmonauts resprayed the mirror of the OST-1. The original surface of the mirror had become contaminated during the first three weeks of use, and apparently the Soviets had planned for the respraying to be an option should it be required. The operation was conducted by Grechko from a remote control panel.

On 7 February the cosmonauts began preparations for their return to Earth as Salyut systems were checked and powered down. The results of the mission experiments were transferred to the Soyuz 17 descent craft. Two days later at 06.08 Soyuz undocked from Salyut and at 11.03 the spacecraft landed safely in the prearranged landing zone in Kazakhstan. After 29 days, the longest Soviet manned spaceflight – and the world's third longest space mis-

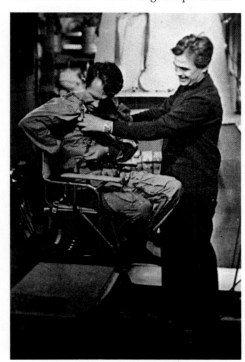

Above: *Makarov (left) and Lazarev working in the Salyut trainer, preparing for their planned two month visit to Salyut 4. The mission failed when their launch vehicle developed problems.*

Above: *Grechko (left) and Gubarev in the Salyut trainer, preparing for their four week mission to Salyut 4. Their's was the longest Soviet mission to that time.*

Left: *Inside the Soyuz assembly building. At the left are the first stage core and strap-on cluster with an orbital stage mated to a Soyuz shroud to the right.*

sion to that time – had been successfully completed. After the earlier false starts, the civilian Salyut programme was fulfilling its promise of scientific returns.

Soyuz 18-1: Launch Failure

Following the month-long flight of Soyuz 17, it was expected that the next mission to Salyut 4 would last for two months, especially as Cosmos 613 had proved that a powered-down Soyuz could fly for such a period. There were no significant orbital manoeuvres during the Soyuz 17 mission, and the orbit of Salyut slowly decayed until the evening of 22 March, when the orbit was raised to an announced altitude of 343-356km. During 31 March - 1 April a smaller manoeuvre was carried out which precisely positioned the station's orbit in readiness for a manned launch which took place on 5 April at 11.02. No launch was announced on 5 April, but the next day the following statement was released by Tass:

"The Flight Control Centre reports that on April 5, 1975 a carrier rocket with a manned Soyuz spacecraft was launched from the Soviet Union to continue experiments jointly with the Salyut 4 station. The crew on board consisted of Col Vasiliy Lazarev and Oleg Makarov.

On the third stage stretch the parameters of the carrier rocket's movements deviated from the pre-set values and an automatic device produced the command to discontinue the flight under the programme and detach the spacecraft for return to Earth.

The descent module soft-landed southwest of the town of Gorno-Altaysk in Western Siberia. The search and rescue service brought the cosmonauts back to the cosmodrome. Vasiliy Lazarev and Oleg Makarov are feeling well."

Normally, such a candid announcement would not have been expected from the Soviet Union, but the joint Apollo-Soyuz mission was only three months away and when the Soviet engineers met their American counterparts after the launch failure, their expressions revealed that they had experienced a set-back. Only openness would prevent rumours circulating, and therefore the Soviets decided to come clean about the failure. Slowly, more details emerged. The first phase of the launch had been successfully completed with the four strap-ons of the SL-4 booster having separated. The core stage was shut down ready for separation from the orbital stage. However, only half the explosive bolts fired and thus the core was still attached to the orbital stage when its ignition took place. The launch vehicle was out of control.

By this time, of course, the payload shroud and tower had separated, and the only way of escape was to fire the Soyuz propulsion system to pull the complete spacecraft away from the errant booster. Once this was done, the Soyuz was rotated through 180° to prepare for re-entry and landing. During the ballistic re-entry the cosmonauts were subjected to a deceleration load of 14-15g. A major concern for the cosmonauts was where they would eventually land. They were heading towards the Sino-Soviet border, and they were worried that they would come down on the wrong side! When it came, the landing was only 320km short of the border. A full Soviet account of the landing has not been published, although it must have been many hours before any rescue teams reached the cosmonauts. One may wonder whether their minds went back to Voskhod 2 which had also landed far off course; the crew that time had to retreat to their descent craft while wolves howled in the distance.

In 1981 a book by Glushko revealed some numerical data for the mission. Naming the flight Soyuz 18-1 (rather than the Western Soyuz 18A designation or simply the "April 5 Anomaly"), the spacecraft mass was quoted as 6,830kg. The peak altitude had been 192km and the spacecraft landed 1,574km down-range from the launch site. The flight had lasted for 21 minutes 27 seconds - unintentionally, the longest suborbital flight flown.

The landing window which Soyuz 18-1 should have used extended from 26 May to 7 June, and this implies that a mission lasting for 51-63 days was being planned. If this was the case, the mission would have been completed before the launch of the Apollo-Soyuz mission on 15 July.

Soyuz 18: Two Months Aloft

The landing opportunities relating to Salyut came at intervals of approximately 60 days and therefore a new launch to Salyut 4 for a mission of the same duration as that planned for Soyuz 18-1 could be expected at the end of May or early in June. However, in order to allow a mission to be flown which would land in the middle of the July landing window, and which would not conflict with the planned launch of two Venus probes in early June, the next Soyuz mission was slightly increased in duration by about 4-7 days, thus dictating an earlier launch date than would have been the case if a straight repeat of Soyuz 18-1 was planned.

Between 1 April and mid-May the orbit of Salyut 4 slowly decayed, but during 12-14 May the orbit was raised. On 22 May a further slight adjustment was made to the orbit ready for a manned launch. On 24 May Soyuz 18 was launched carrying two experienced cosmonauts: Lt-Col P.I. Klimuk (who had previously flown Soyuz 13) and V. I. Sevastyanov (who had flown Soyuz 9). This was the back-up crew for the aborted 18-1 flight. Their mission was to continue the work begun by Soyuz 17.

The initial orbit was announced as 193-247km; within twelve hours it had been raised to an announced 322-384km, somewhat higher than that of Salyut (announced as 344-356km). After being in orbit for a day, Soyuz 18 began an automatic approach to Salyut 4 and at a distance of 100m the cosmonauts took over for the final approach and docking. The docking at 18.44 took place beyond the area of radio contact with the Soviet Union and was not confirmed until the complex emerged from the blackout area and reestablished communications with the mis-

sion controllers. The cosmonauts transferred to Salyut 4 at about midnight on 25-26 May.

The actual docking manoeuvres have been described in some detail. The approach and docking phase began at 18.11 on 25 May when the spacecraft had a relative velocity of 12m/s. At a distance of about 1.5km the automatic system realised that the relative velocity was too high and the rate of approach was slightly reduced. At 18.21 the two craft were 800m apart and they were entering the Earth's shadow. At a distance of 100m the crew took over manual control and established an approach rate of 0.3m/s. The cosmonauts successfully docked in darkness and with no assistance from the controllers. Once inside Salyut, Klimuk and Sevastyanov began readying it for manned operations again. They completed a full check of the Salyut systems and experimental apparatus.

The mission of Soyuz 18 lasted for 63 days, and the activities for each day are too numerous to be described in detail here. The table provides a summary of the areas of research conducted during the mission, compared with the figures for Soyuz 17. The Western prediction that the Soyuz 18 mission would last for about two months was confirmed when, on 14 July, it was announced that the cosmonauts would return during the last ten days of July; this would be after the recovery of the Soviet craft that was participating in the Apollo-Soyuz Test Project (ASTP).

On 15 July Soyuz 19 was launched as the Soviet half of the Apollo-Soyuz mission, and for the first time the Soviet Union had to control two independent space missions (previous Soviet joint flights had been of spacecraft involved in related missions). The differing orbits of Salyut/Soyuz 18 and Apollo/Soyuz 19 meant that the opportunities for conversations between the two Soviet craft were limited to once every two days or so. On 16 July one such conversation between the two spacecraft was reported.

During 24-25 July the Soyuz 18 propulsion system was used to raise the orbit of the complex, an event that served the double purpose of conserving Salyut's propellant load and proving the viability of the Soyuz craft in anticipation of the landing. The new orbit was announced as 349-369km. With the closedown of Salyut 4 complete, at 10.56 on 26 July Soyuz 18 undocked from the station and after retrofire the spacecraft landed just over three

Above: *Following the Soyuz launch abort, Sevastyanov (left) and Klimuk – here posing in front of the Salyut simulator – were assigned to the replacement mission, launched six weeks later.*

hours later. The second longest manned space mission to that time (nearly 63 days) had been successfully completed, more than doubling the record for the longest Soviet manned flight.

In retrospect, the Soviets have hinted that there were problems with Salyut's environmental system which was being asked to support cosmonauts for two months longer than scheduled (if the Soyuz 18-1 mission had not failed, manned operations would have ended by early June). It has been reported that towards the end of the Soyuz 18 mission it was impossible to see out of the Salyut windows and that the interior of the station was covered with mould. Soviet statements, however, do not confirm this serious situation.

It is surprising that no spacewalks took place during either the Soyuz 17 or Soyuz 18 visits. It was stated in 1977 that plans were drawn up for spacewalks to be undertaken on (unspecified) Salyut missions, but that the pressure of the workload meant that these plans had to be cancelled. Comparing the design of the military Salyut with Salyut 4 leads one logically to suppose that the spacewalks were planned for Salyut 4 missions only. The military Salyut design did not include an EVA hatch while Salyut 4 did incorporate such a hatch on the front transfer compartment. Speculating further, the spacewalk(s) might have been planned to take place

SOYUZ 17 AND 18 EXPERIMENTAL WORK		
RESEARCH PROGRAMME	**DAYS OF WORK**	
	Soyuz 17	Soyuz 18
Solar	3	6
X-ray	3	8
Infra-red	2	–
Atmospheric	4	2
Natural resources	3	11
Biomedical	5	9
Technical experiments	3	6

Notes: This Table is based upon Figure 17 of Johnson's volume 'Handbook Of Soviet Manned Spaceflight'.

Above: *In orbit, Sevastyanov and Klimuk undertook the heaviest experimental programme scheduled for a Soviet manned mission up to that time.*

during the April 1975 Soyuz mission, but when the replacement mission was launched in May the Salyut environmental control system was by then too depleted to support re-pressurisation.

Soyuz 20

Following the recovery of Soyuz 18 some commentators predicted that a further manned flight to Salyut 4 would occur (at this time the problems with the environmental system were unknown), but as time went on this became increasingly unlikely. At the end of September an unmanned Soyuz identified as Cosmos 772 was launched, but since its orbital plane was about 100° away from that of Salyut, it could not have been directly related to the Salyut 4 programme.

The orbit of Salyut continued to decay, and at the end of October a mission progress report gave the altitude as 338-358km. During 3-4 November, a manoeuvre was completed which raised the orbit of Salyut, suggesting that it was being readied for another manned flight. Two weeks later a new Soyuz was indeed launched to Salyut 4, but it did not carry a human crew. Soyuz 20 was placed into an initial orbit with the announced parameters 199.7-263.5km, quoted with an unusual accuracy by the Soviets, on 17 November and two days later at 19.20 it docked at the front of Salyut 4. The pro-

gramme scheduled for this unmanned mission related to the testing of the on-board systems during a long space mission, as well as biological investigations. After docking, the orbit of the complex was announced as 343-367km.

The main technological objective of the Soyuz 20 mission was to validate a fully automatic rendezvous and docking system for Salyut and another unmanned craft. In this, Soyuz 20 was related to the development of the unmanned Progress cargo carriers which would be introduced when Salyut 6 was in orbit in 1978. It also proved that it would be possible to rescue a crew if their Soyuz failed; presumably, the defunct spacecraft could be cast off from Salyut and an unmanned craft launched to return the crew.

Further details of the Soyuz 20 mission were announced after the successful launch of Cosmos 782 on 25 November, this being an international biological satellite with a scheduled flight time of nearly three weeks. It was then disclosed that Soyuz 20 was carrying biological specimens, including turtles and plants. The effects of the weightless environment of these specimens would be studied on their

return to Earth. Little further in-flight information was given about the Soyuz 20 mission. On 16 February 1976 it was announced that Soyuz 20 had undocked from Salyut and been recovered; no further details were given. Later, it became known that the undocking had happened on 15 February at 23.04. The flight of Soyuz 20 increased the proven operating time of a Soyuz in Earth orbit from 63 days to about 90 days, although immediate advantage of this capability was not taken by the Soviet mission planners.

Salyut 4 Ends Its Mission

Following the recovery of Soyuz 20, Salyut 4 was abandoned as far as visiting missions of any kind were concerned. The manoeuvre immediately prior to the Soyuz 20 launch was the last that Salyut made until it was de-orbited on 2 February 1977. Despite the Soyuz 18-1 launch abort and the problems experienced on Soyuz 18 when the Salyut systems were being operated beyond their designed lifetime, the mission of Salyut 4 could be counted a total success. Previously, Soviet manned flights had only lasted up to 18 days, but now this record had been extended to 63 days, and Soyuz was proved for flights lasting up to three months. However, the overall duration record was still held by the final crew of the American Skylab, who had completed an 84 day mission.

In all, there had been three launches in this phase of the civilian Salyut programme, if the July 1972 Salyut which failed to reach orbit was not a military vehicle. In addition to the launch failure, another Salyut, identified only as Cosmos 557, was lost in orbit after a successful launch. However, Salyut 4 hosted missions for 28 and 62 days (plus one day on each mission to reach Salyut), more than equalling the time spent by crew on the two manned military Salyuts.

With the demise of Salyut 4, the civilian Salyut programme went into hibernation until September 1977 when the new generation Salyut 6 was launched. In 1976 and early 1977 operations centred on the military Salyut 5 station, discussed in the previous chapter, while in September 1976 a solo Soyuz mission flew in order to prove equipment ready for the next generation of Salyut. During the mid-1970s, the Soviets had three different manned programmes operating almost simultaneously: the military Salyuts, the civilian Salyuts and the solo Soyuz flights. It is with the latter missions that the next chapter is concerned.

FURTHER READING

There have been no books published which deal specifically with the civilian Salyut programme. As with the military Salyuts, the day-to-day activities for the civilian missions have been detailed in the contemporary issues of *Spaceflight* by Gordon Hooper.

Details of the problems said to have been encountered during the Salyut 4/Soyuz 18 visit are given in:

Oberg, James E., *Red Star In Orbit*, Random House, New York, 1981, pp137-138.

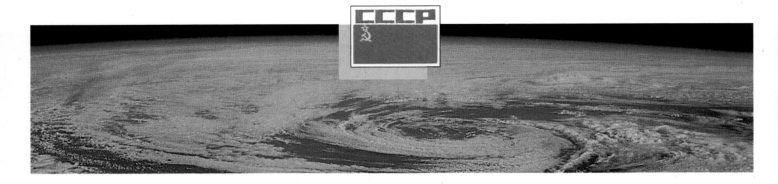

Chapter 9: **The Solo Soyuz Flights**

While the military and civilian Salyut programmes forged ahead, a short series of flights which were independent of the contemporary Salyut programme was also undertaken. As well as some unmanned Cosmos flights, the solo missions of Soyuz 13, Soyuz 16, Soyuz 19 and Soyuz 22 were flown, and these missions are discussed in this chapter. Although Soyuz 12 was flown as a "solo" mission, it was directly connected with the testing of the new Soyuz ferry vehicle and was therefore considered within the context of the Salyut Programme in the previous chapter.

With the exception of Soyuz 13, all the missions within the solo Soyuz programme were directly connected with the Apollo-Soyuz Test Project (ASTP), which, as an expression of *detente* between the superpowers, called for a manned Soyuz to dock in orbit with a manned American Apollo spacecraft. Soyuz 16 was a manned dress rehearsal for the joint flight; Soyuz 19 was the Soviet half of ASTP; and Soyuz 22 used the back-up ASTP Soyuz craft. Soyuz 13 and Soyuz 22 were also related to the Salyut programme in that Soyuz 13 carried the Orion and Oazis experiments – which had first flown on Salyut 1 – and Soyuz 22 carried the MKF-6 multi-spectral camera to qualify the camera prior to its utilisation on Salyut 6, launched in 1977.

Background to ASTP

During the 1960s hopes were expressed that the Soviet Union and the United States could join forces in the manned exploration of space, but in practice the two space programmes advanced with no regard to co-operative ventures. In 1969, as Apollo 11 was preparing for the first manned lunar landing, American enthusiasm for spaceflight was waning; it was even suggested in some quarters that since Apollo 10 had proved theoretically that a manned lunar landing was possible, there was no need to continue with an actual landing itself!

The Apollo programme had been conceived as a reply to an unspoken Soviet challenge to land men on the Moon, and once this goal was achieved the programme had attained its target. Plans did exist to fly lunar landing missions up to Apollo 20, as well as extensive space station missions within the Apollo Applications Programme (AAP). However, the lack of public enthusiasm resulted in the

Above: *The androgynous docking system developed for the Apollo-Soyuz mission is shown in the foreground, with a mock-up of the docking module behind it. The docking module was the one major piece of new equipment developed especially for the ASTP mission.*

lunar programme being cut back to missions up to Apollo 17, and the AAP cut back to only four launches within the Skylab Programme (there was the possibility of a fifth launch if a crew were unable to return to Earth from Skylab and a rescue ferry were required).

The final Apollo lunar mission was scheduled for 1972, with the Skylab programme slated for 1973-1974. After that, American manned space operations would cease until the partially-reuseable Space Shuttle system began operations in about 1979 (actually, the first flight slipped into 1981). A five year lull in American manned spaceflight seemed likely.

As the 1960s drew to a close, the Soviet space programme relaxed slightly its normal policy of strict secrecy and concrete proposals for joint missions were discussed between NASA officials and representatives of the USSR Academy of Sciences. In 1970 the possibility of a joint Soyuz-Apollo mission was mooted, although it was realised that there would be technical difficulties to overcome. The major problem was that the Soviet spacecrft used a normal atmosphere, while the American craft used oxygen at low pressure – some kind of intermediate module would be needed to allow inter-craft crew transfers.

In 1971 the first Salyut was orbited and successfully manned (although, as we have already seen, the Soyuz 11 mission ended in tragedy), and the plans for a joint mission switched to a possible Soyuz-Salyut-Apollo mission, the Salyut being modified to carry a rear docking unit. Soyuz would dock at the back and Apollo at the front. One may wonder what relationship this Salyut design may have had with the later Salyut 6 and Salyut 7 modifications which had both front and rear docking units. However, such plans were quickly abandoned, and the studies reverted to a more simple Apollo and Soyuz docking. After many technical meetings in both the United States and the Soviet Union, an agreement was signed by President Richard Nixon and Premier Aleksey Kosygin on 24 May 1972 which undertook to fly a joint Soyuz-Apollo flight in 1975.

At the time of signing, the penultimate Apollo lunar landing mission had just been successfully flown, while it was less than a year since the last Soviet manned space mission had ended with the deaths of the three cosmonauts. The planned Soviet resumption of manned space missions would be delayed, not by problems with Soyuz, but with major failures in the Salyut programme, and a further 16 months would elapse before Soyuz 12 would take cosmonauts back into orbit.

The Solo Soyuz

The solo Soyuz spacecraft chosen for ASTP incorporated features of both the ferry craft, which was used in the Salyut Programme starting in 1973, and the original Soyuz which had flown until 1971. Like the ferry Soyuz, the descent module was re-designed to allow two cosmonauts

in pressure suits to be carried at launch and recovery, the place of the third cosmonaut having been taken by additional life support systems in case of another sudden de-pressurization as had occurred on Soyuz 11. Of the various missions which took place within the solo Soyuz programme, the spacecraft for Soyuz 19 is the one that has been most widely described. Its measurements were as follows:

Instrument module: length 2.3m, diameter 2.2m, flaring to 2.72m (to interface with the launch vehicle's orbital stage), mass 2,654kg – including propellant.

Descent module: length 2.2m, maximum diameter 2.2m, mass 2,802kg.

Orbital module: length 2.65m, diameter 2.25m, mass 1,224kg.

These figures give a total spacecraft mass of 6,680kg, but at launch Soyuz 19's mass was 6,790kg; most probably the former figure excludes the crew sitting inside the spacecraft. The length of the basic ASTP spacecraft was 7.13m, but with the docking system it measured to 7.48m. The span of the two solar cell wings was 8.37m.

All the four manned missions flown in the solo Soyuz programme carried different assemblies at the front of the orbital module: Soyuz 13 carried the Orion-2 astronomical telescope package; Soyuz 16 had the ASTP docking unit with a second docking ring to simulate dockings in orbit; Soyuz 19 had the docking system alone; and Soyuz 22 had the MKF-6 multi-spectral camera. Some separate details have been published for Soyuz 22 which used the back-up spacecraft built for the Apollo-Soyuz mission: instrument module: total mass 2,600kg, including 500kg propellant; descent module: mass 2,750kg; orbital module: mass 900kg; MKF-6 camera system: diameter 1.3m, mass 250kg; total spacecraft length 7.6m; total mass was

Left: *The view from the American Apollo of the Soviet Soyuz 19 as the craft approached each other. It was as a result of the ASTP mission that the first detailed inflight photographs of the Soyuz spacecraft appeared.*

LAUNCHES WITHIN THE SOLO SOYUZ PROGRAMME

LAUNCH DATE AND TIME	RECOVERY DATE AND TIME	SATELLITE	CREW	MASS (kg)	ORBITAL EPOCH	INCL (deg)	PERIOD (min)	ALTITUDE (km)
1973 18 Dec 11.35	26 Dec 08.50	Soyuz 13	P. I. Klimuk V. V. Lebedev	6,560	18.61 Dec 18.86 Dec 19.29 Dec	51.56 51.58 51.58	88.85 88.91 89.30	189-247 186-255 223-256
1974 3 Apr 07.31*	13 Apr 05.05*	Cosmos 638	–	6,575 ?	3.43 Apr 5.60 Apr 6.66 Apr 7.65 Apr	51.77 51.73 51.73 51.77	89.49 89.06 89.91 89.83	190-309 190-266 240-300 258-274
12 Aug 06.24*	18 Aug 05.02*	Cosmos 672	–	6,575 ?	12.45 Aug 12.51 Aug 14.27 Aug 14.42 Aug 14.54 Aug	51.79 51.76 51.76 51.76 51.75	89.51 88.65 88.96 89.12 89.13	195-305 195-221 223-223 231-231 227-237
2 Dec 09.40	8 Dec 08.04	Soyuz 16	A. V. Filipchenko N. N. Rukavishnikov	6,800	2.58 Dec 3.26 Dec 3.56 Dec 4.12 Dec 7.45 Dec.	51.77 51.77 51.75 51.77 51.75	89.25 88.40 88.97 88.97 88.57	184-291 182-209 191-257 219-229 193-215
1975 15 Jul 12.20	21 Jul 10.51	Soyuz 19	A. A. Leonov V. N. Kubasov	6,790	15.57 Jul 15.75 Jul 15.94 Jul 16.37 Jul 16.68 Jul	51.76 51.76 51.76 51.76 51.76	88.59 88.75 88.70 88.68 88.98	191-218 185-240 198-223 190-229 218-231
1976 15 Sep 09.48	23 Sep 07.42	Soyuz 22	V. F. Bykovsky V. V. Aksyonov	6,510	15.53 Sep 15.78 Sep	64.75 64.78	89.30 89.59	184-296 249-260

Notes: This is a complete listing of the launches in the Solo Soyuz programme which was directly related to the Apollo-Soyuz project (as well as carrying some Salyut-related experiments in the cases of Soyuz 13 and Soyuz 22). It is possible that Cosmos 496 should be included here. Launch times and landing times for the Cosmos flights are calculated using the Two-Line Orbital Elements, and are marked *. All of the orbital data are derived from the Two-Lines.

about 6,500kg. The mass of Soyuz 22 has been refined to 6,510kg.

Although the solo Soyuz programme only included four manned flights, the Soyuz 19 spacecraft at least was assembled in accordance with the safety requirements of the United States, and therefore the programme introduced the Soviet space planners to the design controls operated by the USA. It was noticeable that following this mission, Soviet manned missions and equipment became more flexible and reliable, which suggests that lessons had been learned from American technicians. It is also probable that the best aspects of the ASTP Soyuz design were borne in mind as the design of the modified Soyuz-T spacecraft was being finalised.

Once the technicians had decided upon the actual mission plan for the Apollo-Soyuz flight, crews were assigned to the mission. The first crews to be identified were the American prime and back-up astronauts, the mission assignments being announced on 30 January 1973.

Prime commander	Thomas P. Stafford
Prime CMP	Vance D. Brand
Prime DMP	Donald K. Slayton
Back-up	Alan L. Bean, Ronald E. Evans, Jack R. Lousma
(CMP – Command Module Pilot, DMP – Docking Module Pilot)	

The Apollo was to be launched after the Soyuz had reached orbit. This sequence was dictated by orbital mechanics, but it also meant that if the Soviet mission was delayed, then the Apollo would not be launched. This was especially important since the Soviets were preparing two candidate spacecraft for the July 1975 launch, while there was only one Apollo available. The Apollo would be launched by a Saturn-1B, with the docking module carried beneath the manned module. The manned Command and Service Module (CSM) would separate from the final stage of the launch vehicle and after the four "petals" atop the S-4B stage separated, the CSM would dock with the Docking Module (DM) and pull it away from the rocket stage. The DM would act as an airlock module to allow crew transfer between the different atmospheric conditions in the Apollo and Soyuz spacecraft.

The Paris Air Show was held from 25 May to 3 June 1973, and the Soviet and American space planners arranged for a joint exhibit to be displayed there of full-scale Apollo and Soyuz spacecraft in a flight configuration, including a mock-up of the DM. Western technology was to the forefront, as Skylab was being rescued and the Anglo-French Concorde was displayed during this show. Soviet technological programmes, however, were suffering badly, as two Salyuts had just failed (Salyut 2 and Cosmos 557), while the Soviet technological nadir came as their supersonic airline, the Tupolev Tu-144, crashed while trying to emulate the spectacular display of Concorde at Paris.

While these events were receiving vast publicity, on 24 May the Soviet Union announced its crews for the Apollo-Soyuz mission. Surprisingly, they announced four teams each of two cosmonauts, four of whom were rookies. Since there were due to be two Soyuz spacecraft ready for launch in support of the flight, each had its own prime and back-up crews assigned:

	First Spacecraft	Second Spacecraft
Prime crew	Cdr A. A. Leonov, FE V. N. Kubasov	A. V. Filipchenko, N. N. Rukavishnikov
Back-up crew	Cdr V. A. Dzhanibekov, FE B. D. Andreyev	Y. V. Romanenko, A. S. Ivanchekov
(Cdr – Commander, FE – Flight Engineer)		

It would seem probable that the four cosmonauts in the prime teams had recently been re-assigned from the aborted Salyut/Cosmos 557 mission.

As time went on, it became clear that the Soviet Union was ready to commit more than two spacecraft to the joint mission. Already one unmanned test flight had taken place, and a second was scheduled; in addition, a manned dress rehearsal was planned prior to the joint flight itself. Prior to the beginning of ASTP tests, the Soviets flew the solo Soyuz 12 ferry missions (as part of the Salyut programme) and this was followed by Soyuz 13, which seems to have

Below: *Cosmonauts, astronauts and engineers at the Gagarin Training Centre. Front: Slayton (p), Evans (b), Leonov (p), Cernan, Stafford (p), Kubasov (p) and Brand (p). Standing: Ivanchenkov (b), Romanenko (b), Flannery, Rukavishnikov (p), Overmyer (s), Filipchenko (p), Lousma (b), Bean (b), Bobko (s), Dzhanibekov (b), Andreyev (b) and Forisenko. The men noted as "p" were prime crew members, "b" back-up crews and "s" support crews. Cernan was the US Deputy Director for ASTP.*

been a flight test of elements of the ASTP variant of the Soyuz craft.

Soyuz 13

In September 1973 Soyuz 12 had carried two men in orbit on a test mission designed to validate the new Soyuz ferry variant, and further manned activity was not expected until a new Salyut was in orbit. Meanwhile, in November 1973 the final US crew had been launched to the Skylab space station, and they completed a flight lasting for 84 days 1 hour 16 minutes, so setting a new duration record which would last for four years.

It was a surprise when Soyuz 13 was launched on 18 December, manned by two cosmonuauts, Pyotr Klimuk and Valentin Lebedev. There were some erroneous western reports that Cosmos 613, launched the previous month, was a failed Salyut, and it was speculated that Soyuz 13 might be a rescue mission. However, Soyuz 13 was unconnected with any other currently-orbiting spacecraft. The programme envisaged for the Soyuz 13 flight included astro-physical observation of stars in the ultra-violet range by means of the Orion-2 system of telescopes installed in the orbital module; spectrozonal survey of separate sections of the Earth's surface, with the objective of obtaining data for economic benefits; continuation of the comprehensive verification and checking of systems on the Soyuz spacecraft; further testing of processes of manual and automatic control of the spacecraft; testing of methods of navigation in various flight conditions.

By the fifth circuit of the Earth the cos-

Above: *The Orion-2 assembly of telescopes carried by the Soyuz 13 manned spacecraft. They were placed outside the orbital module, where the docking system was normally located.*

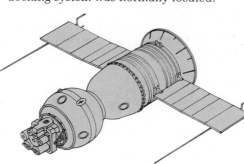

monauts had performed a course correction, and the new orbit was announced as 51.6°, 89.22min, altitude 225-272km. On the second day of the flight the cosmonauts activated the Oazis-2 experiment, which was used for research into the special features of the growth and cultivation of individual biological specimens in a weightless condition. Operations with Orion-2 began soon afterwards. The inclusion of Orion-2 and Oazis-2 experiments raised the possibility that Soyuz 13 was being flown simply as a stop-gap mission because of the delays in the Salyut programme. It is also probable that the opportunity was being taken to fly a development spacecraft for the Apollo-Soyuz programme.

Soyuz 13 returned to Earth on 26 December, having completed a successful eight day mission. This had been the first dedicated science mission to be flown in the Soviet manned space programme, and showed the versatility of the Soyuz spacecraft since it was possible to reconfigure the basic design for different missions.

Soyuz 13
Soyuz 13 undertook the first manned flight of the Soyuz variant which would be flown on the ASTP mission. Externally, Soyuz 13 was almost identical with Soyuz 19 (ASTP), but the docking system was replaced at the front of the orbital module by the Orion-2 telescopes. It has been speculated that Soyuz 13 was flown as a stop-gap measure, using equipment which should have been carried on Salyut missions.

Certainly, Orion-1 and Oazis-1 were flown on Salyut 1, apparently confirming such speculation. A year later a further variant of the solo Soyuz flew the ASTP rehearsal mission, followed by ASTP itself, and the final flight in the series was Soyuz 22 in 1976. Most probably, ASTP was the reason for the solo Soyuz programme since all other Soviet manned flights were almost certainly Salyut oriented.

Above: *Klimuk and Lebedev in orbit during the Soyuz 13 mission. Their flight lasted for eight days, and they were the first Soviets to fly while Americans were in orbit (Skylab was manned at the time).*

Right: *An unusual view of the Soyuz 19 spacecraft in the assembly building. The metal strips outside the descent module carry extra wiring. The white payload shroud is to the right of the picture.*

The Soviet Union flew two unmanned tests for ASTP within the Cosmos programme. Cosmos 638 had been launched on 3 April 1973 – before the Paris Air Show – although its mission profile did not coincide with ASTP itself. ASTP called for a six day Soyuz mission, with docking taking place in a 225km circular oribit. Cosmos 638 remained in orbit for nearly ten days and although it manouevred extensively, none of its tracked orbits matched those planned for ASTP. It is possible that this was simply a shake-down mission for the spacecraft itself, with no attempt made to follow the mission profile, but the possibility also exists that the flight encountered some unacknowledged problems.

Nearly four months later Cosmos 672 was launched on a flight which would closely duplicate the ASTP mission. The initial orbit had an apogee higher than ASTP required, but within hours apogee had been reduced to that scheduled for ASTP. Two days later, Cosmos 672 was tracked in an orbit close to the circular 225km one in which the Soyuz would dock with Apollo. Six days after its launch, Cosmos 672 was returned to Earth, with Soviet controllers having proved that they could accomplish the ASTP goals by means of automatic systems, should the manual systems cause problems.

Soyuz 16: A Final Rehearsal

It was well-known that the Soviet Union would launch a manned Soyuz to conduct a full dress-rehearsal for the ASTP mis-

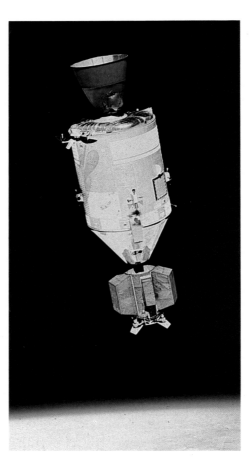

Above: *The last Apollo spacecraft in orbit. The Docking Module is in the lower part of the picture, with the petals of the docking system ready to receive Soyuz 19. The development of the DM overcame one of the mission's major compatability problems.*

sion, and Soviet planners were willing to give NASA advanced knowledge of the planned flight details on the understanding that they would not be given to the press. However, NASA officials were unable to give such an undertaking. As a result, the first information about the flight came when American officials were awoken early in the morning of 2 December 1974 with the news that Soyuz 16 had been launched. It was manned by Anatoly Filipchenko and Nikolai Rukavishnikov, who were the prime crew for the back-up ASTP spacecraft.

A summary of the mission's timetable is given in the accompanying table. It was a surprise in some quarters that the launch took place so early – 09.40 GMT – because the joint flight called for a launch of the Soyuz at 12.20 GMT. This can easily be explained by the Soyuz landing constraints, which were partially governed by the time of sunset at the landing site. A summer ASTP launch meant that sunset would be late in the evening, local time, and therefore for a given mission duration the launch could come late in the day. However, the mid-winter launch of Soyuz 16 meant that sunset occurred much earlier in the day, and in order to retain approximately the same landing conditions an earlier launch was required.

The initial orbit of Soyuz 16 had a far higher apogee than ASTP called for; in fact the initial orbit was never announced by the Soviet Union. A correction was made to the orbit, showing that in the event of an inaccurate oribtal injection, the spacecraft could easily manoeuvre to the planned orbit – so long as the inaccuracy was not too high, of course. Soyuz 16 carried a small added ring, mated with the new

TIMETABLE FOR SOYUZ 16				
GROUND ELAPSED TIME				**EVENT**
hr	min	hr	min	
−2	30			Crew enter spacecraft, check spacecraft status.
0	00			Launch, 2 December at 09.40 GMT.
0	14			After orbital injection, crew raise pressure suit visors: spacecraft pressure 760mm of mercury.
1	30			Crew enter orbital module: check pressure and atmospheric content.
6	43	8	28	Reduce pressure in orbital and descent modules to 540mm of mercury.
10	40	18	50	Sleep.
28	37	28	53	Further venting of cabin pressure, to 510mm of mercury.
34	30	42	20	Sleep.
58	20	66	00	Sleep.
82	00	89	40	Sleep.
102	21			Raise spacecraft pressure to 830mm of mercury.
105	50	113	25	Sleep.
119	20			Jettison APDS simulation ring.
123	30			Reduce pressure to 760mm of mercury.
130	00	137	00	Sleep.
137	40			Don pressure suits, ready for re-entry.
138	10			Crew return to descent module.
141	53			Spacecraft splits into three modules.
142	24			Landing, 8 December.

Notes: This table presents a summary of the events during the Soyuz 16 mission; it is adapted from the table on pages 268-269 of *The Partnership: A History of the Apollo-Soyuz Test Project.* APDS is the Androgynous Peripheral Docking System.

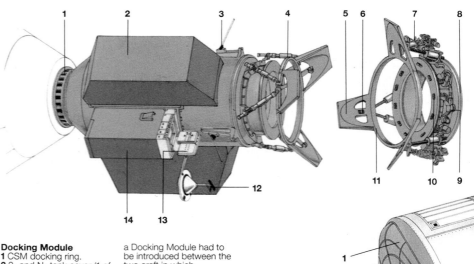

Docking Module
1 CSM docking ring.
2 O_2 and N_2 tank cover (1 of 2).
3 VHF-FM antenna (1 of 3).
4 Attenuators.
5 Guide (1 of 3).
6 Capture latch (1 of 3).
7 Attenuator (1 of 6).
8 Tunnel interface.
9 Cable retract system motors and gearbox.
10 Structural latches.
11 Guide ring (extended).
12 Soyuz docking target.
13 Mutiple operation door.
14 UV spectrometer.

Making Apollo and Soyuz compatible for the space docking involved many modifications. In particular

a Docking Module had to be introduced between the two craft in which cosmonauts and astronauts could acclimatise to cabin atmospheres of different composition and pressures to avoid the "bends" familiar to deep sea divers ascending too rapidly from the ocean depths.

Technical Data
Length of cylindrical chamber: 3.15m.
Maximum diameter: 1.42m.
Total weight with experiments, stowage, fluids and docking system: 5,907kg.

androgynous docking system. In order to simulate the undocking and docking procedures, this ring could be retracted or extended as required. This experiment (which presumably had been previously tested on the unmanned Cosmos flights) proved that the new docking system would work without any problems.

Soyuz 16 remained in orbit for six days, matching the ASTP duration to within ten minutes. Retro-fire was initiated on 8 December at a height of 210km at 07.26 12.7 seconds and lasted for 166.5 seconds. The spacecraft split into its three separate modules at 07.36 at an altitude of 150km.

On the day that Soyuz 16 was launched, a Tass correspondent was promising a "hot space summer of 1975". In the West it was not known that Salyut 4 was undergoing its final checks, prior to its launch at the end of the month. The Soviet Union was planning to have men in orbit for half of the first six months of 1975, and then to undertake ASTP in the latter half of the year.

The Joint Flight

The flight of Soyuz 19 marked the beginning of a new policy of openness in the Soviet space programme, something which gradually developed as information was regularly released about the Salyut 4 missions in the first half of 1975. Any flight with the Americans had to be conducted in the full glare of world publicity and for the first time the Soviet Union broadcast live coverage of events from space to the outside world. American personnel were at the Soviet Kaliningrad Mission Control for the launch, while a rookie cosmonaut, Illarianov, was part of the Soviet team at Houston Mission Control.

American astronauts had visited the Soviet Tyuratam launch site in May 1975 to see the ASTP Soyuz in assembly; this marking the first time that non-Soviet astronauts had visited the launch site. Following the assembly of the spacecraft and its booster, at about 2.00 GMT on 12 July

Above: *The historic first handshake in orbit rehearsed on the ground as Apollo commander Tom Stafford greets Soyuz commander Alexei Leonov; Valery Kubasov looks on as Deke Slayton takes the photograph. This picture was taken at the Johnson Space Center, Houston.*

ASTP Spacecraft in Docked Configuration
1 Apollo Service Module (SM) with bell-shaped engine nozzle. Contained propellants, fuel cells, oxygen and other supplies and equipment.
2 Reaction control quad.
3 Pitch control thrusters.
4 Apollo Command Module (CM) with accommodation for three astronauts. Cabin atmosphere pure oxygen at 0.35kg/cm².
5 Control consoles.
6 Docking Module (DM) which cosmonauts/astronauts entered to acclimatise to different atmospheres before entering the other ship.
7 Apollo VHF antennas (frequency 121.75MHz).
8 Docking target.
9 Guide (1 of 3).

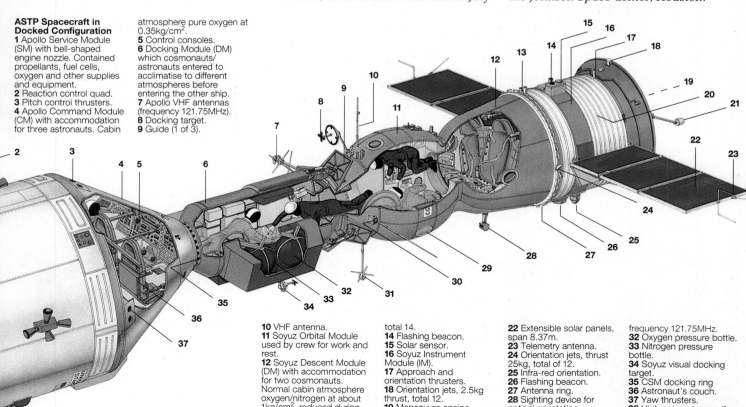

10 VHF antenna.
11 Soyuz Orbital Module used by crew for work and rest.
12 Soyuz Descent Module (DM) with accommodation for two cosmonauts. Normal cabin atmosphere oxygen/nitrogen at about 1kg/cm², reduced during period of docking to 0.7kg/cm².
13 Soft-docking and orientation jets, thrust 10kg,
total 14.
14 Flashing beacon.
15 Solar sensor.
16 Soyuz Instrument Module (IM).
17 Approach and orientation thrusters.
18 Orientation jets, 2.5kg thrust, total 12.
19 Manoeuvre engine, 300kg thrust.
20 Thermal control system radiator.
21 Rendezvous antenna.

22 Extensible solar panels, span 8.37m.
23 Telemetry antenna.
24 Orientation jets, thrust 25kg, total of 12.
25 Infra-red orientation.
26 Flashing beacon.
27 Antenna ring.
28 Sighting device for optical orientation.
29 TV antenna (1 of 2).
30 TV camera and orientation light (red).
31 Apollo VHF antenna,
frequency 121.75MHz.
32 Oxygen pressure bottle.
33 Nitrogen pressure bottle.
34 Soyuz visual docking target.
35 CSM docking ring.
36 Astronaut's couch.
37 Yaw thrusters.
38 High-gain antennas (for communications with ground via satellite ATS-6, frequencies 2256 kMHz out, 2077 kMHz in).

1975 the prime launch vehicle for the Soviet half of the ASTP mission began its journey to the launch pad – the same pad that had been used for the launches of Sputnik 1 and Vostok 1. The back-up Soyuz vehicle was later transported to the second Soyuz launch pad, some 32km distant. Six of the eight cosmonauts who had trained for the mission had arrived at Tyuratam for the launch: Leonov, Kubasov, Filipchenko, Rukavishnikov, Romanenko and Ivanchenkov. This represented the prime crew for the first spacecraft and both the prime and back-ups for the second spacecraft. Unless a serious illness overcame either Leonov or Kubasov, it seems likely that the Soviets would have delayed the launch if either of them fell ill rather than assigning one of the other crews to the mission. The remaining two ASTP cosmonauts, Dzhanibekov and Andreyev, were working in the ground control network, monitoring the flight.

On 15 July Leonov and Kubasov suited up, ready for their flight, in full view of television audiences and boarded the cosmonaut bus to the launch pad. After entering the space craft through the orbital module and dropping into the descent module, the orbital module hatch was sealed at 9.45, ready for the countdown and launch. Two hours before launch, the crew went through their flight check lists.

Television pictures were beamed "live" to the world from Tyurtam, starting at 9.58 GMT, and they continued until Soyuz was safely in orbit. At midday the cosmonauts

fastened their harnesses and put on the their gloves; 15 minutes later they lowered the visors of their pressure suits. Launch in full view of the world's television viewers came at 12.20 GMT. At 120 seconds into the flight the four strap-ons of the launch vehicle had separated and the core second stage was still firing; at 160 seconds after launch the payload tower and shroud were separated; 300 seconds saw the core stage shut down and the ignition of the third,

Below: *The rear of Soyuz 19, as seen from Apollo. This clearly shows the spacecraft's main propulsion system, with boxes of electronics. The spacecraft was covered in a green thermal blanket.*

Bottom: *Looking down inside the Soyuz orbital module, as the two Soviet cosmonauts work at the orbital module's control panel. Leonov is half-way through the hatch to the descent module.*

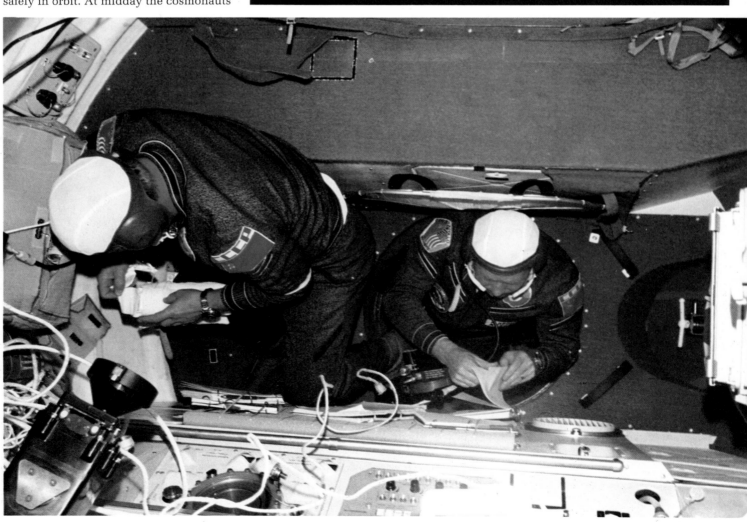

orbital stage; finally, 530 seconds after launch, the third stage shut down, with spacecraft separation quickly following.

After orbital injection, Soyuz had an announced orbit of 51.8°, 186.35-220.35km, almost a perfect match with the planned parameters. In an operation starting at 17.37 and lasting for 2h 34min the pressure in Soyuz was reduced from 867mm to 539mm. Later that evening the cosmonauts began to operate the Biokat-M biological experiment. The only problem that marred the first part of the Soyuz flight was that the spacecraft television camera failed to operate during the first few hours of the mission.

The Apollo crew were asleep as their colleagues were launched into orbit. While Soyuz 19 was completing its fourth orbit, the crew of Stafford, Brand and Slayton climbed aboard Apollo (officially, the spacecraft was simply "Apollo", but it is often listed as "Apollo 18"). At 19.50 GMT the final Saturn booster was launched from the Kennedy Space Center at Cape Canaveral, carrying the last throw-away manned spacecraft which the Americans planned to launch. The mass of the combined CSM was 12,905kg and the DM mass was 2,006kg.

It took Apollo 18 about ten minutes to reach a 155-173km orbit, after which the transposition and docking manoeuvre to link Apollo with the docking module was successfully completed (although Stafford did have some initial difficulties in aligning Apollo with the DM). Two orbital corrections were then completed by Apollo, resulting in orbits of 165-167km and then 169-233km.

A day after the launch, the Soyuz propulsion system fired to place the spacecraft in its "assembly oribt" 51.8°, 222.7-225.4km. It had now to wait for Apollo to arrive. A further pressure reduction in the Soyuz reduced the cabin pressure to about 500mm. On 17 July the world was treated to live television pictures of Soyuz 19 as Apollo 18 approached for the docking. Contact between the spacecraft was made at 16.09 GMT and the final docking was accomplished three minutes later, with Apollo operating as the "active" spacecraft. The crews checked the integrities of their two spacecraft. The cosmonauts then took off their pressure suits which had been worn for safety during the docking, and after checking their modules they entered the orbital module to check the pressure in the DM. Pressure in the DM was adjusted to 250mm of mercury, ready for the first crew transfer. At 19.13 the Soyuz orbital module hatch leading to the DM was opened and four minutes later the DM hatch opened. The world saw Stafford — watched by Slayton in the DM — shake hands with Leonov. Although the flight was far from completed, this symbolic handshake marked the culmination of the ASTP mission.

The spacecraft remained docked for about two days, with various crew transfers taking place. An American always remained in the Apollo and a Soviet in Soyuz, as if an emergency had arisen the astronauts and cosmonauts would return to Earth without any transfers to their

"home" spacecraft. On 19 July at 12.02 the spacecraft undocked for the first time and Apollo aligned itself with the Sun as seen from Soyuz, for an artificial solar eclipse experiment. This occultation completed, the Soyuz operated as the active spacecraft for a second docking with Apollo, which was completed at 12.20. The final undocking took place at 15.26 and Soyuz 19 and Apollo 18 separated to complete their independent missions.

On 20 July, Soyuz 19 was manoeuvred ready for its recovery; the orbit was announced as 51.78°, 88.8min, 216-219km. At 10.10 GMT the Soyuz retro-rocket was fired and 12 minutes later the spacecraft split into its three modules. At 10.28, at a height of about 80km, the spacecraft entered the period of radio black-out, contact being regained at an altitude of 30km. The parachute opened at 10.37 and the descent module landed at 10.51 GMT, again in full

Above: *Stafford and Slayton inside the Soyuz orbital module prepare for a Soviet meal. They are holding containers of bortsch (beet soup) over which vodka labels have been stuck.*

Below: *At the end of their mission, Soviet cosmonauts autograph the charred exterior of their Soyuz descent module: Leonov (right) has already done this and watches as Kubasov adds his signature.*

view of the world's television viewers. The Soviet half of ASTP was complete.

Apollo 18 remained in orbit until late on 24 July. American researchers wanted to get as much as possible out of the mission, since it would be their last involvement in a manned spaceflight until at least the end of the decade. At 21.18 the Command Module splashed down about 430km west of Hawaii close to the recovery ship *New Orleans*, this event marking the final planned water landing of a manned spacecraft in the American space programme. During the descent, gases from the CM's attitude control thrusters had accidentally been drawn into the Command Module and the crew were suffering badly by the time they clambered out of their spacecraft on the *New Orleans*.

However successful the actual flights, it must be acknowledged that the ASTP missions were part of a dead-end project, since Apollo would never fly again and there were no American plans to incorporate docking systems on the Space Shuttle orbiters to allow dockings with Soviet orbital stations or other craft. Furthermore, no plans existed to place American space stations in Earth orbit with which cosmonauts might dock in case of an emergency. ASTP was born in a period of *detente* between the Soviet Union and the United States, and even as the mission was taking place the foundations of the *detente* policy were beginning to shake.

Soyuz 22: Earth Observation

On 24 August Soyuz 21 returned to Earth from the Salyut 5 orbital station, its mission prematurely curtailed. The nominal landing opportunity would have been the first week of September. It therefore came as a great surprise when Soyuz 22 was launched on 15 September 1976, and even

Above: *The end of an era. After the last Apollo splashdown, the three astronauts (left to right, Stafford, Slayton and Brand) speak with President Ford from the USS New Orleans recovery ship.*

more of a surprise when it became clear that the mission was clearly not targeted towards Salyut 5. An initial orbit does not seem to have been announced, but the orbital inclination was 64.8° – never before seen in the Soyuz programme. The day after launch, Soyuz 22 was maneouvred into a 250-280km orbit. It was then revealed that after the successful conclusion of ASTP, the back-up Soyuz spacecraft had been taken back to the assembly shops, refurbished, and was now being flown as Soyuz 22. The commander was Valery Bykovsky who had previously

flown Vostok 5 and had been scheduled to fly the cancelled Soyuz 2 docking mission with the ill-fated Soyuz 1; his flight engineer was Vladimir Aksyonov, on his first space mission.

The orbital inclination of Soyuz 22 was chosen so as to obtain the best orbital passes for the main experiment: in place of the Soyuz docking system, there protruded the MKF-6 multi-spectral camera, which had been built by the Carl Zeiss works in the German Democratic Republic. The camera allowed six simultaneous photographs of the Earth to be taken for

Below left: *Aksyonov and Bykovsky inside their Soyuz 22 spacecraft and* **(right)** *Aksyonov at the controls inside the Soyuz orbital module. They flew a mission using the back-up ASTP Soyuz.*

Above: *A photograph of the Earth's limb and the crescent Moon, taken from the orbiting Soyuz 22. Photography was the major experiment for the mission.*

Left: *The major rationale for Soyuz 22 was photography of the Earth, such as this, using the East German MKF-6 multi-spectral camera. After being tested on Soyuz 22, the MKF-6 system would become a major Salyut experiment.*

Earth resources work; four exposures were in the visible light waveband and two were in the infra-red range of the electro-magnetic spectrum.

While Soyuz 22 was in orbit, NATO forces were undertaking a major European training exercise, and some reports circulated in the western press that Soyuz 22

The Soyuz 22 Orbital Module
In place of the docking system the spacecraft carried a camera shroud (**1**) which was separated in orbit, and the MKF-6 camera system (**2**). Film cassettes could be loaded from inside the orbital module and then returned to Earth at the end of the mission. The orbital module control panel (**3**) included the camera's control system.

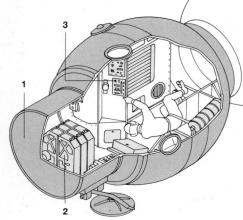

was primarily a manned photo-reconnaissance mission. However, the inclusion of the MKF-6 camera as the main experiment indicates that reconnaissance work was not a major part of the mission plan – if indeed it was a part of it at all. Soyuz 22 remained in orbit for eight days and returned to Earth safely on 23 September. The significance of the mission and the MKF-6 camera would not become clear until Salyut 6 was operational in 1978, when an MKF-6M camera system turned out to be one of the major experiments carried on board.

If one looks back at all the missions discussed in this chapter, it seems very probable that if the Apollo-Soyuz Test Project had not been planned for 1975, the solo Soyuz programme would not have taken place. However, Soyuz 13 was able to conduct important astronomical observations and Soyuz 22 validated a major experimental instrument for the next generation of orbital stations. Both Soyuz 16 and Soyuz 19, as integral parts of ASTP, taught Soviet planners the methodology operated by American planners. In retrospect, one must be thankful that the American tendency to cancel important space missions or abandon proven flight hardware did not rub off on the Soviet Union's space administrators.

FURTHER READING

There are many books available which relate to the ASTP mission of Soyuz 19 and Apollo (18). The following are only a selection:

Ezell, E.C. and L.N., *The Partnership – A History of the Apollo-Soyuz Test Project*, NASA SP-4209, Washington, 1978.

Froehlich, W., *Apollo-Soyuz*, NASA EP-109, Washington, 1976.

Lebedev, L. and Romanov, A., *Rendezvous in Space: Soyuz-Apollo*, Progress Publishers, Moscow, 1979.

NASA Release 74-19, *Apollo-Soyuz Test Project Fact Sheet*, 1974.

NASA Information for the Press, *Apollo-Soyuz Test Project*, 1975.

Rebrov, M.F. and Gilberg, L.A., *Soyuz-Apollo*, Mashinosroyeniye Press, Moscow, 1975 – pre-flight – and 1976 – post-flight.

Various writers, *Soyuz and Apollo*, Politizdat Publishers, Moscow, 1978.

A single book has apparently been dedicated to the Soyuz 22 mission:

Various writers, *Soyuz 22 – Erforscht Die Erde*, Academie der Wissenschaften der DDR and Akademie der Wissenschaften der UdSSR, Published simultaneously in Moscow and Berlin, 1980.

Chapter 10: New Spacecraft to Support Salyut

With the completion of the Salyut 5 programme in February 1977, Soviet space planners were ready to move to operations with what they would describe as a second generation space station. Two flights in this series would be launched: Salyut 6, which hosted crews between 1977 and 1981, and Salyut 7, hosting crews between 1982 and 1986. The major modification for these new stations was the addition of a docking port at the rear of the station. One may speculate that the rear docking system and the resulting modified propulsion system (compared with the civilian Salyut 1 and Salyut 4 which are known in detail) had previously seen service on the military Salyuts, although further modifications were probably also required for the second generation Salyuts.

In support of the Salyut 6 orbital station, the Soviets continued to use the Soyuz ferry craft which was introduced on the manned Soyuz 12 mission in 1973. However, during Salyut 6's lifetime three new spacecraft were brought to operational status to support the manned missions. They were the Progress unmanned cargo ferry, the Soyuz-T manned ferry, and the Heavy Cosmos space station module. To preserve the continuity of the next two chapters, these three spacecraft will be described here.

The Progress Ferry

As a group, the Soviet Union classified Salyuts 1-5 as their first generation space stations, although in the West they are sub-divided into three groups already described. A major drawback with the first generation of Salyut was that it had to be launched with virtually all of the crews' supplies on board. The Soyuz crew ferry could take some extra supplies into orbit, but these were limited.

The first generation of Salyuts had operated at altitudes generally between 250km and 350km. Left to themselves, Salyuts at 250km would decay out of orbit and re-enter the atmosphere within about 20 days of launch (averaging out the atmospheric properties, which vary with the eleven year solar cycle), while a station at 350km would remain in orbit for about a year. Of course, it was possible continually to raise the orbit, but to maintain a 250km orbit would require 4.75 tonnes of propellant each year. At 350km this requirement dropped to about 600kg. Of course, the

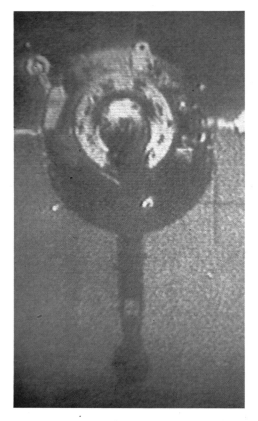

Above: *The docking of Progress 1, as seen from the rear of Salyut 6. Of all the Progress missions launched to the end of May 1988, none had docking problems, while there were still occasional problems with manned missions.*

Soviets allowed the Salyuts to decay to lower orbits and then raised them again in large manoeuvres; this was more economical in terms of fuel expenditure. All in all, to maintain a fully functioning Salyut in orbit for two years with a crew permanently on board required about 20 tonnes of consumables. It was, of course, impractical to launch all these supplies with the station – even if a Soviet booster had the necessary lifting power, which none had at the time.

The second generation of Salyut was planned to support much longer missions than the 63 day flight of Soyuz 18 to Salyut 4, and therefore it was essential that some means be developed for re-supplying a crew in orbit. The first step in this plan was the inclusion of a second docking port on the new Salyut stations (see the descrip-

tion of Salyut 6 in the next chapter). In addition, design work on an unmanned cargo-carrier began in 1973, and the designers chose a Soyuz to be the basis of the spacecraft. Like Soyuz, Progress (as the ferry was called) consisted of three modules. The instrument module, containing all the automatic flight systems and the propulsion system, was modified from Soyuz to accommodate systems which would normally have been in the descent module, and the orbital module at the front of the craft was transformed into a cargo hold. On the outside of the orbital module two television cameras were mounted to allow ground controllers a stereo view of Salyut as the automatic docking was being completed; a standard Soyuz only carries a single camera. The central module, however, involved a total re-design of the Soyuz descent module.

As well as carrying supplies to the cosmonauts on Salyut, Progress would be used to refuel the Salyut station, and the central Progress module consisted of a propellant storage area. This was a sealed unit, and could not be entered by the cosmonauts. It contained two tanks each for the Salyut fuel and oxydizer, together with pressurant tanks and the automatic control system. Like the Soyuz ferry, Progress did not rely on solar cells for power; chemical batteries were carried to supply electrical power during the flight to the Salyut and for the period following undocking, and these could be re-charged while Progress was docked with the orbital station.

The Refuelling Sequence

In order to allow the second generation Salyut stations (and later the Mir station core) to continue orbital manoeuvres, unmanned Progress freighters were capable of carrying about one tonne of propellant for transfer to the Salyut tanks. This was carried in four tanks: two tanks for the nitrogen tetroxide (oxidiser) and two tanks for the UDMH (fuel). These tanks were situated in the space where a descent module would be carried on a Soyuz mission, together with a bottle of high pressure nitrogen gas (see illustration).

The propellant tanks on Salyut are basically spherical with an internal membrane which divides the interior into two parts: one part carries the propellant and the other contains pressurising gas which is added to the tank as the propellant is used on manoeuvres. Salyut carried three tanks for fuel and three for oxidiser, as well as a supply of nitrogen gas which is kept at a high pressure.

The refuelling of Salyut can be conducted either by the crew on the station or automatically under the control of the ground crew. Once Progress has docked at the back of Salyut, the propellant lines and connections are checked for integrity. This done, a compressor slowly reduces the nitrogen pressure in the propellant tanks. The nitrogen is pumped back into its storage bottles ready for the refuelling operation itself. The fuel and oxidiser are transferred at different times for safety reasons, although the procedure is the same for each propellant component.

The UDMH is transferred first to Salyut. This is accomplished by pressure-feeding the nitrogen into the Progress fuel tanks, this forcing the fuel down the connecting pipe system into Salyut. As the amount of fuel in the Salyut tanks increases, more nitrogen pressurant is forced into the Salyut storage tanks. Once this has been accomplished, the nitrogen tetroxide oxidiser is transferred from Progress to Salyut in the same manner.

After all the propellant has been transferred from Progress to Salyut, the connecting lines are purged using high pressure nitrogen (to prevent any contamination or spillage should the propellant transfer lines not be closed when Progress undocks from Salyut) and the gas and any residual propellant are vented into open space.

According to the Soviet record, they have never had a failure during an attempt to transfer propellant from a Progress to Salyut, although in 1983 there was a leak as the transfer was taking place (see details of the Soyuz-T 9 mission to Salyut 7 in Chapter 12).

The Progress spacecraft is about the same length as a Soyuz. It is launched by the basic SL-4 Soyuz booster, although the payload shroud does not include the four aerodynamic flaps which are carried on manned flights as part of the emergency abort system. Television pictures of the Progress 1 launch showed this new configuration. Numerical details have only been released for Progress 1, although all the flights have had similar performances. Progress 1 had a mass of 7,020kg, and could carry 1,300kg of cargo in the orbital module plus 1,000kg of propellant for Salyut. Subsequent details indicate that Progress can carry up to 1,400kg of supplies in the orbital module. Cumulative spacecraft masses have been announced for Progress 1 to Progress 11 (77,160kg) and Progress 1 to Progress 23 (161,400kg). Since the mass of Progress 1 is known, the mean mass for the Progress 2 – Progress 11 series was 7,014kg, while for Progress 12 to Progress 23 the figure is 7,020kg.

The Progress craft took about two days to reach Salyut, and always docked at the rear port, where the Salyut re-fuelling system was situated, whether a refuelling mission was being flown or not. After undocking from Salyut, the Progress could remain in free flight for a further two days, after which the spacecraft was de-orbited and allowed to burn up in the atmosphere over the Pacific Ocean. When Progress 1 was launched, a nominal mission duration of a month was announced, although many missions have lasted for two months or more. As well as being used for transporting cargo to Salyut, the Progress was often used as a "space tug", its on-board engine being fired to raise the orbit of the

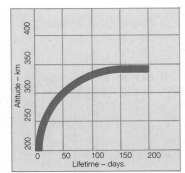

The Progress/Salyut Refuelling Operation
In the Progress propellant cargo bay a nitrogen gas tank (1) provides pressurant to force the propellant out of the oxidiser and fuel tanks (2). The propellant flows in pipes (3) outside the Progress orbital module. Connectors in the docking unit carry the propellant to the Salyut oxidiser and fuel tanks (4) with nitrogen pressurant being forced back into a storage tank (5). The rear of Salyut (6) was extensively modified to permit the refuelling. A modified Soyuz orbital module (7) is carried by Progress with cargo, while a modified Soyuz instrument module (8) manoeuvres the Progress in orbit.

Essential Supplies
The problems of space station missions without resupply facilities are shown in these graphs. The far left diagram shows how much propellant is required each year to maintain a Salyut in orbit, while if there are no manoeuvres periodically to raise the orbit the right hand graph shows how quickly the spacecraft will decay out of orbit and re-enter the atmosphere. Clearly a regular supply of propellant is essential.

Progress
This cutaway of the Progress re-supply craft clearly shows its Soyuz derivation.
1 Short range radar transponder.
2 Modified Soyuz orbital module with cargo for the Salyut crew.
3 Long range radar transponder.
4 Antenna.
5 Soyuz instrument module.
6 Soyuz KTDU-35 propulsion system.
7 Equipment for the automatic control of Progress.
8 Tanks for the propellant to be transferred to Salyut and nitrogen pressurant gas.
9 Docking probe.

Operations with Progress have been automated as much as possible. When a docking is taking place with a Salyut, the cosmonauts on the station need take no part in the operation – ground controllers do all the work. Dockings are always at the rear of the station, where the propellant lines connecting with Salyut's tanks are located. Initially the refuelling of Salyut was controlled by the cosmonauts, but later the operation was conducted under the control of the ground, leaving the crew free for their programme of experiments. The public record shows a 100 per cent success rate in the Progress programme.

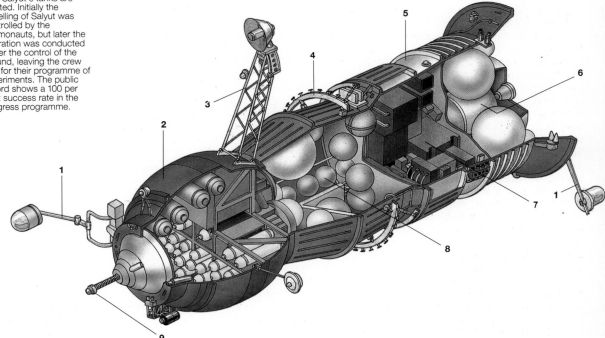

Soyuz-Salyut-Progress orbital complex. This saved fuel on the manned ferry and also on the station.

There does not seem to have been any dedicated test flights of Progress within the Cosmos programme. Soyuz 20 was almost certainly used to test the Progress automatic rendezvous and docking system, and it is possible that the unexplained Soyuz-type flights of Cosmos 670 and Cosmos 772 might have been related to some kind of Progress development. However, both these craft were recovered, while no Progress has ever carried a re-entry vehicle. A promised modification of Progress is expected to carry a re-entry vehicle – for the return of experimental results and data – but this had not appeared by early 1988, and such a modification would seem unlikely after ten years of operations.

Details of the individual Progress flights are given in subsequent chapters within the descriptions of the manned visits which they supported. The first ten Progress missions delivered about 20 tonnes of cargo to Salyut 6, more than the mass of the station itself when it was launched.

Heavy Cosmos Modules

In conjunction with operations of the second generation Salyuts, the Soviets wanted to conduct tests relating to the expansion of orbital complexes, and in support of this work the Heavy Cosmos module was developed. The name "Heavy Cosmos" is apparently a Western designation, since the flights have been within the Cosmos programme, although operational craft launched to the Mir complex may have been given another programme name.

The first flight of a Heavy Cosmos began on 17 July 1977 at about 08.57 GMT when Cosmos 929 was launched. No special announcement was made about the flight, and it was not until 1983 that this was identified as a prototype space station module. The initial orbit was announced as 51.6°, 89.4min, 221-298km, akin to that used for a Salyut. This prompted reports that this mission might possibly be a failed Salyut. However, after its launch the spacecraft manoeuvred in orbit, and its transmissions showed that apparently it was functioning.

After the initial manoeuvres, the western aerospace press reported that a descent craft had returned to Earth on 16 August (estimated landing time 21.22) or 17 August (20.59), although no comment was made by the Soviets. Between 19 August and the end of the month, many manoeuvres were completed with the satellite attaining a 313-328km orbit – akin to some of the orbits used on the Soyuz-T test flights within the Cosmos programme. However, visual observations of the object still in orbit showed that it was far larger than a Soyuz and that unlike the Soyuz ferries then flying, Cosmos 929 carried solar panels.

For more than two months – while the Soviets had expected to begin operations with Salyut 6 (the first mission, Soyuz 25, failed) – the orbit of Cosmos 929 slowly decayed. At some time between 4 December and 9 December the orbit was again raised slightly. 10 December saw the

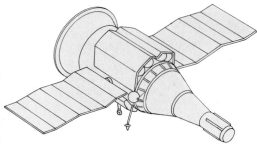

Heavy Cosmos
The Heavy Cosmos module, first flown as Cosmos 929, is believed to be a direct descendant of the Military Salyut stations. The main body of the spacecraft uses the smaller work compartment of Salyut but with only two rotatable solar panels. At the back the cylinder flares to match the diameter of the Proton launch vehicle, the docking probe being located at this end. Propellant tanks are mounted outside the cylinder with solar cells covering one set and a protective shroud the other set. At the front (right) of the satellite is the cone-cylinder of the descent module and its retro-rocket package.

launch of Soyuz 26 to Salyut 6, and once the crew were safely on board the station, Soviet controllers briefly returned to the Cosmos 929 experiment. Between 12 December and 15 December, and again on 18 December or 19 December two further manoeuvres took place, with Cosmos 929 being placed into a high near-circular orbit, about 440km above the Earth's surface. It looked then as if the Cosmos 929 mission might be over, but at the end of January 1978 it began to manoeuvre again. A manoeuvre changed its orbital apogee to 437km, the new perigee being about 335km. On 2 February the satellite was finally de-orbited, presumably to burn up in the atmosphere over the Pacific Ocean.

MANOEUVRES OF COSMOS 929			
PRE MANOEUVRE ORBIT		POST MANOEUVRE ORBIT	
Epoch (1977)	Altitude (km)	Epoch (1977)	Altitude (km)
Initial orbit		17.61 Jul	214-261
17.61 Jul	214-261	17.74 Jul	215-279
29.89 Jul	207-261	30.44 Jul	208-264
31.74 Jul	208-260	2.35 Aug	209-267
Descent craft recovered, 16 or 17 August			
		16.13 Aug	194.228
17.48 Aug	192-222	18.10 Aug	219-232
18.10 Aug	219-232	19.08 Aug	303-327
19.08 Aug	303-327	19.40 Aug	312-318
19.40 Aug	312-318	19.59 Aug	311-319
20.54 Aug	311-318	23.31 Aug	312-319
24.45 Aug	312-319	26.84 Aug	314-325
28.36 Aug	315-324	29.49 Aug	313-326
31.26 Aug	313-325	31.89 Aug	313-328
4.51 Dec	284-294	9.09 Dec	290-301
12.41 Dec	288-300	15.36 Dec	286-305
18.75 Dec	285-303	19.44 Dec	439-447
(1978)		(1978)	
30.28 Jan	437-448	1.23 Feb	335-437
1.23 Feb	335-437	2.12 Feb	337-438
2.12 Feb	337-438	2 Feb	De-orbited

Notes: This table lists all of the manoeuvres completed by Cosmos 929; only the initial orbit was announced by the Soviets and the descent craft recovery was reported by Western sources rather than the Soviet authorities. The orbital manoeuvres listed here are based upon the Two-Line Orbital Elements. In 1983 the Soviets confirmed that it had remained in orbit for 200 days.

A break of more than three years ensued before another flight in the series (Cosmos 1267) took place, and few details of the spacecraft were released. Details of the Heavy Cosmos module were finally published in 1983 when Cosmos 1443 docked with Salyut 7 and was manned by the Soyuz-T 9 crew. The design of the Heavy Cosmos module utilised the central section of the Salyut station. It was 3.8m long and 2.9m in diameter. At one end the cylinder flared to a diameter of 4.15m, providing an interface to the third stage of the Proton SL-13 launch vehicle, and then tapered inwards to a probe docking unit. Overall, this module was about 6m long. At the other end there was an adapter which could be used to interface with different payloads; either descent vehicles or experimental packages, as necessitated by mission requirements.

Of the first four missions (Cosmos 929, Cosmos 1267, Cosmos 1443 and Cosmos 1686), three carried large re-entry vehicles. Only an overall representation of the descent vehicle has been released, revealing a vehicle with the general shape of an American Gemini spacecraft. The "nose" of the descent vehicle seems to be the retro-rocket system, which is used only for the de-orbit manoeuvre. The heat shield must have a hatch in it, to allow the cosmonauts to enter and allow the transfer of equipment. The fourth of the initial flights, Cosmos 1686, has not been depicted or described in detail. It has simply been disclosed that it utilised the same cylindrical unit as the previous flights, and that the descent craft was replaced by a package of telescopes.

When Cosmos 1267 was in orbit, the American publication *Aviation Week & Space Technology* suggested that it was some form of orbital battle station, with anti-satellite weapons mounted externally. In fact, no anti-satellite missions were associated with the Cosmos 1267 mission, and it was only when the Soviets published the first sketches of the Heavy Cosmos that the basis for the story became clear. A completely new engine arrangement was being used, with cylindrical tanks mounted externally around the main cylindrical section of the module, and two nozzles of the main propulsion system protruded outside the module at the end which carried the descent module. The nozzles were mounted in such a way that there was no chance that their exhaust gases could damage the descent craft (or science payload).

The mass of the Heavy Cosmos module was about 20 tonnes, and its total length was about 13m. After the descent craft had been recovered, Cosmos 1267 was described as having more than twice the mass of a Soyuz craft, suggesting a figure of 13.8 tonnes or more. This implies that the descent craft has a mass of 5-6 tonnes.

The Heavy Cosmos can be used in many forms. On its first manned application, it was said that the craft could deliver two and a half times the cargo load that a Progress could carry (depending on whether the Progress propellant for Salyut is included, this suggests a cargo load of either 3.25 tonnes or 5.75 tonnes), and this would be carried in the main module. The descent craft could return with 500kg of cargo. The habitable volume of the Heavy Cosmos is about 50m³, compared with about 90-95m³ for Salyut itself. Two Salyut 6-type solar panels were carried, having an area of 40m².

Four main roles were envisaged for the Heavy Cosmos craft:

a large space station re-supply craft; a space tug, to manoeuvre a space station complex; an add-on module, to extend the volume of the working area available to a Salyut crew; and an autonomous module, to conduct operations independently of Salyut missions.

One further possible role should be considered, even though the Soviets did not explicitly mention it. When the descent craft of Cosmos 1267 returned to Earth (the Soviets did not even announce that a descent craft was being carried), American sources suggested that it was large enough to carry 4-6 men. Therefore, one may wonder whether a future use of the Heavy Cosmos will be to house an emergency escape capsule on a space station complex, in case the normal manned spacecraft should fail to operate and there is not sufficient time for a replacement craft to be launched.

It has long been thought in the West that the design of the Heavy Cosmos is derived from the military Salyut; the Salyut and the Heavy Cosmos share descent craft for experiments and a docking port at the "back" of the vehicle for a Soyuz craft – although on the Salyut the docking unit would have to be a drogue to accept the Soyuz probe. This feeling was strengthened when the Soviet Union stated that Salyut 2, Salyut 3 and Salyut 5, together with the Heavy Cosmos modules, had been designed by the bureau headed by V.N. Chelomei. Although the Heavy Cosmos has seen limited use to the end of 1987, its potential has still to be fully exploited.

Above: *One of the few views available of manned operations with a Heavy Cosmos. Here, Lyakhov and Alexandrov have just entered Cosmos 1443.*

The Manned Soyuz-T Ferry

During the 1970s a series of tests of Soyuz-derived craft were undertaken under the Cosmos identity (see table), and it was not until the end of 1979 that it became clear that a new generation of Soyuz – one adapted to permit the resumption of three-man missions – was ready to begin manned operations.

Following the loss of the Soyuz 11 crew, Soyuz was modified to carry only two men, but the Soviets wished to resume three man missions as soon as possible. In 1976 the cosmonaut Pavel Popovich gave

an interview to the magazine *Flug Revue*, and he confirmed that there were plans afoot to modify Soyuz to carry three men once more. However, he predicted that a crew of three would not be carried on every mission:

"The manning of the spacecraft depends upon the flight programme. We try, however, to keep it as small as possible, even though the work involved [for each man] increases. As it is at the moment a cosmonaut has to know all aspects in various fields: in astronomy, technical matters, biology, medicine, etc."

Some details of the Soyuz-T spacecraft have appeared in Feoktistov's book *Cosmic Apparatus*. The design is derived from the basic Soyuz spacecraft, but significant improvements have been made. The mass

Soyuz-T

The New Soyuz-T space station ferry was identified by the Soviets at the end of 1979 and first flew with men the following year. Although it retains the basic Soyuz design, it has been greatly modified.

1 Orbital module, which appears to be little modified from the earlier Soyuz variants.
2 A pair of solar panels, reinstated to the Soyuz ferry after not being used for eight years.
3 Re-designed instrument module, with the forward section containing much of the electrical equipment.
4 A new unified propulsion system, which allows the attitude control engines to be operated from the same propellant supply as the new main propulsion system; it seems probable that a pair of the Soyuz-T main engines were used for the Salyut 6, Salyut 7 and Mir main propulsion systems.
5 Re-designed descent module, which allows a return to three-person crews. For the first time there is sufficient room for three people to sit in this module wearing pressure suits.

The Soviets proudly announced that the Soyuz-T spacecraft carried a new, improved computer system, but this seems only to match the capabilities which American manned spacecraft possessed more than a decade previously. The spacecraft propellants were changed to match those now used for Salyut: nitrogen tetroxide and UDMH. Although Soyuz-T could carry three people, many of the crews comprised only two. Soyuz-T became operational with the Salyut 7 missions.

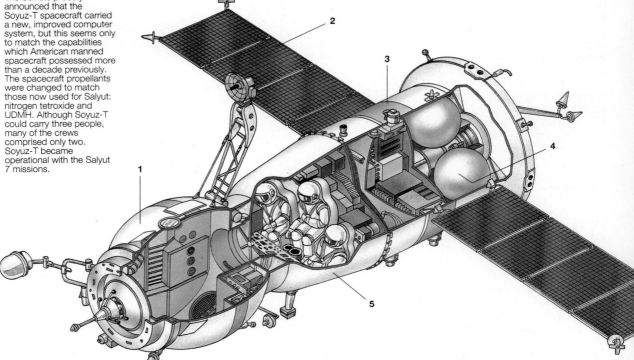

of the spacecraft was quoted as 6,850kg (but see below); this could be split into the orbital module – 1.1 tonnes, descent module – 3.0 tonnes and the instrument module – 2.75 tonnes. The spacecraft length was 6.98m, excluding the extended docking probe; the modules were the same size as a standard Soyuz. The span of the two solar panels which supplied electrical power to the spacecraft systems was 10.6m.

The Soyuz-T included a new computer system and was claimed to be more auto-mated than the earlier Soyuz variants; however, in flight the cosmonauts often had to take over manual control when the automatic systems apparently malfunc-tioned during docking manoeuvres. Soyuz-T was capable of up to four days' independent flight, an improvement on the standard Soyuz which could only man-age two days. Thus, in the event of a dock-ing failure, it was not essential to power down the spacecraft to only the bare mini-mum levels required for the cosmonauts' safety.

Soyuz-T's propulsion system was of a totally new design, using a fully integrated system whereby the attitude control thrus-ters drew on the same propellant supply as the main propulsion system. The thrust of the main engine was reduced to 315kg; in addition there were a further 14 thrusters of 14kg each and 12 thrusters of 2.5kg each. Apparently some of the docking failures of earlier Soyuz missions had occurred because propellant could not be trans-ferred between the attitude control system and the main propulsion system. The Soyuz-T design overcame this problem.

In 1982 a Soyuz-T spacecraft was used for an international mission, carrying a French cosmonaut (or "spationaut", as the French called him), and as a result more details of the spacecraft came into the public domain. The spacecraft carried 700kg of propellant (presumably the standard nitrogen tetroxide and UDMH); this allowed for a nominal 350kg to be used during the rendezvous and docking manoeuvres and a minimum of 150kg was held for the de-orbit manoeuvre. Each solar panel was 1.4m wide and 4.14m long,

Above: *A spectacular view of Salyut 7 docked with Soyuz-T 14, taken from the departing Soyuz-T 13 spacecraft. The configuration of the improved Soyuz-T vehicle, with solar panels reinstated, is clearly seen in this photograph.*

giving a total surface area of about 11.5m^2.

Western observers noted that during Soyuz-T missions the orbital module was separated in orbit prior to the retro-fire manoeuvre, a stratagem designed to save fuel. In some quarters it was speculated that a "train" of orbital modules might be built onto a Salyut to provide more living space, although of course this would have required a heavy docking system to be in-corporated at each end of the orbital module. If ever the Soviets had considered this option, it was never demonstrated in flight.

Most Soviet documentation gives the mass of a Soyuz-T as 6,850kg, as noted

PRESUMED SOYUZ-T TESTS WITHIN THE COSMOS PROGRAMME					
LAUNCH DATE/TIME RECOVERY DATE/TIME	**SATELLITE**	**ORBITAL EPOCH**	**INCL** (deg)	**PERIOD** (min)	**ALTITUDE** (km)
1974 6 Aug 00.01 8 Aug 23.59	Cosmos 670	6.12 Aug	50.57	89.51	209-292
1975 29 Sep 04.18 2 Oct 04.10	Cosmos 772	29.54 Sep 30.41 Sep 1.47 Oct 1.59 Oct	51.74 51.78 51.80 51.80	89.14 89.45 89.75 89.46	193-270 195-300 196-328 196-300
1976 29 Nov 16.00 17 Dec 10.31	Cosmos 869	29.75 Nov 4.00 Dec 4.93 Dec 5.88 Dec 6.69 Dec 9.35 Dec 12.70 Dec	51.78 51.76 51.77 51.78 51.79 51.78 51.78	89.36 89.75 90.46 90.58 90.85 91.12 90.62	196-290 187-335 259-335 260-345 265-368 267-391 300-310
1978 4 Apr 15.07 15 Apr 12.02	Cosmos 1001	4.70 Apr 5.97 Apr 10.81 Apr 11.57 Apr	51.62 51.64 51.60 51.61	88.83 89.37 90.80 90.78	202-231 195-291 306-322 308-318
1979 31 Jan 09.00 1 Apr 10.09	Cosmos 1074	31,55 Jan 2.96 Feb 3.21 Feb 3.84 Feb 7.31 Feb 7.43 Feb 7.63 Feb	51.60 51.67 51.66 51.66 51.64 51.65 51.63	88.87 90.03 90.22 90.83 90.88 92.09 92.02	197-240 255-297 264-306 309-321 279-357 352-402 363-384

Notes: Cosmos 869, Cosmos 1001 and Cosmos 1074 are all believed to have been tests of the spacecraft which would become Soyuz-T. Whether Cosmos 670 and Cosmos 772 were also Soyuz-T development tests is less certain. All of the launch and recovery times are estimated, and are derived from the Two-Line Orbital Elements, as are the orbital data.

Above: *Inside the Soyuz-T 2 descent module trainer. The flight engineer uses the left-hand seat, the commander the centre seat and – for a three-person crew – the researcher the right hand seat.*

Left: *The launch of Soyuz-T 2 with its 2-man crew – Aksyonov and Malyshev. This was to be a four day mission, to man-rate the new spacecraft on the simplest test flight possible.*

The three other flights listed in the table are almost certainly Soyuz-T tests. Cosmos 869 was in orbit for nearly 18 days (matching the Soyuz 9 duration !), and manoeuvred extensively. Cosmos 1001 completed an eleven day mission with manoeuvres. Cosmos 1074 was in orbit for two months, again manoeuvring extensively. The latter mission would qualify the spacecraft for a mission lasting from one Salyut landing "window" to the next.

The next step in the programme came with launch of the unmanned Soyuz-T (un-numbered by the Soviets) in December 1979; since it docked with Salyut 6 for most of its mission, this flight is discussed in the next chapter.

Of the three spacecraft types discussed in this chapter, the Progress was the first to see operational use, with a launch in early 1978 to Salyut 6. Soyuz-T began manned missions in 1980, although it was probably not used operationally until 1982, after Salyut 7 had been launched. The role of the Heavy Cosmos spacecraft is more difficult to assess. Cosmos 1267 docked with Salyut 6, but only after the final crew had abandoned the station; Cosmos 1443 was occupied by the Soyuz-T 9 crew when they docked with Salyut 7; and Cosmos 1686 docked with Salyut 7 while the Soyuz-T 14 crew were on board, and was later used by the Soyuz-T 15 crew. However, on all of these missions it was described as being an experimental craft. Its operational use in the manned programme will probably be revealed as operations with the third generation Mir orbital station complex unfold.

above, but cumulative payload masses published by Glushko do not confirm these figures. The combined masses of Soyuz-T and Soyuz-T 2 is reported 13.58 tonnes, giving a mean value of 6,790kg; cumulative figures for the first twelve orbital missions give the same mean value. Of course, some missions could have reached 6,850kg and others fallen correspondingly short of this figure.

The in-flight development of Soyuz-T seems to have been achieved over a period of at least three years and possibly five years. Cosmos 670, the first of the presumed test flights launched in August 1974, came during a busy period for the Soviets as they were readying the ASTP test, Cosmos 672, and the manned Soyuz 15 mission to Salyut 3. A routine test, one would expect, could have been delayed until a quieter period. The orbital inclination of 50.6° is unique for a Tyuratam launch, suggesting that the Soyuz booster had a slightly greater lifting capability than normal. The spacecraft does not seem to have manoeuvred and it returned to Earth after three days.

The mission lifetime of three days supports a connection with Cosmos 772 which flew at the end of September 1975. However, Cosmos 772 was launched into a normal 51.8° inclination orbit and performed a series of orbital manoeuvres prior to its recovery. Cosmos 772 is often equated with a test prior to Soyuz 20 flying its unmanned mission to Salyut 4, and perhaps the new rendezvous system was being tested on this Cosmos flight.

Repeating Orbits

At a first glance, the orbits chosen for Soviet manned missions may seem to be at random altitudes, but calculations show that they are often chosen quite carefully. One of the major experiments in the Salyut programme has been the observation of the Earth (whether for military or scientific/economic reasons) and therefore the orbits of the Salyuts are chosen to allow the stations to fly over the same areas at specific intervals.

The time between a particular ground track being repeated (retraced) by a satellite depends upon three factors to a first approximation: the average orbital altitude, the orbital "eccentricity" (a measure of how much the orbit deviates from a perfect circle) and the orbital inclination. The table provides a summary of some of the repeating orbits which are found at an inclination of 51.6°.

Using the military Salyut orbital repeating pattern, the ground track will be repeated after 63 circuits with an orbital period of 89.85 minutes – that is, 99.6 minutes earlier every fourth day. Similarly, the standard civilian Salyut groundtrack will repeat 48.2 minutes earlier every alternate day. These patterns, of course, govern the launch opportunities to the orbital stations because a launch can only be made as the groundtrack is passing through the launch site.

Naturally, after being placed in a repeating orbit, the orbit slowly decays because of atmospheric friction, and therefore the patterns of the groundtracks are not perfect. However, the orbital manoeuvres conducted by the orbital station complexes can counter this effect.

In the case of Salyut 7 when the Soyuz-T ferry craft was being used to carry both 2- and 3- manned crews into orbit, it was noted that (apparently because of propellant limits on the Soyuz-T) a 2-manned mission was normally launched with Salyut in a 31 circuit repeater, while a 3-manned mission would be launched with Salyut operating in a lower 78 circuit repeater.

REPEATING ORBITS AT 51.6 DEGREES			
CIRCUITS TO REPEAT	ORBITAL PERIOD (min)	ALTITUDE (km)	APPLICATIONS
63	89.85	260-274	Military Salyut
78	90.74	304-318	Cosmos 929 Soyuz-T tests
31	91.35	334-347	Civilian Salyut
46	92.37	384-398	Salyut 6/Progress 7
91	93.42	435-449	Cosmos 929

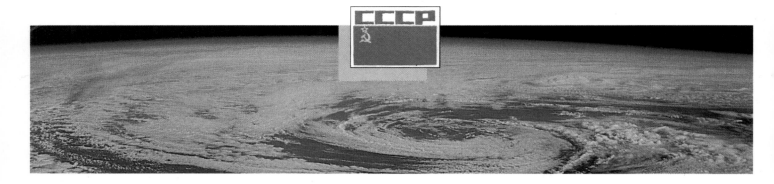

Chapter 11: **The Second Generation Salyut 6**

In retrospect, the Soviet Union identified the first five named Salyut stations as constituting the "first generation" Salyuts; the first of the second generation stations was Salyut 6, launched in September 1977. While retaining the same basic shape as Salyut 4, Salyut 6 represented a more advanced station, capable of supporting not only more complex missions but also far longer missions than the earlier stations.

It was with Salyut 6 that the Soviet space programme reached the peaks of achievement comparable with the American Apollo programme, although the Soviet missions received far less publicity as the operations became more routine. Salyut 6, however, was a major forward step towards the Soviet goal of permanent manned occupation in space which would be realised a decade later.

After its launch, the Salyut 6 programme stumbled as the first resident crew failed to transfer to the station and the mission was aborted after only two days. However, starting with the Soyuz 26 crew, the manned endurance record was extended four times in four resident missions to a maximum of six months (although the fourth mission was not internationally recognised as a record because the previous duration had not been exceeded by the required 10 per cent margin). In addition to extending the duration records, Salyut 6 was visited by a series of guest cosmonauts from socialist states during 1978-1981 and was regularly supplied by unmanned Progress cargo carriers over the same period. The Soviet space programme had come of age, and for the first time in many years the American "lead" in the manned space programme would be under serious threat.

Salyut 6 Description

Outwardly, Salyut 6 was almost identical with the Salyut 4 civilian Salyut, although this was not revealed until more than two months after its launch. Its basic configuration was that of three cylinders comprising a forward transfer compartment and two work compartments of differing diameters. However, gone was the cylindrical propulsion module at the back of Salyut 4 because Salyut 6 carried a rear docking unit as well as a forward unit; instead the propulsion system engines were mounted towards the edge of the rear work module.

According to Soviet accounts, the design of the Proton booster imposes limits on the size of the payload which can be carried. The three stage version, the SL-13, can carry a payload with a maximum diameter of 4.15m (matching the diameter of the third stage of the booster) and a maximum length of about 13.5m in order that the structural loads on the vehicle during the ascent may be kept within safe limits.

Design work on Salyut 6 began in 1973 – the same time as the design of the Progress ferry began – and it was four years before the station was launched. It has previously been noted (in connection with the early Apollo-Soyuz discussions) that a Salyut with front and rear docking ports was considered for a joint Soyuz-Salyut-Apollo mission, but this option was dropped; possibly because of design problems.

Salyut 6 retained the three steerable solar cell arrays which were used to gain power from the Sun, and these were mounted on the exterior of the narrow work compartment (as with Salyut 4). The Soviets described the station as being divided into five main sections:

Forward transfer compartment: length 3.5m, diameter 2m.
First work compartment: length 3.5m, diameter 2.9m.
Connecting frustrum: length 1.2m, diameter 2.9m to 4.15m.
Second work compartment: length 2.7m, diameter 4.15m.
Service compartment: length 2.2m, diameter 4.15.

The forward transfer compartment car-

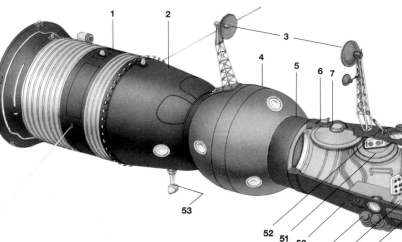

Docking Manoeuvres
1 Soyuz and Salyut 6 at the beginning of docking procedures. Soyuz uses its IGLA system to acquire the station.
2 Salyut orientates itself on Soyuz to align IGLA

antennas. Soyuz fires its main engine to adjust its orbit and approaches Salyut.
3 Soyuz makes a transposition manoeuvre and uses its main engine to reduce speed.

4 Soyuz executes a final transposition and approaches Salyut gently using low-thrust motors for the critical course corrections, guided in to eventual docking by radio commands.

The use of computerised systems in the new 3-man Soyuz-T emphasises the significance the Soviets attach to automatic flight control which may lead to the assembly of large structures in Earth orbit.

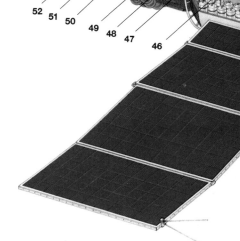

ried the primary docking drogue at its front end, together with a large rendezvous antenna and a smaller search antenna. On the side was a hatch door which opened to allow the cosmonaut crew to perform spacewalks as required. Outside the compartment – and along the exterior of Salyut – were hand-holds to assist the cosmonauts moving outside the station during EVAs. Experiments carried outside the transfer compartment were a micrometeorite detector and a container for various materials which were being tested for reactions to long exposure to space conditions. Inside the transfer compartment was a storage area where the spacesuits were carried, ready for use on spacewalks. It also housed an astronomical photography camera, a sextant port and a station for an experiment to study the Earth's horizon. The connecting hatch between the forward transfer compartment and the first work compartment had a door, allowing the transfer compartment to be isolated from the rest of the station, and thus act as an airlock when a spacewalk was being performed.

On entering the first work compartment from the forward transfer compartment, one immediately passed the main control consoles of the station – in the same position as those of Salyut 1 and Salyut 4. On the "ceiling" and on each wall of the compartment was a single drive for the solar

"GUEST COSMONAUT" SELECTIONS FOR SALYUT 6		
SELECTION DATE	**COUNTRY**	**TRAINEE COSMONAUTS**
1976 September	Czechoslovakia Poland GDR	V. Remek, O. Pelczak M. Hermaszewski, Z. Jankowski S. Jähn, E. Koellner
1978 March	Bulgaria Hungary Cuba Mongolia Romania	G. Ivanov, A. Alexandrov B. Farkas, B. Magyari A. Tamayo-Mendez, J. Lopez-Falcon J. Gurragcha, M. Ganzorig D. Prunariu, L. Dediu
1979 April	Vietnam	Pham Tuan, Bui Thanh Liem

Notes: This table lists all the "guest cosmonaut" researchers who trained for Soyuz visiting missions to Salyut 6. With the exception of the Vietnamese mission (which flew between the Hungarian and Cuban missions) the list reflects the order in which the flights took place. The trainee-cosmonaut who actually flew is shown first, followed by his back-up; this ordering ignores some of the switches of prime and back-up cosmonauts which have been rumoured to have taken place just before the launches. All of the men were air force officers and – with the exception of the Mongolians (who were engineers) – all were military pilots.

panel which extended from each of the three positions. There were two seats at the main control console, one for each of the two cosmonauts who would make up the standard "resident" crew for the station. The walls of the compartment contained cupboards and storage space for the cosmonauts' use and for experiments; one wall had a table which could be lowered into position when required. The ceiling had two pieces of biomedical equipment: the cycle exerciser and the body mass meter. Along the floor of the compartment

was the container for the Rodnik drinking water supply and the viewport for the KATE-140 (also called KT-140) topological camera system.

The frustrum connecting the two work compartments contained the MKF-6M Earth resources camera on the floor. The second, larger, work compartment was dominated by the equipment for the BST-1M sub-millimetre telescope, a conical apparatus which replaced the larger conical unit which had carried the OST-1 on Salyut 4; on top of the BST-1M was the Yelena gamma radiation detector. In front of this, and to one side, was the shower unit which was kept in a stored position, nearly flush against the ceiling when not in use. On the other side of the compartment from the shower was a treadmill, another part of the exercise equipment which the station carried. Three sleeping bags were stowed on the wall, near the ceiling. On the ceiling were two waste disposal airlocks; one of these airlocks later contained the Splav-01 materials processing furnace, delivered to the station by a Progress mission. Food storage lockers were mounted on either side of the far end of the compartment. At the centre of the rear of the second work compartment was another transfer tunnel, leading to a second docking port. Above the inner hatch were dust filters and below was the toilet system (below the general level of the floor); a pri-

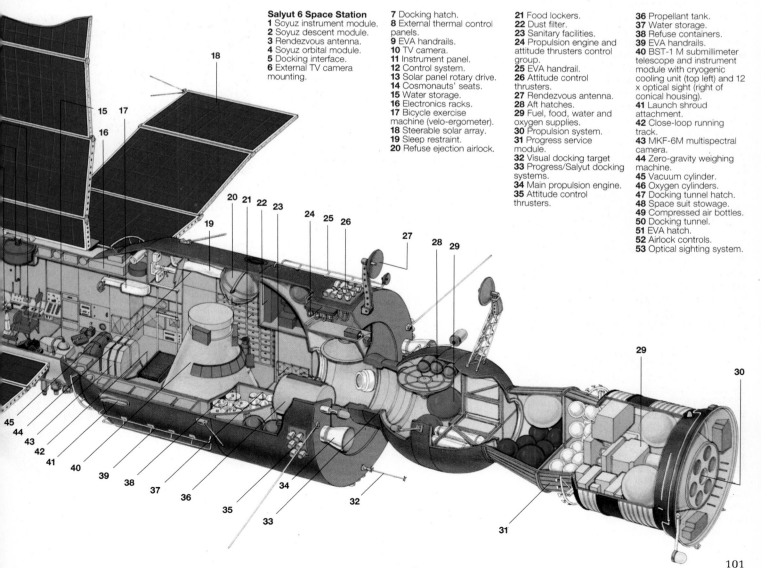

Salyut 6 Space Station
1 Soyuz instrument module.
2 Soyuz descent module.
3 Rendezvous antenna.
4 Soyuz orbital module.
5 Docking interface.
6 External TV camera mounting.
7 Docking hatch.
8 External thermal control panels.
9 EVA handrails.
10 TV camera.
11 Instrument panel.
12 Control system.
13 Solar panel rotary drive.
14 Cosmonauts' seats.
15 Water storage.
16 Electronics racks.
17 Bicycle exercise machine (velo-ergometer).
18 Steerable solar array.
19 Sleep restraint.
20 Refuse ejection airlock.
21 Food lockers.
22 Dust filter.
23 Sanitary facilities.
24 Propulsion engine and attitude thrusters control group.
25 EVA handrail.
26 Attitude control thrusters.
27 Rendezvous antenna.
28 Aft hatches.
29 Fuel, food, water and oxygen supplies.
30 Propulsion system.
31 Progress service module.
32 Visual docking target
33 Progress/Salyut docking systems.
34 Main propulsion engine.
35 Attitude control thrusters.
36 Propellant tank.
37 Water storage.
38 Refuse containers.
39 EVA handrails.
40 BST-1 M submillimeter telescope and instrument module with cryogenic cooling unit (top left) and 12 x optical sight (right of conical housing).
41 Launch shroud attachment.
42 Close-loop running track.
43 MKF-6M multispectral camera.
44 Zero-gravity weighing machine.
45 Vacuum cylinder.
46 Oxygen cylinders.
47 Docking tunnel hatch.
48 Space suit stowage.
49 Compressed air bottles.
50 Docking tunnel.
51 EVA hatch.
52 Airlock controls.
53 Optical sighting system.

vacy curtain was carried for the toilet.

At the back of Salyut was the service compartment, which replaced the Soyuz-derived propulsion system carried on Salyut 1 and Salyut 4. Along the centre of the cylinder ran the rear transfer tunnel, which allowed access to the rear docking port. Around the tunnel – and inaccessible to the cosmonauts – were the propellant tanks and propulsion system: six propellant tanks were carried, three each for UDMH and nitrogen tetroxide. Outside the service compartment were two sets of yaw attitude control thrusters and two sets of pitch and roll attitude control thrusters. On the back of the compartment were the two nozzles of the main propulsion system: the performance of this has not been described in detail, but each of the two engines has a thrust of 300kg – the same as the Soyuz-T propulsion system, suggesting that a pair of Soyuz-T engines might also power Salyut 6. Like Soyuz-T, Salyut 6 is operated as part of a unified control system, with the attitude control propellant being drawn from the main propulsion system's supply. Each of the two main engine nozzles had a cover, which would be closed to prevent outgassing when the system was not in use. The rear of Salyut also carried a second rendezvous radar antenna and search antenna, solar attitude sensors and handrails for use during spacewalks. The docking unit at the back of Salyut was almost identical to that at the front, except that the rim included pipes which allowed propellant to be transferred from Progress cargo craft to the Salyut propellant tanks. A single piece of experimental equipment was contained in the transfer tunnel: a control panel for the Kristall materials processing furnace which was delivered to Salyut 6 by a Progress mission.

The Control System

Salyut 6 contained an "Orientation And Motion Control System", identified by the acronym SOUD, which performed the following functions: automatic orientation of the station for the performance of scientific observations or experiments; automatic raising or lowering of the orbit of the station when preparing for a rendezvous with a transport craft; automatic orientation of the station during the rendezvous and docking manoeuvres with approaching spacecraft; and orientation of the station when under the manual control of the crew.

The SOUD included a number of instruments: gyroscopes, angular velocity gauges, longitudinal acceleration integrators, infra-red plotters of the local vertical, solar and ionic sensors and optical orientation instruments. In addition, the SOUD included the radio rendezvous equipment which, jointly with approaching spacecraft, provided measurements of the relative approach parameters.

The Kaskad orientation system, which was tested on Salyut 4, was included as part of SOUD, as was the Delta navigation system.

On-board Experiments

Salyut 6 was launched with a number of large pieces of experimental equipment already installed, while other important experiments were carried into orbit on Progress cargo missions. Here, brief details of the major experiments which Salyut 6 itself carried into orbit will be noted, while details of the other experiments will be given in the review of the resident mission during which they were launched into orbit.

The **MKF-6M** Earth resources camera system was a modification of the camera which had been carried into orbit during the Soyuz 22 mission in September 1976. The camera enabled photographs of the Earth to be taken simultaneously in six spectral bands. The camera was modified from the MKF-6 version because the earlier camera had been designed to operate in

orbit for only about a week, while the Salyut 6 experiment was expected to last for many years. Therefore, a more robust camera system was carried on Salyut 6 (the second "M" in the MKF-6M stands for "modified"). The film cassette system was changed to allow for easy replacement (each cassette carried 1,200 exposures and had a mass of 13kg). Fresh cassettes were carried into orbit by Progress craft because exposure to the space environment might have led to the film being affected by radiation. They were returned in manned Soyuz craft. The ground resolution of the camera

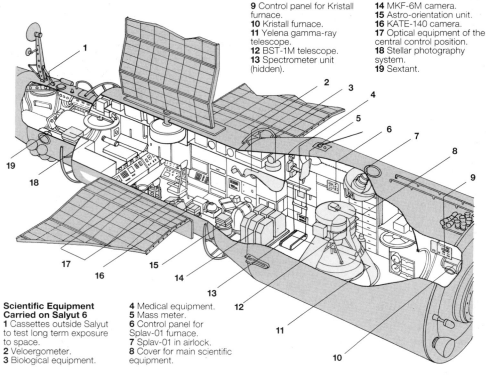

Above: *Bykovsky with experimental equipment aboard the Soyuz 31 visiting mission to Salyut 6 in 1978. Launched with an East German cosmonaut, the mission involved a spacecraft switch.*

Scientific Equipment Carried on Salyut 6
1 Cassettes outside Salyut to test long term exposure to space.
2 Veloergometer.
3 Biological equipment.
4 Medical equipment.
5 Mass meter.
6 Control panel for Splav-01 furnace.
7 Splav-01 in airlock.
8 Cover for main scientific equipment.
9 Control panel for Kristall furnace.
10 Kristall furnace.
11 Yelena gamma-ray telescope.
12 BST-1M telescope.
13 Spectrometer unit (hidden).
14 MKF-6M camera.
15 Astro-orientation unit.
16 KATE-140 camera.
17 Optical equipment of the central control position.
18 Stellar photography system.
19 Sextant.

was probably about 20m, half that of the MKF-6 because of the differing operating altitudes.

The **KATE-140** system was a wide angle, stereographic, topographical camera, used for making contour maps. The camera could supply either single or strip photographs (as required): two film cassettes (each with 600 exposures) might be used at once, and the camera could be operated either by the crew or by automatic ground control.

The **BST-1M** telescope was used for recording atmospheric data in the infrared, ultra-violet and sub-millimetre spectral ranges: it was the largest single instrument on Salyut 6, with a total mass of 650kg. It had cryogenically cooled receivers which required calibration each time the instrument was used. Because of its high level of power consumption, it was apparently used more sparingly than other major instruments like the MKF-6M. As well as being used to study the Earth's atmosphere, the BST-1M was used to study planets (Venus, Mars and Jupiter), stars (specifically Sirius and Beta Centauri), galaxies and the interstellar medium. There were also some observations made of the Moon during lunar eclipses.

Salyut 6 Missions

The Soviet Union planned to use Salyut 6 to support manned space missions lasting more than twice the length of the 1977 manned duration record (84 days, held by the Skylab 4 astronauts). This posed a problem for the Soviet mission planners, in that the Soyuz spacecraft was only man-rated for flights of about two-and-a-half months. In order to fly missions longer than two months it was necessary for the

Left: *The cosmonaut training centre at Zvezdny Gorodok has a large hydrotank where cosmonauts can practise spacewalking operations in simulated weightless conditions.*

Above: *Dzhanibekov and Gurragcha training for the Soyuz 39 mission. Although all Soviet missions are planned to return to solid ground, crews routinely train in the Black Sea for water landings.*

Soyuz spacecraft to be replaced during the residency.

As a result, the Soviets planned for two types of manned mission to be flown to Salyut 6: the main missions were designated *"Ekspeditsya Osnovnoi"* (Principal Expedition) or "EO", while the shorter missions which would sometimes result in spacecraft replacements were *"Ekspeditsya Poseshchenya"* (Visiting Expedition) or "EP". Following the launch of Salyut 6, the Soyuz containing the crew for the first main mission, EO-1, would be launched 10-16 days later. The preliminary flight plan called for three main expeditions to be flown: EO-1, lasting about 90 days;

Below: *Cosmonaut engineers involved with the Salyut 6 programme. Left to right: Makarov, Lebedev, Feoktistov (designer), Aksyonov, Sevastyanov, Yeliseyev and Kubasov.*

EO-2, lasting 120-140 days; EO-3, lasting about 180 days.

Originally, it seems that the timing of the visiting missions was not decided when the main expeditions were being scheduled; a Soviet description of the pre-launch plans for Salyut 6 only lists EP-1 as scheduled to spend 7 days on Salyut during EO-1.

A study of the cosmonaut crew assignments shows that there were two different groups of cosmonauts training. One group comprised the EO cosmonauts; the back-up crew for one EO mission would normally be assigned as the prime crew of the next EO mission. The second group were the EP cosmonauts, where the back-up commander for one mission would normally be assigned as the prime commander three EP missions later. Halfway through the Salyut 6 operations, a third group of cosmonauts was established: the men who would fly the test missions of the new Soyuz-T ferry. After 1979, it seems that when cosmonauts had completed their prime mission assignments with the original Soyuz ferry craft, they were transferred to Soyuz-T training.

Manned Intercosmos Flights

Intercosmos is an organisation set up in November 1965 by the Soviet Union and eight other socialist states: Bulgaria, Czechoslovakia, Cuba, the German Democratic Republic (GDR), Hungary, Mongolia, Poland and Romania. In 1979 Vietnam also joined. Intercosmos organised collaborative space missions launched by the Soviet Union, but with instruments and experiments supplied by its member states. As time went on, Intercosmos encouraged collaboration with other non-socialist countries who were not members of the organisation itself.

On 13 July 1976 the nine original members of Intercosmos signed a fresh agreement which called for co-operation in areas relating to launching of space objects of scientific and applied significance; the design of equipment for space exploration; experiments on geophysical and meteorological rockets; mutual scientific-technical assistance; the exchange of scientific documentation and information. The themes of the new Intercosmos space programme would be the study of the physical properties of outer space, space meteorology, space medicine, space biology and space communications.

On 16 July it was agreed within Intercosmos that the Soviet Union would conduct a series of manned space missions which would involve cosmonauts from the Intercosmos states; the manned flights would involve "Soviet spaceships and stations". On 15 September, Lt-Gen Shatalov, the former cosmonaut who was head of the Soviet cosmonaut training programme, said that it would take about two-and-a-half years to train the crews, providing there were no language problems. The mission commander for the joint flights would be a Soviet cosmonaut, while the "flight engineer and research engineer will be from other countries". A further statement on 7 October indicated that the joint flights would take place in the period 1978-1983.

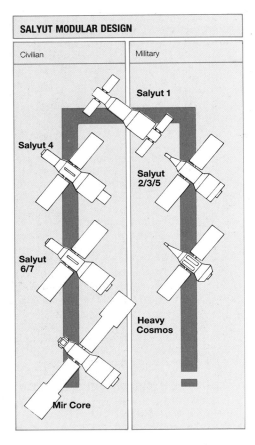

SALYUT MODULAR DESIGN

Civilian	Military
Salyut 4	Salyut 1
Salyut 6/7	Salyut 2/3/5
Mir Core	Heavy Cosmos

Modular Design
The diagram reveals the commonality of Salyut and Heavy Cosmos module designs. Salyut 1 was used as the basis of all of the civilian Salyuts. The military

Salyuts took the smaller work compartment of the Salyut 1 design and added a recoverable capsule and a rear docking port. The Heavy Cosmos is derived from Salyut 2/3/5.

The first group of six international cosmonauts arrived at Zvezdny Gorodok in December 1976 to begin training (see table), and comprised two representatives from Czechoslovakia, Poland and the GDR. The basic training programme was completed in May 1977 and the trainee cosmonauts were then paired with their scheduled commanders:

Czechoslovakia: prime commander Gubarev, back-up Rukavishnikov.

Poland: prime commander Klimuk, back-up Kubasov.

GDR: prime commander Dzhanibekov, back-up Makarov.

The first mission to Salyut 6 failed, and as a result the cosmonaut assignments were altered. The GDR commanders were paired as a two-man Soyuz crew, and were replaced by Bykovsky and Gorbatko. If the original pairings had been retained, it is interesting to note that the GDR mission would have been commanded by a rookie cosmonaut (Dzhanibekov had trained as a back-up commander within the Apollo-Soyuz project). Additionally, the original selection was the first time that civilian engineers were assigned to command a Soyuz mission.

In March 1978, as the manned Czechoslovakian mission was underway, the second group of socialist cosmonaut trainees arrived to begin their training programme and only a year later their first representative was launched into space. In April 1979 Vietnam joined the Intercosmos organisation and two Vietnamese trainee cosmonauts began work which would lead to a joint flight in 1980.

The Launch of Salyut 6

The final mission to Salyut 5 had been completed in February 1977, and in July the Cosmos 929 space station module had begun its tests on a "solo" mission. October would mark the twentieth anniversary of the Sputnik 1 launch and it was expected that the Soviet Union was planning a major space experiment to celebrate this fact.

On 29 September at 06.50 a Proton SL-13 booster was launched from Tyuratam and soon afterwards a new orbital station was placed in its parking orbit. The launch announcement gave the initial orbit as 51.6°, 89.1', 219-275km. Over the next few days the station manoeuvred to a higher orbit which indicated that it was part of the civilian Salyut programme, probably a fol-

Above: *The unlucky first crew to be launched to Salyut 6. Kovalyonok and Ryumin docked with the station, but a fault with their Soyuz docking unit prevented entry into Salyut.*

low-on to Salyut 4 (see table). Later, the mass of Salyut 6, immediately after orbital injection, was said to be 19,825kg, although most Soviet documentation repeated the "standard" Salyut mass of 18,900kg. The Soviet flight schedule called for the EO-1 manned mission to be launched between 10-16 days after Salyut.

On 9 October 1977 the first manned mission to Salyut 6 was launched; commanded by Lt-Col Kovalyonok with the flight engineer Ryumin, this was scheduled to be the mission designated EO-1, lasting for about 90 days in space. From a landing window point of view, the first recovery opportunity would be 16-29 November with the next being 15-26 January the following year. Therefore, it would appear that the actual duration assigned to Soyuz 25 would be close to 100 days.

During the first day in orbit, Soyuz 25 manoeuvred to bring itself close to Salyut 6, approaching the front docking unit (at this time the Soviet Union had not announced that Salyut 6 carried two docking units, although this was suspected by Western observers). However, the docking attempt ran into serious problems. A press announcement ran,

"At 07.09 Moscow time today [10 October] the automatic rendezvous of the Soyuz 25 ship and the Salyut 6 station was begun. From a distance of 120 metres, the vehicles performed a docking manoeuvre.

Due to deviations from the planned procedure for docking, the link-up was

The launch of Salyut 6 had, by this time, been scheduled for the third quarter of 1977 and the Soviet Union had decided on a novel way of flying missions which would include the replacing of an old Soyuz by a new one. The new Soyuz could have been launched unmanned or simply with a commander on board. However, an unmanned mission would not help alleviate the main Salyut crew's sense of isolation from other people, and a one-manned Soyuz mission would leave a seat empty. Therefore, it was decided to fill the otherwise empty seat with a cosmonaut from a friendly socialist state. Salyut 6 was expected to operate for up to five years in orbit, and the announced time-span of the international manned missions more than covered the life of Salyut.

CREW ASSIGNMENTS FOR THE "GUEST COSMONAUT" MISSIONS

COUNTRY	SOYUZ	PRIME CREW	BACK-UP CREW
Czechoslovakia	28	A. N. Gubarev, V. Remek	N. N. Rukavishnikov, O. Pelczak
Poland	30	P. I. Klimuk, M. Hermaszewski	V. N. Kubasov, Z. Jankowski
GDR	31	V. F. Bykovsky, S. Jähn	V. V. Gorbatko, E. Koellner
Bulgaria	33	N. N. Rukavishnikov, G. Ivanov	Y. V. Romanenko, A. Alexandrov
Hungary	36	V. N. Kubasov, B. Farkas	V. A. Dzhanibekov, B. Magyari
Vietnam	37	V. V. Gorbatko, Pham Tuan	V. F. Bykovsky, Bui Thanh Liem
Cuba	38	Y. V. Romanenko, A. Tamayo-Mendez	Y. V. Khrunov, J. Lopez-Falcon
Mongolia	39	V. A. Dzhanibekov, J. Gurragcha	V. A. Lyakhov, M. Ganzorig
Romania	40	L. I. Popov, D. Prunariu	Y. V. Romanenko, L. Dediu

Notes: This table shows for each mission the prime commander and researcher (guest cosmonaut), and the back-up commander and researcher. The original pattern of a Soviet back-up commander being assigned to the prime crew three missions later is clearly shown, although the late insertion of the Vietnamese mission and the transfer of cosmonauts to the Soyuz-T training groups in 1980 meant that later missions did not follow the pattern in every case. For example, Khrunov refused the opportunity of flying the Soyuz 40 mission, and two commanders from 1980 Salyut 6 missions were assigned to that flight.

called off. The crew has begun making preparations for a return to Earth.''

The Soviet media had given the launch of Soyuz 25 a high profile, specifically noting that it had been launched from the same pad from which Sputnik 1 and Vostok 1 had been launched. The docking failure was thus a serious blow to the Soviet plans. At least three docking attempts had been made by Soyuz 25 (from monitoring the cosmonauts' communications with the ground controllers the Kettering Group has reported that there were four attempts), and it is thought that these resulted in at least one "soft docking" but not the final "hard docking" which would have left the spacecraft firmly attached with the electrical systems connected.

Soyuz 25 returned to Earth only two days after its launch, and unusually the two cosmonauts were only awarded an Order of Lenin without being made Heroes of the Soviet Union: perhaps they were partially to blame for the aborted mission? Had the docking succeeded, it would seem that on about 8 November a Soyuz 26 would have been launched with the Czechoslovakian manned mission, and this would have been traded for the older Soyuz 25 spacecraft. Kovalyonok and Ryumin would have switched the new Soyuz from the back of Salyut to the front and then received the first Progress supply craft before their return to Earth in January 1978.

The failure of the Soyuz 25 mission resulted in a major change in the Soviet crewing policy which is still being adhered to more than a decade later. Never again would an all-rookie crew be launched on a space mission; there would always be at least one experienced cosmonaut inside a spacecraft at launch. As a result, the Soyuz 25 back-up crew of Romanenko and Ivanchenkov, both of whom were rookies, was split and paired with cosmonauts who had already flown in space: Romanenko with Grechko and Ivanchenkov with Kovalyonok.

The Soyuz 26 Residency

The failure of Soyuz 25 left the Soviet mission planners with a serious problem. Was the front docking port of Salyut 6 malfunctioning and if so, how could the mission be salvaged with two fully-functional docking ports being a prerequisite for successful operations? Georgi Grechko, an experienced spacecraft designer and veteran of the Soyuz 17 mission to Salyut 4, was drafted onto the next Soyuz crew; he would be able to give the Salyut systems a thorough check, as well as undertaking the scheduled EO-1 long duration mission.

From landing window considerations, the conditions of the Soyuz 25 launch would be closely paralleled on 8 December 1977, and a new Soyuz was expected to fly about that date. In fact, Soyuz 26 was launched on 10 December with Lt-Col Romanenko and Grechko. Two days later at 03.02 the manned ferry successfully docked with Salyut, and for the first time the Soviets admitted that the station had a second docking port:

"The Salyut 6 orbital station is equipped with two docking systems.

Above: *Georgi Grechko on board Salyut 6. He was the replacement flight engineer for the Soyuz 26 mission when, after the failure of Soyuz 25, it was decided not to fly any more "all-rookie" crews.*

Below: *The first group of guest cosmonauts to begin training for flights in December 1976. Left to right: Jähn, Koellner, Remek, Hermaszewski, Pelczak and Jankowski.*

The first is installed on the station's transfer compartment and the second on the opposite side, in the equipment bay. The two docking systems make it possible for two spacecraft to service manned stations. Unlike the Soyuz 25 craft, which – in October this year – approached the station from the transfer compartment side, the Soyuz 26 docked with the station's second docking assembly.''

Once inside Salyut 6, the Soyuz 26 crew began to de-mothball the station and its equipment, although many experiments were still awaiting launch on a Progress ferry. Even so, the mass of scientific equipment on the station was in excess of two tonnes. Since this was due to become the longest manned flight attempted to that time (ignoring the aborted Soyuz 25 mission), an emphasis was placed on the medical experiments on board Salyut.

However, the Soyuz 25 docking failure still endangered the overall Salyut 6 mission. The problem might involve the Salyut docking unit – in which case the mission might not have been salvaged – or it might have been caused by Soyuz 25's docking unit, which was destroyed as the spacecraft re-entered the atmosphere and so not available for examination. The only way to resolve this uncertainty was to go outside Salyut and inspect the front docking cone.

On 19 December, Romanenko and Grechko began to prepare for a short spacewalk. At 21.36 the side hatch of Salyut

opened and Grechko floated outside; this was the first Soviet spacewalk for nearly nine years and only the third in the Soviet space programme. Grechko examined the cone of the docking unit, noting that it had not even been scratched by the Soyuz 25 docking attempt. He then used some tools to check that the docking unit was functioning properly. The result was that the docking unit was given a clean bill of health, and the Salyut 6 programme was rescued.

The 88-minute spacewalk was apparently not without its drama, however. It has been reported that only Grechko was scheduled to leave Salyut, with Romanenko present only to assist him. However, it is claimed that Romanenko fancied a taste of spacewalking experience and slowly left the transfer compartment. Since he did not have a line connecting him to the station, he began to float away and Grechko had to rescue him.

The next problem to overome was the fact that Soyuz 26 was docked at the rear of the station and the ground controllers did not wish to risk undocking the Soyuz from the rear of Salyut and redocking it at the front end – a manoeuvre which would later become standard when one Soyuz was replacing another. Additionally, they did not wish to commit the international Czechoslovakian mission to docking at the front of Salyut. As a result, a new, short visiting mission was inserted into the standard EO-1 flight plan. This called for the launch of a second all-Soviet crew who would dock at the front of Salyut in a new Soyuz ferry to verify the integrity of the docking unit and who would return to Earth in the older Soyuz, thus freeing the rear docking unit of Salyut.

Following the Soyuz 26 spacewalk, the next landing opportunity for a Soyuz would be 15-26 January, and the launch of a new Soyuz was expected shortly before that date. In fact, Soyuz 27 was launched on 10 January 1978 with a crew which comprised the original prime and back-up commanders for the scheduled GDR Intercosmos manned mission: Lt-Col Dzhanibekov and Oleg Makarov, who had only been re-assigned following the Soyuz 25 docking failure. In terms of experience, this crew was a parallel to the Soyuz 26 crew, in that both Romanenko and Dzhanibekov had trained as back-up commanders for Apollo-Soyuz and both Grechko and Makarov were the prime flight engineers for Salyut 4 (in fact, Makarov's mission was the aborted Soyuz 18-1 flight).

On 12 January at 14.06 Soyuz 27 docked at the front of Salyut 6, forming the first orbital complex of three independently-launched spacecraft. During the approach of Soyuz 27, Romanenko and Grechko retreated to their Soyuz 26 spacecraft in case there were problems with the new docking which might endanger the integrity of Salyut 6. Although not specifically intended as such, Soyuz 27 was used to ferry mail and newspapers to the Soyuz 26 crew. Together with the Soyuz 27 crew, Romanenko and Grechko conducted the French "Cytos" experiment, to study the kinematics of the cell division in microorganisms, on Salyut. The results of the experiments would be returned to Earth by the Soyuz 27 crew in containers marked "Biotherm-8".

"Resonance"

While the three spacecraft were docked, the cosmonauts completed an experiment named "Resonance" which simply called for the cosmonauts to jump up and down in the station, while equipment measured the stresses which were set up in the complex's structure. This proved that larger space structures than previously assembled could be constructed in Earth orbit with safety. After this, it became standard when a new docking configuration was being tested for the first time for a similar "Resonance" experiment to be conducted.

During the joint flight the four cosmonauts switched their personal couches in the Soyuz descent modules from those in which they were launched to those in which the crews would return to Earth. On 16 January at 08.05 Soyuz 26 containing Dzhanibekov and Makarov undocked from the Salyut 6/Soyuz 27 complex and after being de-orbited, it was recovered at 11.25. Unusually, the undocking time and landing time were not announced by the Soviet Union immediately, although the latter could be calculated from the Salyut groundtrack and the former usually came 3h 20m (±2 minutes) before the landing: in fact, when the Soviets finally announced the actual times, they tallied with the estimated figures to the nearest minute.

After the return of Soyuz 26, Romanenko and Grechko continued their work in orbit, although they would soon be receiving a new visitor. On 20 January at 08.25 the first unmanned cargo craft was launched into orbit; it was Progress 1. As previously described, this was derived from the Soyuz spacecraft, retaining the orbital and instrument modules but with the descent module replaced by tanks containing the propellant to be transferred to Salyut 6. After an approach taking two days (like that of the unmanned Soyuz 20 to Salyut 4, compared with the normal one day approach for a manned spacecraft), Progress 1 docked with the orbital complex at the rear port on 22 January at 10.12. The Salyut cosmonauts carefully monitored the automatic docking of Progress, but this time they did not retreat to the safety of the Soyuz spacecraft during the final docking operation.

Soon after Progress 1 had docked with the orbital complex, Romanenko and Grechko opened the connecting hatches and began to unload the freighter which was carrying propellant for Salyut 6, oxygen for the air supply, water, foodstuffs, films, equipment for the station and towels for the cosmonauts. The cosmonauts were generating 20-30kg of waste material each day, and this would be loaded into Progress, and destroyed when it burned up in the atmosphere.

The First Refuelling

On 24 January the cosmonauts began to prepare for the refuelling of Salyut, checking to ensure that the propellant and gas supplies were in order. The following day, the cosmonauts began the unloading of Progress in earnest; the spacecraft had brought them air-purification filters, carbon dioxide absorber units and air circulation fans which would replace older units which had been inside Salyut at launch. Additionally, the cargo included new gravity suits, equipment for "technological experiments", safety belts for seats and extra life-support systems.

26 January saw continued preparations for the refuelling of Salyut, with the oxidation tanks being readied for the operation. The slow preparations for the operation

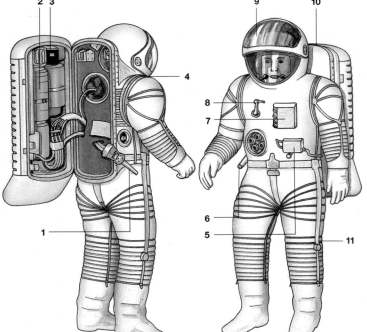

The EVA Suit Introduced with Salyut 6
(The same semi-rigid suit was designed to fit all cosmonauts.)
1 Entry hatch release handle.
2 Replaceable canisters.
3 Life support system on the open suit entry hatch.
4 Hermetic joint.
5 Pressure regulator.
6 Umbilical connector.
7 Suit master control panel.
8 Reserve oxygen source valve.
9 Sun filter visor.
10 Rigid helmet, back-pack and torso frame.
11 Arm and leg length adjustment straps for universal unit.

A cosmonaut enters the single-piece suit through an access hatch which is revealed when the back-pack with its life support system is swung open. The arm and leg lengths can be adjusted for individual cosmonauts. In use, a white overall covers the suit as shown above. Cosmonauts have shown that they can work outside Salyut for 5 hours or more wearing this suit.

Above: *Romanenko wearing the Salyut EVA suit, prior to the spacewalk in December 1978. He is inside the transfer compartment. The EVA cleared the way to continue the Salyut 6 programme.*

continued during the next four days, and on 30 January the crew monitored the pumping of compressed gas from the oxidiser tanks. Two days later the actual refuelling operations were completed – the first time that one spacecraft had refuelled another in orbit (there are unconfirmed reports that during one of the Cosmos docking exercises in 1967 and 1968 a transfer of propellant took place, but the geometry of the Soyuz spacecraft really makes such an operation doubtful: certainly, the transfer could not have been to a propulsion *system*). The Soviet controllers were well-pleased with this experiment, which marked another important milestone in proving the modifications introduced with Salyut 6. If the refuelling had proved impossible, it would have proved essential for a spacecraft docked with Salyut to perform all the necessary orbital manoeuvres and attitude control changes during the manned visits.

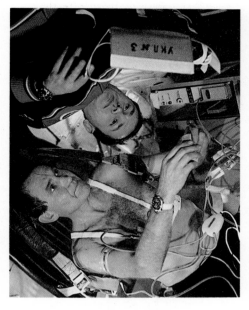

Below: *Cosmonauts preparing to enter the EVA suits for a training session in the hydrotank at Zvezdny Gorodok. Inside the suits, the men are gently lowered using a lift device into the tank.*

Above: *Dzhanibekov (top) and Makarov (bottom) during their short visit to Salyut 6. Makarov is using the station's medical monitoring equipment to test his adaptation to weightlessness.*

With this major operation complete, the unloading of the Progress supplies continued and subsequently the waste material generated during the Soyuz 26 mission was loaded into the orbital module of Progress, ready for destruction. On 6 February at 05.53 Progress undocked from the rear of Salyut and moved away slowly. However, its programme of work with Salyut was not yet complete. During the first orbit after undocking Progress was scheduled to drift some 12-15km away from the station: the freighter's back-up automatic search and approach system was to be tested to prove that it would work if necessary. It was later said that the maximum separation during these tests was 13km, and an actual re-docking was not planned. Once Salyut had been re-acquired and the automatic system began to manoeuvre back to Salyut, the programme was interrupted and Progress continued its independent flight.

On 8 February the propulsion system of Progress was switched on at 02.39 and the spacecraft was de-orbited. Re-entry came over the Pacific Ocean so that any debris which survived re-entry would simply fall into the Ocean harmlessly.

SOYUZ 26 RESIDENCY EXPERIMENTS	
EXPERIMENT	**DESCRIPTION**
Resonance	Crew jump on running track and the resulting vibrations are measured.
Splav-01	Smelting kiln, with the following programme: study of crystallisation process, production of alloys of high degree of uniformity, obtaining crystals from the vapour phase, obtaining monocrystals from liquid phase, obtaining of epitaxial layers from metal alloys, obtaining of metal alloys which will not combine in the liquid state on Earth.
IFS	Four greenhouses with algae, for checking the rate of growth.
Biotherm-2M	Reproduction of dropsophilia fly; radiological study with seeds of arabidopsis and crepis plants; study of vestibular system of aquarium fish in zero gravity.
Oazis	Growth of plants in weightlessness from seeds.
Beta-3	Produce electro- and seismo- grams and register breathing rates.
Reograph-3	Make reographs to determine the amount of blood passing through the heart.
Amak-3	Taking blood samples for later analysis.
Chibis	Spacesuits which fit over lower torso and subject it to low pressures, making the heart work harder.
Running track	For general exercise.
VEA-3	Velo-ergometer.
Tonus-2	Muscle stimulator.
Pingvin and Atlet	Suits worn in orbit to force muscles to work and maintain normal body stature.

CZECHOSLOVAKIAN EXPERIMENTS	
Questionaire	Question cosmonauts about their reaction to weightless environment.
EDC-1K	Study of cosmonaut's body heat exchange.
Kislorodomer	Use of an oxygen meter to test ability of human tissues to absorb oxygen during flight.
Comfort	Use of Chibis suit to lessen orthostatic disturbances; also used Polinom-2M, Reograph-2. Beta-3, Amak-2M biomedical experiments.

Notes: This table, and the corresponding tables for other Salyut 6 residencies, will note the experiments which were conducted for the first time during that residency, particularly noting the experiments which were introduced during the Intercosmos visiting missions. It does not claim to be exhaustive, and many experiments can be further split into individually-named sub-sections. Experiments will be listed only by the residency where they were introduced, and will not be repeated as later crews continued the research.

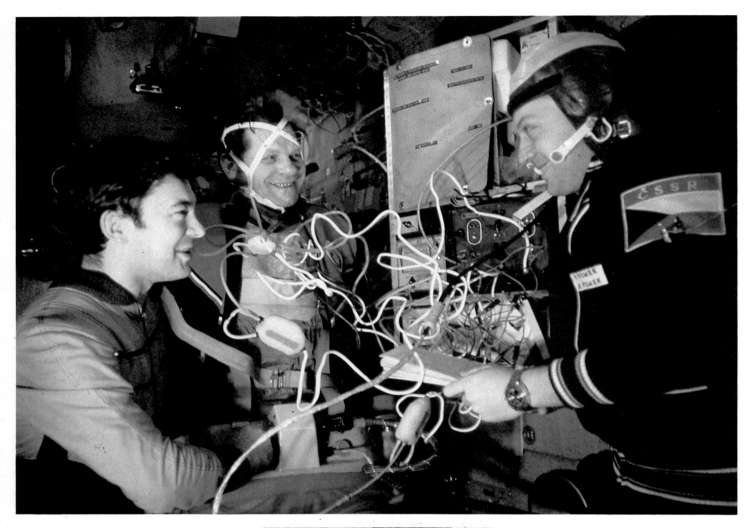

Three days later, Romanenko and Grechko received the congratulations of Klimuk and Sevastyanov for having broken their record for the longest Soviet manned space mission (Soyuz 18 with Salyut 4 in 1975).

Experiments Undertaken

One of the major pieces of equipment carried into orbit was the Splav-01 materials processing furnace. This was designed to operate in a vacuum environment, and the Soyuz 26 crew therefore placed it in one of the two waste disposal airlocks in the larger work compartment of Salyut on 14 February. By this time it had been discovered that if left to itself, Salyut would orient itself in orbit so that the service compartment of the station was pointing towards the Earth and the Soyuz on the transfer compartment was pointing

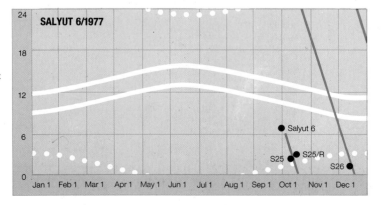

directly away from the Earth. This discovery would allow much valuable attitude control propellant to be conserved.

On 16 February the first experiments with Splav-01 were completed. A capsule with materials for the experiment was placed in the electric furnace. The airlock containing Splav-01 was de-pressurised be

Above: *The first international crew aboard Salyut 6. The Czech Vladimir Remek (left) watches as Gubarev uses the Chibis lower body, negative pressure suit, aided by Romanenko.*

Left: *The first international crew on Salyut, with the resident crew. Left to right: Remek, Gubarev, Grechko and Romanenko. Gubarev and Grechko – here launched separately – had flown Soyuz 17/Salyut 4 together.*

fore the furnace was heated to a temperature exceeding 1,000°C. Since one end of the airlock was open to the vacuum of space, the excess heat generated by the experiment was easily dissipated. The first experiment had involved copper-indium, aluminium-magnesium and indium-antimonide, while a second experiment on 17 February involved aluminium-tungsten, molybdenum-gallium and a semi-conductor. The Splav-01 permitted the Soviets to mix materials which on Earth would not mix uniformly because of the effects of the Earth's gravitation. In the weightlessness of space, near perfect mixing was possible for the first time.

As well as conducting these experiments and making observations of the Earth, planets and other astronomical objects, the cosmonauts continued with a heavy load of medical experiments. As far as mission durations were concerned, they and most of the resident crews which followed them on Salyut 6, Salyut 7 and, a

Launch and Landing Windows
The launch of Salyut 6 at the end of September 1977 was followed within two weeks by an attempt to man the station. The scheduled mission of Soyuz 25 has not been confirmed, but the spacecraft was probably scheduled to return to Earth during the November landing window. Because of problems the mission actually lasted for only two days. A similar duration mission, landing in the January 1978 window, was expected in December 1977, when Soyuz 26 was launched.

SALYUT 6/1977

Salyut 6
S25   S25/R
S26

24 18 12 6 0

Jan 1 Feb 1 Mar 1 Apr 1 May 1 Jun 1 Jul 1 Aug 1 Sep 1 Oct 1 Nov 1 Dec 1

decade later, the Mir complex would be entering "uncharted territory" as longer and longer manned space missions were conducted. With the American Space Shuttle system only capable of short (less than two weeks) missions, all the research into the long-term effects of weightlessness would have to be completed by the Soviet Union.

The first experiments with the BST-1M telescope were conducted on 22 February, although full details were not immediately available of the observations made.

Towards the end of February another landing opportunity was approaching, the actual window being 10-21 March, and the record-breaking flight of Romanenko and Grechko was due to end within this period. However, first they would receive two more visitors from Earth. On 2 March at 15.28 the Soyuz 28 ferry was launched to Salyut, carrying the first international crew to orbit: the flight commander was Col Alexei Gubarev, who had previously flown with Grechko on Soyuz 17 three years earlier, and with him was Vladimir Remek from Czechoslovakia. Afer a successful approach, Soyuz 28 docked at the rear port of Salyut 6 on 3 March at 17.10. The new visiting crew quickly transferred into Salyut 6, joining Romanenko and Grechko.

The following day, Romanenko and Grechko exceeded the duration record of 84 days set by the final crew to visit the American Skylab station. The four cosmonauts began a series of joint experiments, which included the Czechoslovakian experiment "Moravia" with the Splav-01 furnace. Biological and medical experiments were conducted by the Soyuz 28 crew using the Chibis suit and the Polynom-2M recording equipment. On 6 March a further Czechoslovakian experiment, "Ekstinktsiya" was conducted which studied the change in the brightness of stars as they set below the Earth's night horizon; this work allowed the layer of micrometeoritic dust at an altitude of 80-100km to be studied. 9 March saw the completion of the joint experiments, and the visiting crew prepared for their return to Earth.

At 07.17 on 10 March, Gubarev and Remek transferred back to the Soyuz 28 spacecraft and the hatches between it and Salyut were closed two minutes later. Soyuz 28 undocked from Salyut at 10.23, with the spacecraft performing retro-fire at 11.51; landing came at 13.45. The first Intercosmos manned mission was over.

Meanwhile, the Soyuz 26 crew continued to prepare for their own return to Earth by starting to mothball the Salyut 6 equipment, while they increased their exercise regime so that their bodies might be prepared for the return to the Earth's gravity. They used Chibis suits which created a negative pressure in the lower part of the body, which caused the body's systems to react as if they were operating in the normal terrestrial environment. On 16 March at 08.00 Soyuz 27 undocked with Salyut and the spacecraft landed at 11.19. The crew had undertaken a flight of more than 96 days.

Thus, the first Salyut resident mission was successfully completed. Romanenko and Grechko had been launched not knowing whether or not they could save the Salyut mission; on their return Salyut was operating to its full capacity. However, far longer missions were planned before the station's mission would be considered complete.

The Soyuz 29 Residency

Following the return of the Soyuz 26 crew, the orbit of Salyut 6 was allowed to slowly decay until mid-May 1978 when it was boosted to a higher orbit. Nearly a month later, the orbit was again raised after another period of decay, and this heralded the launch of a new manned mission. It will be recalled that the EO-2 freight called for a mission lasting 120-140 days.

On 15 June at 20.17 the Soyuz 29 manned

Above: *Bykovsky (right) assisting in the moving of equipment aboard Salyut, while Ivanchenkov (left) and Kovalyonok (centre) watch. Jähn from the GDR presumably took the picture.*

spacecraft was launched from Tyuratam with a two-manned crew. Quickly being re-cycled was the mission commander, Col Kovalyonok, while the flight engineer Alexander Ivanchenkov was on his first space mission, having earlier trained for Apollo-Soyuz. Docking with Salyut 6 was completed at the front port the following day at 21.58 and the crew transferred to Salyut to begin their long mission. The life support systems were the first to be brought to full operational levels, and the cosmonauts began to wear their "Pingvin" (or Penguin) suits which were especially constructed to force the body muscles to work against the tendency of the suits to make the body curl up in the foetal position in order to provide physical exercise. The cosmonauts were beginning to work as they meant to continue with the exercise equipment.

By 22 June the cosmonauts were ready to begin experiments with the Splav furnace and the following day these experiments began: the programme for 23 June also included the testing of the Kaskad orientation system. Although not announced prior to the event, the cosmonauts were also preparing for their first visitors.

The appropriate Salyut landing window lasted between 30 June-12 July, and the

MANOEUVRES IN SUPPORT OF THE SOYUZ 25 AND SOYUZ 26 RESIDENCIES

PRE MANOEUVRE ORBIT		POST MANOEUVRE ORBIT	
Epoch (1977)	Altitude (km)	Epoch (1977)	Altitude (km)
Initial orbit		29.96 Sep	214-253
29.96 Sep	214-253	1.07 Oct	214-260
2.43 Oct	211-258	4.49 Oct	229-351
6.43 Oct	229-348	7.37 Oct	340-349
Soyuz 25: docking failure 10 October			
		10.42 Oct	341-348
26.51 Nov	332-339	28.79 Nov	340-354
Soyuz 26: docked 11 December			
		11.74 Dec	337-354
(1978)			
Soyuz 27: docked 11 January			
		14.17 Jan	330-352
Soyuz 26: undocked 16 January			
		16.26 Jan	329-350
Progress 1: docked 22 January			
		22.47 Jan	328-349
4.33 Feb	327-345	5.47 Feb	329-346
Progress 1: undocked 6 February			
		8.19 Feb	328-346
23.32 Feb	325-340	24.46 Feb	335-356
Soyuz 28: docked 3 March			
		3.89 Mar	334-353
9.31 Mar	333-352	9.85 Mar	337-351
Soyuz 28: undocked 10 March			
		11.37 Mar	338-350
Soyuz 27: undocked 16 March			
		17.08 Mar	333-348

Notes: This table lists all the Salyut 6 orbital manoeuvres which took place between the station's launch and the return of the first resident crew in March 1978. The various dockings and undockings of spacecraft are shown, as well as the first orbit for which data are available after the event. The orbital data are derived from the Two-Line Orbital Elements, and the same format of table will be used for each of the Salyut residencies.

LAUNCHES IN SUPPORT OF THE SOYUZ 25 AND SOYUZ 26 RESIDENCIES

LAUNCH DATE AND TIME	RE-ENTRY DATE AND TIME	SPACECRAFT	MASS (kg)	CREW
1977 29 Sep 06.50	(1982 29 Jul)	Salyut 6	19,825	
9 Oct 02.40	(1977) 11 Oct 03.25	Soyuz 25	6,860	V. V. Kovalyonok, V. V. Ryumin
10 Dec 01.19	(1078) 16 Jan 11.25	Soyuz 26	6,800 ?	Y. V. Romanenko, G. M. Grechko
1978 10 Jan 12.26	16 Mar 11.19	Soyuz 27	6,800 ?	V. A. Dzhanibekov, O. G. Makarov
20 Jan 08.25	8 Feb 02.39*	Progress 1	7,020	
2 Mar 15.28	10 Mar 10.24	Soyuz 28	6,800 ?	A. N. Gubarev, V. Remek

Notes: All the times are announced by the Soviet Union; the figure for Progress 1 marked with an asterisk is the time of retro-fire. There was a spacecraft/crew switch involving the Soyuz 26 and Soyuz 27 crews; their flight times were: Romanenko and Grechko — 96d 10h, Dzhanibekov and Makarov — 5d 22h 59m. Similar tables, using the same conventions, are prepared for each Salyut residency.

second manned Intercosmos mission was scheduled for this period. On 27 June Soyuz 30 was launched, containing Col Klimuk, on his third space mission, and the Polish cosmonaut-researcher Miroslav Hermaszewski. Following the now-standard approach to Salyut, Soyuz 30 docked at the rear port of the station at 17.08 on 28 June. For the third time, Salyut 6 would be operating as a four-manned space laboratory.

On 29 June the international crew began examinations of their blood circulation system with the Polynom-2M apparatus. The day's programme also included the Polish "Sirena" experiment using the Splav-01 furnace, the operations being scheduled to last for about two days. On 1 July it was reported that the Soyuz 30 crew had tested the "Cardioleader" experiment, which checked their cardiovascular systems. The following day the cosmonauts used the BST-1M and MKF-6M equipment, and filmed the Aurora Borealis ("Northern Lights") from orbit. The mission was successfully completed, with Soyuz 30 containing the international crew undocking from Salyut on 5 July at 10.15 and successfully landing back on Earth.

The Soyuz 29 crew were not to have the rear port of Salyut vacant for long, because Progress 2 was launched on 7 July and docked two days later at 12.59. On 10 July the cosmonauts began to work with Progress 2. As Kovalyonok and Ivanchenkov began to unload the ferry, mission controllers prepared the spacecraft for a transfer of propellant from Progress to Salyut under ground command. On 12 July the controllers began to pump compressed gas from the Salyut 6 propulsion tanks, preparing for the actual propellant transfer. This operation was completed during 19 July. Replacement equipment was carried to Salyut by Progress, including a new "Globus" instrument panel. One piece of

Above: *Ivanchenkov outside Salyut 6 during the Soyuz 29 EVA. His sun visor reflects Kovalyonok taking the picture by the EVA hatch, and the front of Salyut with the Soyuz docked.*

Below: *In the weightlessness of space, Ivanchenkov and Kovalyonok float through Salyut. Their's was the first residency to extend the duration record significantly beyond US experience.*

SOYUZ 29 RESIDENCY EXPERIMENTS	
EXPERIMENT	**DESCRIPTION**
Kristall	Improved furnace for smelting experiments.
Medusa	Mounted outside Salyut: test of biopolymer reaction to space environment.
MMK-1M	Mounted outside Salyut: micrometeorite detector.
POLISH EXPERIMENTS	
Siren	Use of Splav-01: study of connection between alloy uniformity and transfer of mass in weightlessness, study of crystallographic structure, observations of magnitude of crystal growth.
Taste	Check cosmonaut's ability to differentiate between various tastes.
Additional: work with EDC-1K, Questionaire, Kiloxodomer.	
GDR EXPERIMENTS	
Pentakon-6M and Praktika-EE2	Hand-held cameras for Earth resources photography.
Speech	Jähn repeated "226" in German to check his emotional condition (deduced from voice changes).
Berolina	Use of Splav-01 and Kristall furnaces for six experiments: growth of crystals in cylindrical shapes, growth of alloy crystals, controlled setting of crystals obtained from alloys, growth of mono-crystal for use in optical electronics industry, growing of a germanium mono-crystal from the vapour phase and experiments with the hardening and fusing of different kinds of optical lenses.

totally new equipment to be delivered was the Kristall furnace, an improved version of the Splav-01. Kristall was located in the transfer tunnel leading to the rear docking unit.

On 29 July the second spacewalk in the Salyut 6 programme was undertaken. The purpose of the exercise was to replace and return to Earth some equipment which had been mounted outside Salyut prior to launch and which had been deliberately subjected to solar radiation and the general space environment for ten months. The two cosmonauts donned the adjustable spacesuits (only two suits were carried by Salyut for spacewalks, and these could be adjusted to fit most cosmonaut sizes), de-pressurised the forward transfer compartment and opened the EVA hatch at 04.00. Ivanchenkov stepped outside the Salyut and secured himself to a footstand, ready to begin work. Kovalyonok assisted in the operation and operated a colour television camera which returned pictures of the spacewalk to the ground controllers. Ivanchenkov removed one of the micrometeorite detectors, cassettes with polymers, cassettes with biopolymers, optical "and other structural material used in the build-

ing of advanced spacecraft". The complete spacewalk lasted for 125 minutes.

Following the excitement of the spacewalk, Kovalyonok and Ivanchenkov returned to the routine of operations inside Salyut. Re-pressurisation of the station following the spacewalk was completed using the air supplies carried by the Progress 2 freighter. On 31 July work with Progress was complete and the spacecraft was filled with the accumulated garbage generated by the Soyuz 29 crew to that time. On 2 August at 04.57 Progress separated from

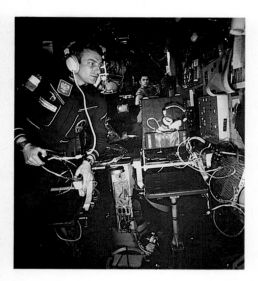

Above: *The first international cosmonaut mission during the Soyuz 29 residency carried Hermaszewski from Poland (foreground) into orbit.*

Salyut 6 and two days later the spacecraft was de-orbited to destruction over the Pacific Ocean.

On 3 August the cosmonauts completed a further Kristall semiconductor experiment, while the next day was devoted to a series of regular biomedical experiments, allowing the mission controllers to check the health of the cosmonauts. 7 August saw an orbital manoeuvre by Salyut, with a further Progress mission being readied. Progress 3 was launched on 7 August (8 August Moscow Time) and it docked with the Salyut 6/Soyuz 29 complex at midnight 9-10 August. On 10 August Kovalyonok and Ivanchenkov began to unload their second ferry craft. An unusual piece of cargo was a guitar for Ivanchenkov, since he was missing this particular pastime. Pastimes aside, the cosmonauts continued with their usual heavy workload, conducting experiments with the Splav-01 and Kristall equipment. Progress 3 was the first of the freighters not to carry a fresh supply of propellant for Salyut, the station's tanks having been filled by Progress 2. Its mission complete, Progress 3 undocked from Salyut on 21 August at 19.29, and the spacecraft was de-orbited and destroyed two days later.

Already, there were plans for a further launch to Salyut 6. On 26 August Soyuz 31 was launched with two more visiting cosmonauts: commander Col Valery Bykovsky and cosmonaut-researcher Sigmund Jähn from the GDR. The flight of an East German cosmonaut was considered to be particularly significant because of the MKF-6M camera on Salyut 6, this having been built by the Carl Zeiss works at Jena. The day after launch, at 16.38, Soyuz 31 docked at the rear docking port of Salyut 6. Although not initially detailed by Soviet sources, East German radio reports indicated that the Soyuz 31 crew would return to Earth in Soyuz 29, and that the Soyuz 29 crew would then undock the new craft from the rear of Salyut and re-dock at the front, thus freeing the rear port for further Progress missions as might be required.

By 29 August the cosmonauts had switched the seats from the original Soyuz craft to those in which the two crews would return to Earth. In addition the four cosmonauts undertook a series of biological experiments the same day. This was followed by a series of photographic sessions using the MKF-6M camera. A special GDR experiment was called "Rech" which required the cosmonauts to repeat vocally a series of numbers during the flight, in order to carefully monitor any changes in cosmonaut speech mannerisms. A further experiment called "Audio" tested the hearing of the cosmonauts during the joint flight, trying to detect changes in that also.

On 3 September Bykovsky and Jähn transferred to Soyuz 29 and undocked from Salyut at 08.23. Retro-fire began at 10.53 and at 11.11 the three modules of the Soyuz separated. Landing came at 11.40. Calculations showed that with this return to Earth the standard recovery procedure had been changed; subsequent landings were to use the new procedure also. Normally, the recovery of civilian Salyut crew had been made on the orbit *following* that which provided a nominal launch opportunity to Salyut, but starting with the Soyuz 29 recovery, the landing occurred during the orbit which provided the nominal launch opportunity. This meant that the landing window would open some 2-3 days earlier than it would have done otherwise.

Homecoming

The September 1978 landing window was 28 August-10 September, and Soyuz 29 returned to Earth roughly in the middle of that period. On 30 August, the East Germans released specific details of the planned switch of Soyuz 31 from the rear to the front docking port. The whole operation would take about 3-4 days. Initially, over a two day period, Kovalyonok and Ivanchenkov would partially mothball the Salyut 6 systems, and then they would enter Soyuz 31. The spacecraft would then undock from Salyut and retreat to a distance of 100-200m. Salyut would then perform "half a somersault", resulting in its front docking port facing Soyuz. Once the station and the ferry were correctly aligned, Soyuz would re-dock at the front port.

Above: *Jähn (left) and Bykovsky suiting up ready for the launch of Soyuz 31. This was the first international mission to involve a spacecraft switch.*

On 7 September at 10.53 Kovalyonok and Ivanchenkov inside Soyuz 31 undocked from Salyut and performed the re-docking exercise as described above, the re-docking being completed at 12.03. The manoeuvre was undertaken during the Salyut landing window, presumably to allow a return to Earth under the nominal flight conditions should the re-docking have failed for any reason. The manoeuvre freed the rear docking port for a fresh Progress mission.

Once back on board Salyut, the crew set to work reactivating the systems which they had closed down in case the re-docking was a failure. The flight settled down to the routine of medical checks, experiments with the Splav and Kristall furnaces and the various telescopes and cameras on the station. The duration record set by the Soyuz 26 cosmonauts earlier in the year was exceeded on 20 September, and there were no signs that the Soyuz 29 intended returning to Earth in the immediate future. The regular exercises which the crew were performing on Salyut ensured that they were physically in better shape than the Soyuz 26 crew had been during their record-breaking trip.

The Salyut 6 Life Support System
1 Device for air regulation.
2 Control panel for air regulation system.
3 Device for regenerating water.
4 Velo-ergometer.
5 Shower (stored).
6 Medical supplies.
7 Mass meter.
8 Sleeping place.
9 Food storage area.
10 Airlock.
11 Anti-dust filter.
12 Storage for underwear and hygiene items.
13 Mirror.
14 Electric toothbrush storage.
15 Docked Soyuz orbital module (or Progress at the rear).
16 Air hose (to assist circulation of air).
17 Hygiene napkins.
18 Collector of liquid waste material.
19 Receptical for solid and liquid waste material.
20 Drinking water storage area.
21 Other waste collection devices.
22 Running track.
23 Vacuum chamber.
24 Rodnik drinking water device.
25 Storage space for EVA suits.

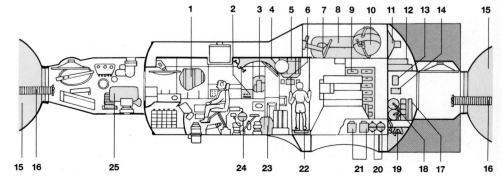

On 3 October Progress 4 was launched towards Salyut on the final re-supply mission for Kovalyonok and Ivanchenkov; docking at the rear of Salyut came on 6 October at 01.00 and the preliminary announcement said that the freighter was carrying "propellant for the combined engine system of the Salyut 6 station, equipment, apparatus, life-support material, material for scientific research and experiments and mail". Later on the day of docking, the cosmonauts opened the connecting hatches leading to the freighter. The next day they began to prepare Salyut for the refuelling operation, and on the same day came the first indications that "the day is not far off" when the crew would begin to mothball the station and return to Earth. Calculations showed that the nominal landing window would be 31 October-12 November, assuming that the new, earlier landing opportunity was used.

The refuelling operation with Progress 4 was completed on 11 October while the cosmonauts continued their flight programme. More and more emphasis was being placed on physical exercise as the landing drew nearer. During 19 and 20

October two manoeuvres of the complex were completed by Progress 4, resulting in the station being placed into a 51.6°, 91.7', 359-376km orbit (as announced on 23 October). This would allow Salyut to remain in a fairly high, though still accessible, orbit following the return of the Soyuz 29 crew. Progress 4 undocked from Salyut on 24 October at 13.07 and was de-orbited two days later at 16.28 (this was the first time that the time of retro-fire was announced immediately following the event; the times for the earlier missions were only published some years after the events).

The cosmonauts continued with their work, although as the recovery day approached the work load was reduced

Above: *The first clear view of Salyut 6 in orbit to be released. This was taken during the re-docking operation as Soyuz 31 was moved from the rear to the front port of the station.*

and the exercise was increased. On 29 October, one of the cosmonauts' regular "days off" the crew were congratulated by mission control on the successful completion of their programme. The non-medical work which would be undertaken would concentrate on the mothballing of the station's equipment until a new crew could be launched.

Finally, Kovalyonok and Ivanchenkov transferred into the Soyuz 31 spacecraft for the last time on 2 November and at 07.46

MANOEUVRES IN SUPPORT OF THE SOYUZ 29 RESIDENCY

PRE MANOEUVRE ORBIT		POST MANOEUVRE ORBIT	
Epoch (1978)	Altitude (km)	Epoch (1978)	Altitude (km)
15.27 May	309-323	16.65 May	321-362
10.30 Jun	313-355	11.18 Jun	340-357
Soyuz 29: docked 16 June			
		16.96 Jun	339-355
Soyuz 30: docked 28 June			
		1.34 Jul	334-343
Soyuz 30: undocked 5 July			
		5.79 Jul	332-340
Progress 2: docked 9 July			
		10.22 Jul	331-338
Progress 2: undocked 2 August			
		2.95 Aug	324-332
5.41 Aug	326-330	5.66 Aug	327-357
Progress 3: docked 10 August			
		10.36 Aug	327-355
17.08 Aug	326-353	17.84 Aug	339-352
Progress 3: undocked 21 August			
		21.96 Aug	335-352
23.36 Aug	336-350	24.88 Aug	338-356
Soyuz 31: docked 27 August			
		28.43 Aug	338-355
Soyuz 29: undocked 3 September			
		3.59 Sep	335-352
Soyuz 31: redocked 7 September			
		8.16 Sep	333-351
Progress 4: docked 6 October			
		6.21 Oct	322-340
6.21 Oct	322-340	7.04 Oct	324-348
19.76 Oct	321-341	20.52 Oct	329-366
20.52 Oct	329-366	20.84 Oct	359-362
Progress 4: undocked 24 October			
		25.36 Oct	358-360
Soyuz 29: undocked 2 November			
		2.57 Nov	356-357

Launch and Landing Windows
Soyuz 27 was launched to allow a short joint flight in January, prior to the recovery of Soyuz 26. Soyuz 27 and Soyuz 28 were both targeted for the March opportunity. When Soyuz 29 returned in September, the recovery came one orbit earlier than previous missions, a practice which became standard; this allowed the nominal landing opportunity to begin a few days earlier. When Soyuz 29 was launched, a mission of about 140 days duration could be confidently predicted.

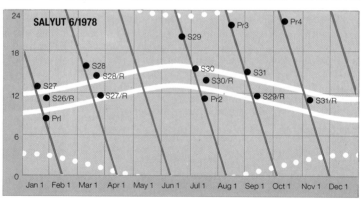

LAUNCHES IN SUPPORT OF THE SOYUZ 29 RESIDENCY

LAUNCH DATE AND TIME	RE-ENTRY DATE AND TIME	SPACECRAFT	MASS (kg)	CREW
1978 15 Jun 20.17	3 Sep 11.40	Soyuz 29	6,800 ?	V. V. Kovalyonok, A. S. Ivanchenkov
27 Jun 15.27	5 Jul 13.30	Soyuz 30	6,800 ?	P. I. Klimuk, M. Hermaszewski
7 Jul 11.26	4 Aug 01.32*	Progress 2	7,000 ?	
7 Aug 22.31	23 Aug 17.30*	Progress 3	7,000 ?	
26 Aug 14.51	2 Nov 11.05	Soyuz 31	6,800 ?	V. F. Bykovsky, S. Jähn
3 Oct 23.09	26 Oct 16.28*	Progress 4	7,000 ?	

Notes: There was a spacecraft/crew switch involving the Soyuz 29 and Soyuz 31 crews; the flight times were: Kovalyonok and Ivanchenkov – 139d 14h 48m, Bykovsky and Jähn – 7d 20h 49m.

they undocked from the station. Following retro-fire and a successful descent, they reached terra firma at 11.05. Their flight had lasted nearly 140 days. The day following the landing, the doctors attending to the cosmonauts announced that the crew felt no worse than the Soyuz 26 crew after they returned from 96 days in orbit. On 4 November the cosmonauts were able to go for walks for the first time since their return to Earth, and specialists indicated that this rapid re-adaptation to the conditions at the Earth's surface was a direct result of the hard exercise programme which the cosmonauts completed during their final days in orbit – as well as the level of exercise throughout their orbital mission. This cleared the way for a further extension of the manned duration record in 1979.

The Soyuz 32 Residency

Following the return of the Soyuz 29 crew, the orbit of Salyut 6 was allowed to decay until a minor adjustment was completed between 20-22 February 1979. Although the Soviets did not state this until later, one of the membranes in the propulsion system tanks had disintegrated and as a result its performance was impaired; only the attitude control system could be relied upon. By the standards set by previous Salyuts, the completion of the Soyuz 29 mission Salyut 6 should have signalled the end of manned operations; in reality the station would continue to host cosmonauts for a further two years. The environmental problems which plagued Salyut 4, resulting in reports that the walls of the station were covered in mould by the time that the Soyuz 18 mission ended, were not repeated in Salyut 6.

On 25 February, Soyuz 32 was launched towards Salyut 6, the objective being to complete the mission designated EO-3, and planned to last six months in orbit. The crew comprised Lt-Col Vladimir Lyakhov, on his first space mission, and Valery Ryumin who had been recycled following the Soyuz 25 failure. It was planned that in addition to the regular Progress supply missions, two visiting missions involving a Bulgarian and a Hungarian cosmonaut-researcher would rendezvous with Salyut. However, these plans were to go awry.

It was noted that Soyuz 32 had been launched during the 22 February – 4 March Salyut landing window, so that if there had been problems with Salyut when the crew docked, then they could return to Earth immediately under the normal landing conditions. However, since the landing windows repeat every two months, the completion of a six month mission automatically dictated that launch must come during a landing window.

Docking with Salyut came on 26 February at 13.30, and Shatalov, the head of the training programme, described the mission as being a "bonus" in the Salyut 6 programme. Naturally, as soon as they boarded Salyut, Lyakhov and Ryumin began to de-mothball the station and prepared a list of equipment which needed replacing so that it could be supplied on the next Progress mission. In fact, the former cosmonaut and spacecraft

Above: *Soyuz 32 is launched with Lyakhov and Ryumin to begin six months in orbit. International missions were planned, but the second was cancelled following the Soyuz 33 failure.*

Below: *Lyakhov (left) and Ryumin in the Salyut trainer, as seen through the EVA hatch on the side of the station. Behind Lyakhov is the hatch to Salyut's main work compartment.*

designer, Konstantin Feoktistov, said that during the first week in orbit the cosmonauts would be carefully checking the station's systems for malfunctions, although apparently most of the major systems on the station seemed to be working properly. On 1 March the Soyuz 32 propulsion system was used to raise the orbit of Salyut slightly, although it was not known in the West that Salyut was in fact having problems with its own propulsion system.

While the Salyut was being checked, the cosmonauts were beginning their exercise programme, since it was anticipated that they would be in orbit longer than the Soyuz 29 crew. Certainly, it was considered that the high level of exercise completed by the Soyuz 29 crew compared

with the Soyuz 26 crew was a major factor in their good health and rapid re-adaptation to normal living conditions.

By 5 March, a week after the cosmonauts had entered Salyut, they had re-activated all of the systems, sending a list of requirements to the ground controllers. They had conducted biological experiments and loaded the MKF-6M camera, conducting some tests of its performance. The following day Lyakhov and Ryumin repaired a faulty videotape recorder using a soldering iron – the first time that such repair equipment had been used in orbit – and changed worn-out lamps and ventilators. To conserve propellant, on 7 March the Salyut/Soyuz complex was put into a gravity-stabilised attitude, with the rear docking port facing the Earth and the Soyuz pointing away from it. By 10 March Salyut had been completely de-mothballed and the cosmonauts were awaiting a supply freighter.

On 12 March, Progress 5 was launched from Tyuratam and two days later at 07.20 it docked at the rear port of Salyut. The contents of Progress 5 were described thus:

"The cargo craft has brought supplies of propellant, water, food and clothes both for work and exercise, a linen drier and shampoo . . . The Progress has brought a standby storage battery. Specialists have also provided for additional means of security at the work areas of the cosmonauts . . . some additional equipment to be used in any contingency. Six signalling devices, which are to be placed in different parts of the station, are capable of detecting even a minute content of carbon dioxide.

New things brought in to space for the first time include a black-and-white television set for viewing both programmes of Moscow TV and special programmes. The TV channel will make it possible to retransmit to cosmonauts blueprints and copies of pictures previously received on Earth from them. The cosmonauts received a new tape recorder, a "ring" system for wireless communications between crew members, a technologically improved design of the Kristall furnace, sets of films and materials for technological experiments. Progress has also supplied the cosmonauts with personal parcels. . . ."

On 16 March, when reporting the unloading of Progress 5, the Soviet authorities noted for the first time that there were problems with the propulsion system of Salyut:

" . . . preventative maintenance and repair operations were started with the propulsion system of the joint power unit of the station and its preparation for refuelling . . . by the cargo craft. At the end of their work aboard the station, cosmonauts Kovalyonok and Ivanchenkov noted some deviation in the control parameters in the main pneumatic line of the engine installation's supercharge system. These deviations did not affect the overall functioning of the power unit. Technical experiments with the station during its flight in an automatic mode

and an analysis showed that the reason for these deviations was damage to the mobile membrane dividing liquid fuel and gaseous nitrogen in one of the three fuel tanks. During a prolonged stay in space this could produce unstable functioning of the regulating valves of the fuel system and a subsequent breakdown in the normal functioning of the power unit. For this reason the decision was taken not to use the tank with the defective membrane in the future work with the station, to exclude it from the general circuit of the power system and transfer its remaining fuel to the two other serviceable tanks."

That day, the cosmonauts began operations which would help the propulsion system problem. They slowly spun the Salyut/Soyuz/Progress complex on its transverse axis, allowing the separation of the fuel and nitrogen in the defective tank by centrifugal force. Most of the fuel was then transferred to another fuel tank, and the residual fuel and the nitrogen gas were transferred to an empty container on Progress 5. Then a valve leading to the vacuum of space was opened to allow the fuel tank and fuel lines to clear any residual fuel. During the next week the tanks and lines were further purged, ready for the transfer of fresh propellant from Progress 5.

Following the repair work on 16 March the two cosmonauts were allowed to take a rest for the next two days, although they did complete some minor repair work. On 19 March, they continued to unload Progress. They removed the original Kristall furnace and replaced it with the improved version which Progress had delivered. A new command and signalling unit and a clock were installed on the station's control desk and a scientific instrumentation panel was replaced. Additional storage batteries were installed to help with the power shortage which the cosmonauts were experiencing when all the equipment was operated.

A milestone (quite a major one for space station crews) was reached on 24 March when the cosmonauts installed their television monitor and were able to have a two-way television link with the ground controllers. It had been standard practice for television pictures to be beamed down to the mission controllers, but for the first time the cosmonauts could receive televi-sion pictures. Setting aside the importance that this would have in the exchange of information between Salyut and the controllers, it was important from a morale point-of-view since the cosmonauts were able to regularly *see* their families and friends, rather than simply hear disembodied voices. As missions became longer and longer, this facility came to assume even greater psychological significance.

By 30 March the propulsion system of Salyut had been purged, and the new supply brought by Progress was pumped into Salyut's tanks. The Progress engine was used to raise the orbit of the complex to an announced altitude of 284-357km. A further manoeuvre was completed by Progress on 2 April, resulting in the complex once more being placed in its normal "two-day repeater" orbit. The following day at 16.10 Progress 5 undocked from Salyut and on 5 April it was de-orbited.

Soyuz 33

Another Salyut landing window was approaching, this time lasting between 18-27 April, and consequently a new manned Intercosmos mission was anticipated. Observers were not disappointed when on 10 April the two-manned Soyuz 33 was launched into orbit. In command was Nikolai N. Rukavishnikov, the first civilian to command a Soviet space mission (it will be recalled that he had been the back-up commander for Soyuz 28), and Georgi Ivanov from Bulgaria. It was expected that Soyuz 33 would dock with Salyut and the crew would return in Soyuz 32, leaving the Soyuz 32 cosmonauts with the job of switching the new Soyuz from the rear to the front docking port. However, this was not to be.

At first the Soyuz 33 mission seemed to be going well, but problems arose during the final approach manoeuvres. Late on 10 April it was said that the Soyuz 33 propulsion system had been used twice to correct its orbit. On its 13th circuit of the Earth the orbit was announced as 273-330km, and about four circuits later the orbit was due to be raised to match that of Salyut. According to a *Tass* announcement early on 12 April:

"In accordance with the programme of flight of the international crew, on 11 April at 21.54 Moscow Time [18.54 GMT] the approach of the spacecraft Soyuz 33 to the orbital complex Salyut 6/Soyuz 32 was commenced. During the process of approach there occurred deviations from the regular mode of operation of the approach correcting propulsion unit of the Soyuz 33 spacecraft, and the docking of the craft with the Salyut 6 station was aborted."

This marked the first in-orbit failure of the Soyuz propulsion system, and it was particularly disappointing that it happened during a prestigious international flight. The Soyuz was powered down as the cosmonauts drifted in orbit awaiting their recovery attempt which could not happen for almost a complete further day. Only in 1983 was the true situation described by the Soviets, when it became clear that they were seriously worried about the safe return of the crew. Since the main propulsion system had failed for the first time during a flight, the single-burn (in automatic mode) back-up propulsion system would have to be used. The nominal burn time was 188 seconds; so long as the burn lasted for more than 90 seconds, the crew could manually re-start the engine to compensate, but this would mean that the landing would be very inaccurate. If the burn was less than 90 seconds, then the crew would be stranded in orbit. A burn for longer than 188 seconds could result in unacceptable overloads on the crew during the descent.

On 12 April at 15.47 the back-up engine fired and continued to fire for a total of 213 seconds before Rukavishnikov shut it down manually. As a result, a ballistic re-entry was completed during which the cosmonauts were subjected to overloads of 8-10g instead of the normal 3-4g encountered on a nominal Soyuz re-entry. The crew survived the landing, but on Salyut 6 Lyakhov and Ryumin were none too happy.

The spirits of the Salyut 6 crew had been raised by the launch and impending visit of two more cosmonauts. They were following the approach of Soyuz 33 from the vantage point of Salyut 6, and they saw the Soyuz 33 engine ignite to begin the final approach and docking, followed by an almost immediate shut-down and a glowing in the area of the main engine. It has been reported that when the docking was cancelled, the news was received by the Soyuz 32 crew with a series of grunts, followed by their terminating all voice communications with the mission controllers. The disappointed cosmonauts immediately went to bed and their mood was none too good for some days after the failure.

Still alone in orbit, the Soyuz 32 crew returned to their routine of experiments, observations and exercises. The equipment for the experiments to be conducted during the Bulgarian visit had been delivered to Salyut by Progress 5, so at least they could be completed. The mission planners had to decide how to continue the mission. Soyuz 32 was man-rated for

Below: *The prime and back-up cosmonauts for Soyuz 33. Left to right: Romanenko, Alexandrov, Rukavishnikov and Ivanov. Alexandrov was the back-up, but he flew to Mir in June 1988.*

SOYUZ 32 RESIDENCY EXPERIMENTS	
EXPERIMENT	**DESCRIPTION**
Yelena-F	Gamma radiation telescope.
Kristall	Replacement for first Kristall furnace.
Biogravistat	Biological experiments with centrifuge.
Incubation	Use of quail eggs to study development of embryos.
KRT-10	Radio telescope.
Deformation	Determine possible deformation of Salyut's shape due to solar heat.
BULGARIAN EXPERIMENTS	
Spektr-15	Spectrophotometer for Earth resources.
Sredets	Check influence of spaceflight on the condition of cosmonauts.
Pirin	Metallurgical experiments.

flights of three months at a maximum, and the next Soyuz would not be ready for launching until June. That would mean that a crew returning in the old Soyuz would be using a craft that was nearly four months old. It had been planned that Soyuz 34 would be launched about 6 June with a Soviet-Hungarian crew, and presumably they would have returned in Soyuz 33 had that mission been success-

Above: *Survival training prepares cosmonauts for landings far off course. Here, Rukavishnikov and Ivanov prepare for a possible landing in a forest, lighting flares to assist rescue.*

ful. However, it was decided that the second manned Intercosmos mission of the year would be cancelled and Soyuz 34 would be launched on time, but in unmanned mode. Thus, Soyuz 32 could be returned to Earth in unmanned mode during the June landing window. Interestingly, when the Press Office at the London Hungarian Embassy issued the crew biographies in May 1980 for the joint Soviet-Hungarian mission, they were still dated June 1979, thus confirming Western suspicions that the mission had been delayed for nearly a year.

This decision having been made, the Soviets continued with their scheduled launches. For nearly a month Lyakhov and Ryumin continued to orbit alone, then on 13 May Progress 6 was launched with fresh supplies. Docking came two days later at 06.19 with a full load of cargo for the crew. One can imagine that some of the personal items might have previously been carried aboard Soyuz 33. The routine transfer of supplies from Progress to Salyut was completed and during 24-27 May the propellant tanks of Salyut were topped up from the Progress propellant supply.

Progress 6 remained docked with Salyut until 8 June at 08.00, when separation took place and the following day Progress was de-orbited. Already, a new spacecraft had been launched towards Salyut; on 6 June the unmanned Soyuz 34 had been launched, and it took two days to reach Salyut. In what must have been a difficult period for the ground controllers, Soyuz 34 docked at the back of Salyut at 20.02 on 8 June, twelve hours after the Progress undocking. Some supplies were carried by Soyuz 34, but these could not match those carried by a Progress. The cosmonauts

removed their contoured seats from Soyuz 32 and transferred them to Soyuz 34. After a short joint flight, at 09.51 on 13 June Soyuz 32 was undocked from the front of Salyut in unmanned mode and was de-orbited towards a recovery in the standard Soyuz landing zone. The spacecraft was found to be in normal condition after a mission which, unexpectedly, had lasted for slightly more than 108 days. Taking advan-

Above: *Ivanov (facing the camera) and Rukavishnikov in the Soyuz simulator. Although their flight was to last for eight days, an engine failure curtailed it and prevented a Soyuz switch.*

Below: *The Soyuz 33 launch vehicle backing onto the launch pad at Tyuratam in April 1979. The launch complex system is designed to allow rapid re-use — within 24 hours if necessary.*

tage of all of the space in the Soyuz descent module, Lyakhov and Ryumin had filled it with 180kg of photographic film, specimens from the Splav and Kristall furnaces and the results of other scientific experiments. This represented double the experimental load which a single Soyuz would normally have returned because half of the material should have returned in April.

After the recovery of Soyuz 32, the Soviets announced that the Salyut crew would be returning to Earth in Soyuz 34, thus confirming Western estimates of the mission duration. On 14 June at 16.18 the cosmonauts undocked Soyuz 34 from the rear port of Salyut and re-docked at the front port 90 minutes later. All was now ready for the launch of a routine Progress mission.

The launch of Progress 7 came on 28 June, and docking was accomplished two days later at 11.18. There were no indications at this stage of the novel experiment that would be undertaken at the end of the Progress mission. During 3-4 July the Progress propulsion system was used to raise the orbit of the complex to an announced 399-411km – the highest at which a Salyut had been operated. Because there would be no more Progress missions in 1979 and Salyut's propulsion system was considered unreliable, it was decided to boost the station as high as possible before the Soyuz 32 crew returned to Earth. During 13 July Progress refuelled Salyut, so that it would at least have sufficient attitude control propellant to last until the next re-supply mission.

The undocking of Progress 7 came on 18 July at 03.50, but this time it was not a routine operation. A folded radio-telescope, designated KRT-10, had been attached to the docking tunnel at the rear of the station. The outer hatch had remained open, while the tunnel hatch inside Salyut was closed. The folded telescope lay along the tunnel and inside the orbital module of Progress. As the freighter pulled away from the station, the now-exposed telescope dish unfolded and deployed to its full 10m diameter. Television pictures of the operation were beamed to the ground from the retreating Progress. This done, Progress was de-orbited late on 19 July.

The deployment of the KRT-10, of course, meant that the rear docking unit of Salyut could no longer be used, and this was interpreted by some analysts as suggesting that the Salyut 6 programme was drawing to a close. Surprisingly, though, the KRT-10 experiment was destined to be only a short one. The telescope and its supporting structure had a mass of 200kg, and it was used in conjunction with a ground-based radio-telescope with a dish diameter of 70m. It has been suggested that the KRT-10 dish did not deploy correctly, and certainly the data which were returned were of a lower quality than one would have expected from a dish of the announced diameter.

Meanwhile Lyakhov and Ryumin continued their routine work on Salyut, with their exercise periods being increased as their return to Earth drew nearer. On 6 August it was announced that the following week the cosmonauts would begin to

Above: *While in orbit the cosmonauts like to spend part of their leisure time simply watching the Earth pass more than 300km beneath them. Soyuz 32 introduced two-way television contact with the Mission Control centre.*

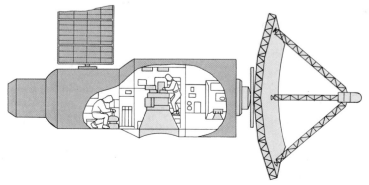

KRT-10
The KRT-10 radio telescope system, deployed from the rear docking port of Salyut 6 while Progress 7 undocked from the station. The telescope was operated from a control system located with the main BST-1M telescope inside Salyut. It is uncertain how successful the experiment was – there have been rumours that it was fouled with Salyut's antennas during deployment, reducing its usefulness. It was discarded during an EVA after a month's operations.

mothball Salyut prior to their return to Earth. The nominal landing window would be 15-27 August, with recovery scheduled close to the middle of the window. However, a further problem needed to be overcome before the crew could return to Earth.

On 9 August the programme with the KRT-10 was concluded, and it was necessary to clear the rear docking port of the station. The telescope should have separated from Salyut cleanly, but instead it became tangled with antennas at the back of Salyut (it has been suggested that it was tangled at deployment, and this prevented the clean separation). As a result, the tired Soyuz 32 cosmonauts had to complete an unscheduled spacewalk to remove the wayward antenna.

On 15 August at 14.16 the EVA hatch on the forward transfer compartment opened and Ryumin followed by Lyakhov emerged into open space. They crawled along the exterior of Salyut to the back and cut the antenna free of the station. This work done, they returned to the area of the EVA hatch, where they took advantage of the unexpected activity to remove some micrometeorite panels and other materials for return to Earth (echoing the work done a year earlier by the Soyuz 29 crew). Lyakhov and Ryumin returned inside Salyut and shut the hatch; their spacewalk had lasted for 83 minutes.

From this point onwards, no further problems arose with the mission, and on 19 August at 09.07 Soyuz 34 with the two cosmonauts undocked from Salyut and returned to Earth, landing at 12.30. Their mission had lasted a few minutes in excess of 175 days. Once more, the cosmonauts were found to be in good health, although they were unable to bear the weight of even a small bunch of flowers being presented to them, because it apparently felt like a giant sheaf of wheat. Some four days after the landing, the crew were considered to be in better health than the Soyuz 29 crew had been at a similar time.

The successful conclusion of the Soyuz 32 mission meant that all three of the EO expeditions scheduled for Salyut 6 had been successfully completed. However, as a result of the repair work completed by Lyakhov and Ryumin, Salyut was now functioning well – with the exception of the main propulsion system – and it was considered that much further use could be made of the station.

Soyuz-T 1

After a break of four months in the flight programme of Salyut 6, another manned mission was expected for December 1979. The landing window was 18-28 December, and since four months had also separated the return of Soyuz 29 and the launch of Soyuz 32 a new Soyuz mission was considered to be a distinct possibility. When it came, though, the mission was quite unlike what anyone had expected.

On 16 December the unmanned Soyuz-T 1 (the Soviets identified it only as Soyuz-T with no number; the next mission in the series was Soyuz-T 2) spacecraft was launched towards Salyut, and docking was achieved three days later at 14.05. The

three day approach and docking profile was considered unusual, and it seems certain that a two-day profile was intended. Geoffrey Perry of the Kettering Group has indicated that Soyuz-T overshot Salyut on its 32nd circuit when the docking should have been completed, and had to be manoeuvred back to a docking the following day.

Further apparent confirmation of the failure of the Salyut propulsion system came on 25 December when Soyuz-T was used to manoeuvre the complex to a higher orbit. On 8 January 1980 it was announced that there would be a further manoeuvre of the complex using the Soyuz-T propulsion system, but this does not seem to have taken place. The continued operations with Soyuz-T indicate that, unlike earlier long unmanned missions, it was not powered down while it was docked with Salyut. On 23 March at 21.04 Soyuz-T undocked from Salyut, and after an independent flight lasting for two days its descent module was returned to Earth.

The return of Soyuz-T signalled a change in procedure for the recovery. On previous Soyuz missions the whole spacecraft had been de-orbited, but to save propellant on Soyuz-T missions the orbital module was separated in orbit prior to retro-fire. This meant that the extra fuel used in de-orbiting the orbital module could in future be used for in-orbit manoeuvres.

With the benefit of hindsight, it would appear that Soyuz-T spacecraft were never

Above: *Ryumin (left) and Lyakhov recline in special couches on return to Earth after their six months in orbit. The "hatch" in the Soyuz descent module's side is the attachment point of the jettisoned periscope.*

originally intended to operate with Salyut 6. Testing had begun in 1976 – possibly with related flights in 1974 and 1975 – and development had been slow compared with the flight testing of the normal Soyuz in the late 1960s. The extended life of Salyut 6 was put to good advantage so that Soyuz-T could be tested in conjunction with the station, but the new ferry was not actually used operationally until the launch of Salyut 7, the next of the Soviet space stations, in 1982.

The landing of Soyuz-T came outside a normal Salyut landing window. The spacecraft had to be flight-rated in excess

MANOEUVRES IN SUPPORT OF SOYUZ-T 1			
PRE MANOEUVRE ORBIT		**POST MANOEUVRE ORBIT**	
Epoch (1979)	Altitude (km)	Epoch (1979)	Altitude (km)
Soyuz-T 1: docked 19 December			
		20.27 Dec	341-345
24.27 Dec	340-346	25.53 Dec	365-378
(1980) 22.46 Mar	336-345		
Soyuz-T 1: undocked 23 March			
		23.90 Mar	344-347

LAUNCHES IN SUPPORT OF THE SOYUZ 32 RESIDENCY AND SOYUZ-T 1				
LAUNCH DATE AND TIME	**RE-ENTRY DATE AND TIME**	**SPACECRAFT**	**MASS** (kg)	**CREW**
1979 25 Feb 11.54	13 Jun 16.18	Soyuz 32	6,800 ?	V. A. Lyakhov, V. V. Ryumin
12 Mar 05.47	5 Apr 01.04*	Progress 5	7,000 ?	
10 Apr 17.34	12 Apr 16.35	Soyuz 33	6,860	N. N. Rukavishnikov, G. Ivanov
13 May 04.17	9 Jun 18.51*	Progress 6	7,000 ?	
6 Jun 18.13	19 Aug 12.30	Soyuz 34	6,800 ?	(unmanned)
28 Jun 09.25	20 Jul 01.57*	Progress 7	7,000 ?	
16 Dec 12.30	1980 25 Mar 21.47	Soyuz-T 1	6,900 ?	(unmanned)

Notes: The Soyuz 32 crew returned to Earth in Soyuz 34, which had been launched unmanned. The crew flight time was 175d 0h 36m. The Soviets simply call Soyuz-T 1 "Soyuz-T".

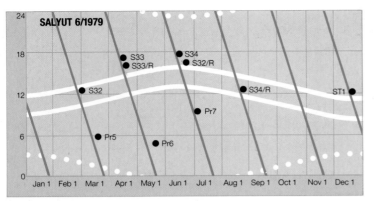

SALYUT 6/1979

(graph with vertical axis 0, 6, 12, 18, 24 and horizontal axis Jan 1, Feb 1, Mar 1, Apr 1, May 1, Jun 1, Jul 1, Aug 1, Sep 1, Oct 1, Nov 1, Dec 1)

Data points: S33, S33/R, S34, S32/R, S32, S34/R, ST1, Pr7, Pr5, Pr6

Launch and Landing Windows
The Salyut 6 landing graph for 1979 shows that the launch of Soyuz 32 was chosen during a landing window so that if Salyut had proved uninhabitable, the crew could have returned to Earth immediately. Also, the six month mission required a launch during a landing window since it would last for three two-month landing window cycles. The manned involvement in Soyuz 34's launch was scrapped after the failure of Soyuz 33's propulsion system.

MANOEUVRES IN SUPPORT OF THE SOYUZ 32 RESIDENCY			
PRE MANOEUVRE ORBIT		POST MANOEUVRE ORBIT	
Epoch (1979)	Altitude (km)	Epoch (1979)	Altitude (km)
20.88 Feb	299-318	22.39 Feb	302-313
Soyuz 32: docked 26 February			
		26.67 Feb	296-309
1.25 Mar	295-305	1.69 Mar	306-334
Progress 5: docked 14 March			
		14.61 Mar	295-324
31.37 Mar	277-306	31.69 Mar	278-340
2.65 Apr	277-335	3.03 Apr	334-348
Progress 5: undocked 3 April			
		4.17 Apr	333-349
6.13 Apr	333-346	6.83 Apr	340-359
Soyuz 33: docking failure 11 April			
		12.74 Apr	338-355
Progress 6: docked 15 May			
		15.44 May	324-339
20.43 May	322-337	23.03 May	333-340
2.16 Jun	328-337	4.75 Jun	328-353
5.57 Jun	327-353		
Progress 6: undocked 8 June			
Soyuz 34: docked 8 June			
		9.52 Jun	351-363
11.30 Jun	353-361	13.34 Jun	354-365
Soyuz 32: undocked 13 June			
		14.04 Jun	356-366
Soyuz 34: redocked 14 June			
		15.31 Jun	357-361
Progress 7: docked 30 June			
		2.05 Jul	353-360
3.35 Jul	352-360	12.43 Jul	395-406
Progress 7: undocked 18 July			
		18.40 Jul	394-405
Soyuz 34: undocked 19 August			
		22.00 Aug	384-393

of the standard two-and-a-half months and the Soviets were wanting to launch a new manned mission during the April 1980 landing window.

The Soyuz 35 Residency

The manned flights during 1980 marked the most successful run of manned and unmanned space missions which the Soviets had flown. After the return of Soyuz-T on 25 March, Progress 8 was launched on 27 March to begin the manned programme for 1980. Docking came on 29 March at 20.01, with the rear Salyut port being used.

Although unmanned, Progress provided Salyut with means of propulsion before the manned mission would begin, and two manoeuvres were completed on 2 April.

The anticipated launch of Soyuz 35 came on 9 April, but the crew was unexpected. The commander was a rookie, Lt-Col Leonid Popov, but the flight engineer was space veteran Valery Ryumin who had only returned from orbit eight months earlier. The reason for this soon became clear. The original flight engineer for the mission had been Valentin Lebedev, who had flown Soyuz 13, but shortly before the flight he had suffered a knee injury and was disqualified from the launch. The back-up crew was apparently Zudov (who had commanded Soyuz 23) and Andreyev (the only un-flown rookie cosmonaut who trained for Apollo-Soyuz), with support for the mission involving Illiarianov (rank unknown), Col V. Titov and G.M. Strekalov. Of these five men, only the commander had spaceflight experience, and therefore Lebedev could not be replaced by his back-up or anyone else in the training group. As a result, Ryumin was drafted into the crew as both a qualified Salyut engineer and as one of the few engineers

not then involved with Soyuz-T training.

Soyuz 35 docked with Salyut 6 at the vacant front docking port on 10 April at 15.16. Once more, another long mission to Salyut was about to begin. It was traditional for one Salyut crew to leave a note of greeting for the next resident crew to read when they arrived; little had Ryumin expected to be reading the note himself as it was written in August 1979. Some two and a half hours after the docking, the cosmonauts entered the station for the first time. Within days, operations had begun the routine with which Ryumin was so familiar. As the Salyut systems were being demothballed, work began in earnest tending the Oazis miniature garden on the station. The cosmonauts unloaded Progress, installing a further extra storage battery to help the power systems. By 15 April it was reported that the unloading of Progress was complete and the cosmonauts had refilled the freighter with garbage. Preparations were being made for the refuelling of Salyut which actually took place on 25 April. On 18 April Popov and Ryumin had loaded film into the MKF-6M and KATE-140 cameras and had checked their operations. Progress undocked from Salyut on 25 April at 08.04 and was deorbited the following day.

The rear port of Salyut was not unoccupied for long, since on 27 April Progress 9 was launched, to dock with Salyut two days later at 08.09. Unloading of Progress began on 2 May, with more than a tonne of cargo to be transferred to Salyut. Preparations for further refuelling of Salyut began on 5 May. The following day the first trans-

Below: *Ryumin (left) and Popov with their pressure suits, worn during the launch, docking and landing of Soyuz. This was Ryumin's second six-month visit to Salyut during 1979 and 1980.*

Above: *The first international crew to be launched in 1980. Farkas (left) from Hungary and Kubasov aboard Salyut with their pressure suits. Their's was the first of two spacecraft switches during the 1980 residency.*

Left: *Kubasov and Farkas during their survival training. Since they should have flown in 1979, they probably trained with the Soyuz 33 crew (compare page 115).*

Above: *Kubasov and Farkas switching their individually tailored couches from the Soyuz 36 spacecraft to Soyuz 35. This was the second spacecraft switch which involved an international crew.*

fer of water was completed between a freighter and Salyut, with an extra 180kg of water having been taken into orbit by Progress 9. The planned refuelling of Salyut was completed on 12 May.

Some minor repair work was undertaken by the crew; the electrical motor of the Biogravistat installation was replaced, as well as one of the gas analyser filters. The video recorder had also required some repair work. Equipment called "Lotos" was being used to make plastic items using a special mould with a quick-setting substance. In addition, the cosmonauts had performed some experiments with the production of polyurethane foam, the hope being that this light, strong material could be used for assembling structures in orbit.

As a new Salyut landing window approached (31 May to 11 June), Progress 9 was undocked on 20 May at 18.51 and was de-orbited two days later. All was ready for the first Intercosmos manned visit to Salyut 6 in 1980.

On 26 May Soyuz 36 was launched into orbit, with two cosmonauts: in command was Valery Kubasov (the second time that a civilian was commanding a manned Soviet mission) and Bertalan Farkas from Hungary. The spacecraft made a successful docking with Salyut 6 at the rear port the

following day at 19.56. Two Hungarian experiments were announced on 28 May: Pille would measure doses of radiation using a miniature thermoluminescent device attached to the cosmonaut's clothing or the walls of the space station, and Interferon studied the formation of interferon in human cells under the conditions of weightlessness. The Hungarian press agency claimed that Farkas had adjusted to the spaceflight environment far quicker that had Kubasov who was on his third flight.

A Spacecraft Switch

The now-standard eight day mission was completed by the international crew, and on 3 June at 11.47 they undocked from Salyut 6 in Soyuz 35, leaving the fresh craft for the use of the resident crew, should they need it to return to Earth. A successful landing was accomplished a few hours later.

On 4 June at 16.39 Soyuz 36 with Popov and Ryumin undocked from the rear port of Salyut and 90 minutes later it re-docked at the front port. This rapid switch of the ferry vehicle, coupled with the Soyuz 36 launch coming at virtually the earliest date to allow a crew recovery in the nominal landing window, raised the question of a

second ferry mission being planned to Salyut during the June landing opportunity. Possibly another international mission would be flown, to make up for time lost following the Soyuz 33 failure?

This speculation was partially confirmed, when a new manned Soyuz was launched on 5 June, although it was not the mission that had been predicted. Soyuz-T 2 was placed in orbit with two cosmonauts on board: Lt-Col Yuri Malyshev, a rookie, and Vladimir Aksyonov who had previously flown on Soyuz 22 in 1976. This was to be a tentative first manned spaceflight of the improved Soyuz ferry variant. Soyuz-T 2 docked at the rear port of Salyut on 6 June at 15.58. During the approach to Salyut, the crew of Soyuz-T 2 had tested the spacecraft in free flight and checked the working of the solar cells which were re-introduced to the Soyuz ferry. The approach to Salyut was completed automatically, while the final 180 metres and the docking itself were accomplished manually. Since Soyuz-T was designed for automatic dockings, it seems probable that there was a partial failure in the automatic approach and docking system; this was a problem which would be regularly repeated during Soyuz-T missions.

Apart from the testing of Soyuz-T, Malyshev and Aksyonov seem to have completed a minimum of scientific experiments in orbit. Their mission was to be a short one, with undocking from Salyut coming on 9 June at 09.20 and recovery following 3h 20m later. Overall, the short mission was judged a success.

Popov and Ryumin once more settled down to a solitary life on Salyut, although they were expecting two more manned visits prior to their return to Earth. On 29 June Progress 10 was placed into orbit around the Earth, and it docked at the rear of Salyut two days later at 05.53. Further

SOYUZ 35 RESIDENCY EXPERIMENTS	
EXPERIMENT	DESCRIPTION
HUNGARIAN EXPERIMENTS	
Biosphere-M	Study of natural phenomena by visual observation: photography of geomorphological objects, photographs of geological structures, study of oceans, wave motions and plankton and observations of meteorological phenomena.
Diagnosticator	Measurement of various physiological parameters, to obtain details of changes in cosmonauts following the landing which could be compared with similar data taken prior to launch.
Metabolism	Studies of protein metabolism; calcium and potassium content in hair, urine and blood; influence of radiation, using urinalysis.
Interferon-1	Effect of spaceflight on interferon production in lymphatic system of the human body.
Interferon-2	Check for any changes in prepared interferon samples due to the conditions of spaceflight.
Interferon-3	Check the effect of interferon on blood samples taken before launch and after recovery.
Work capacity	Check for possible changes in mental activity and capabilities during a spaceflight using the Batalon device.
VIETNAMESE EXPERIMENTS	
Halong and Imitator	Melting of siliceous samples using the Kristall kiln.
Halong	Photography and mapping of Vietnam.
Azolla	Experiments with azolla plant.
CUBAN EXPERIMENTS	
Sugar	Observations of saccharose crystallisation process.
Cortex	Study of brain functions and changes in the brain's electrical activity during a spaceflight and following the return to Earth.
Support	Study of degree of adaptation of Earth conditions after a spaceflight.
Blood Circulation	Cardiograms made before, during and after the flight to determine bio-electrical changes in the heart.
Anthropometry	Study of the effects of spaceflight on muscle size and bone structure.
Stress	Study of changes in the hormone content and metabolism before and after flight.
Co-ordination	Co-ordinograph device used to study the effects of weightlessness on a cosmonaut's motor co-ordination and ability to concentrate.
Metabolism	Study of cosmonaut's water and mineral balance.
Immunity	Study of changes in proteins and mineral compounds aiding the body to ward off illness.
Spectrum	Photography of Earth using Spektr-15 camera.

Notes: The Vietnamese have apparently used the name "Halong" for two different experiments.

replacement equipment was carried by Progress, as well as the normal supplies for the crew: a fresh set of intensifiers for the BST-1M telescope were carried to extend the life of the instrument.

On 17 July at 22.21, after refuelling Salyut, Progress 10 separated from the station, and was de-orbited early in the morning of 19 July. Again, a Salyut landing window was approaching, and a fresh Intercosmos mission was expected. Most observers were expecting that the Cuban mission would be the next to be flown, but they would be proved wrong. Soyuz 37 was launched on 23 July; it was commanded by Col Gorbatko – on his third space mission – who was accompanied by Pham Tuan from Vietnam (the first Asian cosmonaut). The launch was timed to take advantage of the world's media who were in Moscow to cover the 1980 Olympic Games, although of course the launch date was governed by the landing window of 1-13 August.

Another Switch

Soyuz 37 docked at the rear of Salyut at 20.02 after the standard one day approach, and for the third time the Soyuz 35 cosmonauts were playing host in a four-manned space laboratory. There seem to have been few experiments from Vietnam conducted in orbit, although the opportunity was taken for Pham Tuan to photograph his homeland from space. The four cosmonauts switched the seats of the two Soyuz craft and on 31 July at 11.55 the international crew left the orbital station in Soyuz 36, heading towards a successful recovery. The next day at 16.43 Soyuz 37 with Popov and Ryumin undocked from the rear of Salyut and after the station had rotated through 180° they re-docked at the front port at 18.20.

The rapid re-docking of Soyuz 37 sug-

gested the possibility of a further manned visit during the landing window, although this was not to happen. A further surprise was that no Progress craft was launched in the next two months. Progress 10 had carried sufficient supplies into orbit to suffice Popov and Ryumin until their scheduled recovery in October.

The next landing window would last between 2-15 October, and it was expected that prior to that the Cuban international crew would be flown for a return early in the window (implying a launch on 24 September). This would allow the Soyuz 35 crew to mothball Salyut and return in the same window. Once more, what was a reasonable analysis would be proved wrong, but for reasons which were originally obscure.

The launch of Soyuz 38 came on 18 Sep-

Above: *The second guest cosmonaut visit during the Soyuz 35 residency brought Pham Tuan from Vietnam to orbit (left), shown with Popov and Ryumin.*

Below: *Soyuz-T 2 in flight as it approached Salyut 6. This was the first manned flight of the improved Soyuz ferry, carrying two men.*

tember, nearly a full week earlier than the landing window constraints allowed. On board were Col Yuri Romanenko and Arnaldo Tamayo-Mendez from Cuba (the first black cosmonaut). The fact that a Cuban cosmonaut-researcher was carried provided a clue to the early launch for the mission. Although the nominal landing window was the ideal, another more political constraint on the international launches could over-ride the normal landing conditions. It was required that when an international cosmonaut was on Salyut, it should be possible to see Salyut in the night sky from the country which had supplied the guest cosmonaut. For most of the Intercosmos nations this was not a major consideration, since orbital mechanics meant that for countries close to the eastern Soviet Union Salyut would be seen at night automatically when an international mission was launched with a landing targeted for the nominal window. However, for Vietnam and, more especially, Cuba, this rule did not hold, and for inhabitants of Vietnam and Cuba to see Salyut while ''their cosmonaut'' was on board an earlier launch than normal was required. For Vietnam the difference was only about a day, but for Cuba it was nearly a week.

Once in orbit, Soyuz 38 docked with Salyut at 20.49 on the day after the launch, and soon all four cosmonauts were at work inside the Salyut station. During the joint flight, a number of bio-medical experiments – designed by Cuban doctors – were undertaken, as well as photography of Cuba. Little specialist equipment was taken into orbit for the mission, since it was realised that operations with Salyut 6 would soon be ending. Earlier international crews had taken their own equipment into orbit, and this was left for use by the resident Salyut crews, but for their later joint flights as much use was made as possible of existing hardware.

With the impending return of the Soyuz 35 crew, there was no need to switch Soyuz craft again, so the international crew returned to Earth in their Soyuz 38 spacecraft on 26 September.

Unexpected Progress

The launch on 28 September of Progress 11 came as a complete surprise, since it was not anticipated that the returning Soyuz 35 cosmonauts would require any more supplies. Progress docked at the back of Salyut on 30 September at 17.03. The next day some equipment was transferred from Progress to Salyut, although clearly the cosmonauts were ready to return to Earth. On 5 October it was announced that the cosmonauts's exercise programme had been stepped up in anticipation of their impending return.

Progress 11 was left only partially unloaded when Soyuz 37 with Popov and Ryumin undocked from the station on 11 October at 06.30. The cosmonauts were safely back on Earth at 09.50 after extending the duration record once more, this time to 185 days.

The new record was not internationally recognised, however. In order to establish a new record, the IAF demands that the previous mark must have been exceeded

Above: *As the Soyuz 35 residency was drawing to a close, the third international crew to visit the cosmonauts prepared for flight. Romanenko and Tamayo-Mendez are training in the Black Sea for emergency splashdowns.*

LAUNCHES IN SUPPORT OF THE SOYUZ 35 RESIDENCY				
LAUNCH DATE AND TIME	**RE-ENTRY DATE AND TIME**	**SPACECRAFT**	**MASS** (kg)	**CREW**
1980 27 Mar 18.53	26 Apr 06.54*	Progress 8	7,000 ?	
9 Apr 13.38	3 Jun 15.07	Soyuz 35	6,800 ?	L. I. Popov, V. V. Ryumin
27 Apr 06.24	22 May 00.44*	Progress 9	7,000 ?	
26 May 18.21	31 Jul 15.15	Soyuz 36	6,800 ?	V. N. Kubasov, B. Farkas
5 Jun 14.19	9 Jun 12.40	Soyuz-T 2	6,850 ?	Y. V. Malyshev, V. V. Aksyonov
29 Jun 04.41	19 Jul 01.47*	Progress 10	7,000 ?	
23 Jul 18.33	11 Oct 09.50	Soyuz 37	6,800 ?	V. V. Gorbatko, Pham Tuan
18 Sep 19.11	26 Sep 15.54	Soyuz 38	6,800 ?	Y. V. Romanenko, A. Tamayo-Mendez
28 Sep 15.10	11 Dec 14.00*	Progress 11	7,000 ?	

Notes: There were a number of spacecraft/crew switches: the Soyuz 35 crew returned in Soyuz 37, the Soyuz 36 crew returned in Soyuz 35 and the Soyuz 37 crew returned in Soyuz 36. The flight times were: Popov and Ryumin – 184d 20h 12m, Kubasov and Farkas – 7d 20h 46m, Gorbatko and Tuan – 7d 20h 42m. Strictly speaking, Progress 11 was launched to support the Soyuz-T 3 mission in November 1980, but it is included here because it was launched before the Soyuz 35 crew returned to Earth.

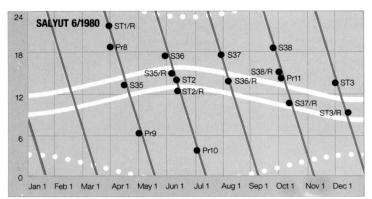

SALYUT 6/1980

(graph with vertical axis 0, 6, 12, 18, 24 and horizontal axis Jan 1, Feb 1, Mar 1, Apr 1, May 1, Jun 1, Jul 1, Aug 1, Sep 1, Oct 1, Nov 1, Dec 1; plotted points labelled ST1/R, Pr8, S36, S37, S38, S35/R, ST2, S38/R, Pr11, ST3, S35, ST2/R, S36/R, S37/R, ST3/R, Pr9, Pr10)

Launch and Landing Windows
Missions to Salyut 6 during 1980 are summarised here. There were two Soyuz spacecraft/crew switches: Soyuz 35/Soyuz 36 and Soyuz 36/Soyuz 37. The launch of Soyuz 38 was much earlier than expected, and this was dictated not by the landing window but by the need for Salyut to be seen during the mission from Cuba, the guest cosmonaut's home. Progress 11 was primarily launched to support the Soyuz-T 3 repair mission that was flown at the end of the year.

MANOEUVRES IN SUPPORT OF THE SOYUZ 35 RESIDENCY			
PRE MANOEUVRE ORBIT		POST MANOEUVRE ORBIT	
Epoch (1980)	Altitude (km)	Epoch (1980)	Altitude (km)
Progress 8: docked 29 March			
		30.46 Mar	340-344
2.50 Apr	338-342	2.88 Apr	342-350
2.88 Apr	342-350	2.94 Apr	342-355
Soyuz 35: docked 10 April			
		10.81 Apr	336-348
24.12 Apr	330-341	24.44 Apr	334-365
Progress 8: undocked 25 April			
		26.40 Apr	334-363
Progress 9: docked 29 April			
		29.90 Apr	333-361
15.25 May	326-355	19.06 May	339-363
Progress 9: undocked 20 May			
		21.41 May	339-362
Soyuz 36: docked 27 May			
		28.20 May	334-355
29.41 May	334-355	29.91 May	338-353
Soyuz 35: undocked 3 June			
		4.18 June	335-352
Soyuz 36: redocked 4 June			
		6.21 Jun	334-350
Soyuz-T 2: docked 6 June			
		7.03 Jun	333-349
Soyuz-T 2: undocked 9 June			
		9.50 Jun	335-348
Progress 10: docked 1 July			
		1.48 Jul	325-340
17.22 Jul	319-334	17.86 Jul	327-342
Progress 10: undocked 17 July			
		18.17 Jul	327-341
21.21 Jul	326-339	21.78 Jul	338-352
Soyuz 37: docked 24 July			
		25.27 Jul	336-351
Soyuz 36: undocked 31 July			
		1.25 Aug	335-349
Soyuz 37: redocked 1 August			
		2.07 Aug	334-348
4.06 Sep	323-338	5.08 Sep	337-350
16.24 Sep	333-345	16.87 Sep	335-352
16.87 Sep	335-352	17.19 Sep	339-352
Soyuz 38: docked 19 September			
		21.25 Sep	338-350
25.18 Sep	337-349	26.01 Sep	332-344
Soyuz 38: undocked 26 September			
		27.27 Sep	334-341
Progress 11: docked 30 September			
		1.26 Oct	332-339
Soyuz 37: undocked 11 October			
		11.64 Oct	318-332

Above: *Posing aboard Salyut for a group picture are (left to right) Ryumin, Tamayo-Mendez, Romanenko and Popov. The space laboratory's walls carry Cuban and Soviet flags for decoration.*

Below: *Although not internationally recognised as a world record, Popov (in couch) and Ryumin (being assisted from Soyuz 37) completed the longest flight to the end of 1980 – 185 days.*

by at least 10 per cent, and therefore for the Soyuz 32 record to be beaten a flight of nearly 193 days was required. During the period February 1979 and October 1980 Ryumin had logged a total of 350 days in orbit; with the aborted Soyuz 25 taken into account Ryumin was by far the most travelled man in space at that time, having logged a total of 352 days.

The leaving of the partially used Progress 11 docked with Salyut apparently made no sense, unless it was simply to ensure that there was a spacecraft docked with Salyut which could manoeuvre the complex since the main propulsion system was not operating. Following the return of the Soyuz 35 cosmonauts, Salyut's orbit was allowed to decay slowly until 17 November when Progress 11 boosted it slightly.

Soyuz-T 3

Once more, a Salyut landing window was looming (4-15 December) and it was suspected that another Soyuz-T test flight would be undertaken, to gain more manned flight experience than the short Soyuz-T 2 mission had given. Therefore, the launch of Soyuz-T 3 on 27 November was not totally unexpected, although the crew size was. For the first time since the Soyuz 11 accident in 1971, a three-man cosmonaut team was in orbit: the commander was Lt-Col Leonid Kizim, the flight engineer was Oleg Makarov – on his fourth space mission and third orbital flight (his second flight was the aborted Soyuz 18-1) – and the research cosmonaut was Gennady Strekalov.

The crewing assignments for the mission are interesting. In 1982 the book *Where All Roads Into Space Begin* was published in English, and the authors (I. Borisenko and A. Romanov) include the following comment (page 94-95):

"We had one more meeting with Konstantin Feoktistov. On that occasion we were interested in him not only as a designer, but as a member of the crew of the forthcoming mission aboard Soyuz-T 3, a spacecraft which he had helped construct. 'You intend to make a second flight, 16 years after the first. Why?'

'If you remember, in 1964 on the Voskhod my duties were those of flight engineer. The chief designer Sergei Korolyov gave me the job, as one of those involved in its design, of testing the Voskhod in flight together with Vladimir Komarov and Boris Yegorov. Actually working on the Voskhod in space provided us with much material for modernising the craft and then for producing the multi-purpose Soyuz and designing the Salyut stations. I think it's high time I went up again to take a look for myself how the systems of the new Soyuz-T function, as well as those of Salyut which has surpassed all expectations in operating so long. . . .'"

Although Feoktistov was disqualified from the Soyuz-T 3 mission on health grounds (he would have been nearly 57 years old), some photographs are available of him doing his sea training with the other Soyuz-T 3 cosmonauts, thus confirming his claim to have expected the flight.

The cosmonauts in Soyuz-T wore new, light-weight pressure suits, and these allowed sufficient room for three space-suited cosmonauts to sit next to each other in the descent module. However, the Soyuz-T could be flown by either two or three men, as the mission required.

Soyuz-T 3 docked at the front of Salyut on 28 November at 15.54 and soon the cos-

Above: *The flight of Soyuz-T 3 at the end of 1980 was the first three-man Soviet mission since the Soyuz 11 accident in 1971. The new crew were Strekalov (left), Makarov (centre) and Kizim (right).*

monauts transferred to Salyut. The mission was not simply a test of Soyuz-T; it was meant to allow the repair of many Salyut systems which were nearing the end of their lifetimes, thus allowing the possibility of a further resident mission to Salyut early in 1981.

The three cosmonauts de-mothballed Salyut, noting areas which required repair or replacement, and on 4 December they dismantled and repaired a faulty electronic unit in the telemetric system. The next day was scheduled for the refuelling of Salyut with supplies carried by Progress 11. 6 December saw the cosmonauts working on the thermo-regulator system which had been exhausted after three years of operations. A great deal of experimental work using the furnaces was completed during the short stay of Soyuz-T 3. On 8 December it was reported that the mission was drawing to a close, and the cosmonauts had begun to tidy and mothball the station once more. Progress 11 was undocked on 9 December at 10.23 and pulled away from the station.

The end of the Soyuz-T 3 mission came on 10 December. Undocking from Salyut was effected at 06.10 and recovery came at 09.26. Although it had been short, the Soyuz-T 3 mission had ensured that a further short resident mission could be conducted on Salyut 6, which had already operated for a year longer than its nominal lifetime. Operations associated with Soyuz-T 3 ended on 11 December when Progress 11 was de-orbited to destruction over the Pacific Ocean.

Soyuz-T 4

As a result of the failed Soyuz 33 mission forcing the cancellation of the Hungarian manned flight in 1979 and the addition of the Vietnamese manned flight to the launch schedule, when the Soyuz 35 mission ended in October 1980 two international crews still awaited launch. They were from Mongolia and Romania. The testing of the new Soyuz-T manned spacecraft was well underway, ready for its operational use in 1982 when a new Salyut was scheduled for launch.

When the Soyuz-T 3 mission cleared Salyut 6 for a further resident mission, the opportunity was taken to conduct a series of flights in order to test further the new Soyuz-T spacecraft for the period between one landing window and the next (approximately two months), and to complete the two outstanding flights in the manned Intercosmos programme.

The two cosmonauts for the manned residency were not paired until late 1980, meaning that they would not be ready for a launch until just before the March 1981 landing window, rather than the earlier February 1981 window. Since Salyut had to be operating with a resident crew before the Mongolian mission was launched and the crew had to remain on the station until after the recovery of the Romanian mission, the residency would have to last for longer than the nominal 60 days interval between landing windows.

The first activity in preparation for the new series of manned flights was the launch on 24 January 1981 of Progress 12. After the now-standard approach, Progress docked at the back of Salyut on 26 January at 15.56. Once docked, Progress completed a series of manoeuvres which raised the orbit of Salyut to that required for the nominal "two-day repeater".

The launch of Soyuz-T 4 came on 12 March, and the crew comprised Col Vladimir Kovalyonok, on his third visit to Salyut 6, and rookie engineer Viktor Savinykh. Docking at the front port of Salyut came a day later at 20.33. The next day the cosmonauts opened the connecting hatch leading to the Progress 12 orbital module and began to unload it. Whilst the station was being de-mothballed, a fresh supply of water was pumped from Progress to the Rodnik water supply system on Salyut. By 18 March, Salyut had been completely reactivated; all the cargo had been transferred from Progress to Salyut, and garbage had been transferred from Salyut to Progress. The cosmonauts continued maintenance work with Salyut by installing a new "unit of solar battery orientation control" and they replaced a pump required for removing condensation in the thermo-regulation system.

Progress 12 undocked from Salyut on 19 March at 18.14 and the following day it was de-orbited. This was the last freighter mission to fly to Salyut 6.

The anticipated international flight began on 22 March with the launch of Soyuz 39; the commander was Col Vladi-

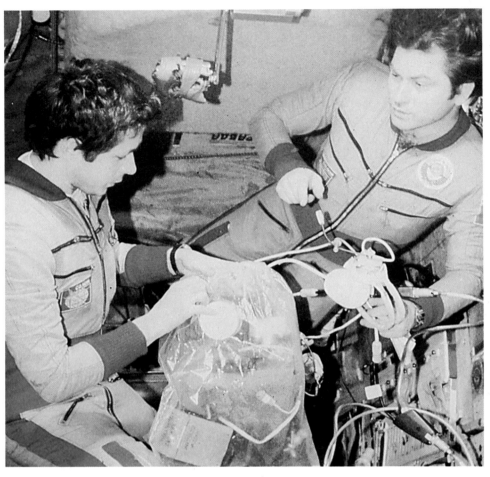

Above: *Savinykh (left) and Kovalyonok aboard Salyut 6 for the final residency. They tested Soyuz-T for more than two months in orbit and received the last two international crews which were planned.*

Left: *The recovery of the Mongolian Gurragcha (left) and Dzhanibekov after the Soyuz 39 visit to Salyut 6. It is widely reported that Gurragcha was incapacitated during the mission.*

Below: *Inside the Soyuz simulator Gurragcha and Dzhanibekov train for their week-long visit to Salyut 6. The mission was given the final go-ahead in late 1980 after the Soyuz-T 3 repairs.*

SOYUZ-T 4 RESIDENCY EXPERIMENTS	
EXPERIMENT	**DESCRIPTION**
Malakhit	Greenhouse experiment for growing flower and plant seeds.
MONGOLIAN EXPERIMENTS	
Collar	Collar device worn, to prevent rapid head movements which could cause spacesickness.
Biosphere-Mon	Observations of Mongolia using the Spektr-15.
Hologram	Three experiments: transmission of hologram to Earth, transmission of hologram from Earth, and use of Fulo apparatus to get images of crystals growing in zero gravity.
ROMANIAN EXPERIMENTS	
Ballisto	Ballistocardiograms used to monitor cosmonaut heart parameters.
REO	Detect changes in the peripheral and central blood circulation in the brain.
Capillary-1	Experiments to obtain crystals by capillary action, using Splav-01 and Kristall furnaces.

mir Dzhanibekov and the cosmonaut-researcher was Jugderdemidiyn Gurragcha from Mongolia. The Soyuz docked at the back of Salyut 6 on 23 March at 16.28 and the cosmonauts transferred to Salyut. Most of the experiments conducted by the four men were medical in nature. It has been widely suggested that Gurragcha was ill in orbit, although this has not been officially confirmed. However, it is certainly notable that only one photograph of Gurragcha on board Salyut 6 has been released. Of course, it is possible that few were taken or the film was not developed properly . . . Soyuz 39 undocked from Salyut on 30 March at about 08.20 and the cosmonauts were safely returned to Earth.

The routine of the Soyuz-T 4 mission continued in the same vein as previous residencies on Salyut, with medical work, exercises, scientific and technical experiments and Earth observations being completed. Although it did not interact with the manned mission on Salyut, the launch of Cosmos 1267 was announced on 25

April into a 51.6°, 89′, 200-278km orbit. Observers of the Soviet space programme noted that the spacecraft was similar to the then-mysterious Cosmos 929 flight in 1977-1978. More will be said about Cosmos 1267 following the end of the description of the Soyuz-T 4 mission.

On 13 May an interesting piece of work was reported by Moscow Radio. It was said that during a television communications session on their 61st day in orbit (12 May) Kovalyonok and Savinykh had explained that one of their jobs in the last few days had been the dismantling of the active docking unit of Soyuz-T 4. This had shown that it was possible to install another unit, turning Soyuz-T 4 into a passive spacecraft with which other craft could dock. This was experimental work for future craft, in case it should prove necessary to rescue cosmonauts in orbit. Possibly this experiment had been prompted by the near-loss of the Soyuz 33 mission in 1979? Certainly, to the end of 1987 there has been no apparent application of this experiment.

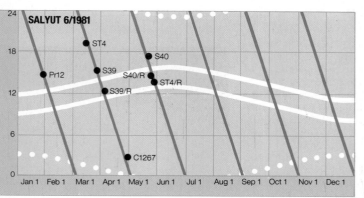

SALYUT 6/1981

MANOEUVRES IN SUPPORT OF THE SOYUZ-T 3 AND THE SOYUZ-T 4 RESIDENCIES

PRE MANOEUVRE ORBIT		POST MANOEUVRE ORBIT	
Epoch (1980)	Altitude (km)	Epoch (1980)	Altitude (km)
17.09 Nov	286-295	17.91 Nov	297-311
Soyuz-T 3: docked 28 November			
		29.40 Nov	285-298
8.29 Dec	278-287	8.60 Dec	287-353
Progress 11: undocked 9 December			
		9.86 Dec	284-355
Soyuz-T 3: undocked 10 December			
		10.43 Dec	286-352
11.57 Dec	283-351	11.82 Dec	284-357
12.33 Dec	284-356	12.58 Dec	316-355
		(1981)	
Progress 12: docked 26 January			
		26.27 Jan	295-321
27.28 Jan	293-319	28.72 Jan	291-355
30.24 Jan	293-352	30.74 Jan	348-360
28.32 Feb	339-348	2.23 Mar	343-356
Soyuz-T 4: docked 13 March			
		14.48 Mar	338-350
17.33 Mar	337-348	17.78 Mar	339-353
Progress 12: undocked 19 March			
		20.26 Mar	338-352
Soyuz 39: docked 23 March			
		24.06 Mar	336-350
27.29 Mar	334-348	29.64 Mar	342-355
Soyuz 39: undocked 30 March			
		31.29 Mar	342-354
1.24 Apr	341-353	1.56 Apr	350-361
7.78 May	330-336	7.85 May	334-353
Soyuz 40: docked 15 May			
		16.22 May	331-346
20.27 May	330-343	20.59 May	339-374
Soyuz 40: undocked 22 May			
		23.20 May	338-372
Soyuz-T 4: undocked 26 May			
		27.20 May	339-372
Cosmos 1267: docked 19 June			
		19.40 Jun	333-363

Notes: The remaining orbital manoeuvres of the Salyut 6/Cosmos 1267 complex are detailed in a separate table.

Launch and Landing Windows
The landing graph for 1981 shows how the launch and landing of Soyuz-T 4 were chosen to encompass the Soyuz 39 and Soyuz 40 international crew visits to Salyut 6. Although Cosmos 1267 was launched before the Soyuz 40 mission, it did not dock with Salyut until June 1981, by which time the final crew had departed. Despite rumours of a further manned visit to Salyut 6 – docked with Cosmos 1267 – during 1981, none took place and probably none was planned.

LAUNCHES IN SUPPORT OF SOYUZ-T 3 VISIT AND THE SOYUZ-T 4 RESIDENCY

LAUNCH DATE AND TIME	RE-ENTRY DATE AND TIME	SPACECRAFT	MASS (kg)	CREW
1980 27 Nov 14.18	10 Dec 09.26	Soyuz-T 3	6,850 ?	L. D. Kizim, O. G. Makarov, G. M. Strekalov
1981 24 Jan 14.18	20 Mar 16.59*	Progress 12	7,000 ?	
12 Mar 19.00	26 May 12.38	Soyuz-T 4	6,850 ?	V. V. Kovalyonok, V. P. Savinykh
22 Mar 14.59	30 Mar 11.42	Soyuz 39	6,800 ?	V. A. Dzhanibekov, J. Gurragcha
25 Apr 02.01	(1982 29 Jul)	Cosmos 1267	20,000 ?	
14 May 17.17	22 May 13.58	Soyuz 40	6,800 ?	L. I. Popov, D. Prunariu

Notes: The launch time of Cosmos 1267 is estimated, and should be correct to ±1 minute. The spacecraft docked with Salyut 6 and remained docked until the complex was de-orbited in July 1982.

On 14 May Soyuz 40 was launched with a crew comprising Col Popov (who had only returned from the Soyuz 35 mission seven months earlier) and the Romanian cosmonaut-researcher, Dumitru Prunariu. The final docking at the rear port of Salyut came the following day at 18.50. The four cosmonauts spent a week on Salyut completing medical, scientific and technological experiments before Soyuz 40 returned to Earth on 22 May at 13.58. At the postflight press conference for the Soyuz 40 mission, Popov confirmed that the flight had been the last intended to use the old Soyuz ferry (something which Western observers had suspected), with Soyuz-T

Left: *The returning Soyuz 40 crew – Popov (left) and Prunariu (right) – sign their descent module before travelling back to the launch site at Tyuratam, on their way to official welcomes in Moscow.*

becoming the standard ferry for future manned missions.

As the Romanian mission was returning to Earth, Kovalyonok and Savinykh were beginning to mothball Salyut 6 for the last time as their own mission drew to a close. The work completed, Soyuz-T 4 undocked from Salyut on 26 May and made a safe landing on Earth. This was the third recovery in the manned space programme in the space of only four days, since it is reported that the descent craft from Cosmos 1267 was recovered on 24 May. The final resident crew on Salyut 6 had remained in orbit for nearly 75 days.

Cosmos 1267

Experiments with Salyut 6 were not over yet. Because the main propulsion system had proved unreliable following the problems encountered towards the end of the Soyuz 29 mission in 1978, it could not be relied upon to perform the de-orbiting of Salyut – and a controlled re-entry and burn-up was the standard end for Salyut missions. The problem could have been overcome by docking a Progress with Salyut and using the Progress engines to perform the de-orbit. However, the Soviets decided to take another route to ensure the safe de-orbiting of Salyut 6.

On 25 April at 02.01 the unmanned Cosmos 1267 had been launched from Tyuratam using the three-stage Proton SL-13 booster, the same vehicle as used for Salyut launches. The Soviets only gave their "standard" launch announcement for the mission, noting the launch date, the orbit and the fact that the scientific mission was proceeding.

Western observers realised that Cosmos 1267 resembled the large Cosmos 929 payload which had been launched in July 1977 on a then-obscure mission. Cosmos 1267 quickly manoeuvred to an orbit which was similar in altitude to the military Salyut stations, and this led to conjecture that it might be an unmanned variant of the military Salyuts.

It was reported in the West that on 24 May Cosmos 1267 detached a recoverable module, and calculations suggest that the recovery time would have been about 13.25 on that day. No Soviet comments were made about this.

The orbital plane of Cosmos 1267 was close to that of Salyut 6, and following the reported capsule recovery Cosmos 1267 began to climb to the Salyut 6 altitude. Following a series of manoeuvres which lasted between 27 May and 19 June, Cosmos 1267 docked with Salyut 6 at the front port on 19 June at 06.52. When reporting the docking of Cosmos 1267 and Salyut 6, the Soviets indicated that the mass of the Cosmos was more than twice that of a Soyuz – that is, about 13,600kg. The launch mass of the Cosmos could have been about 20 tonnes; bearing in mind the amount of propellant which had been expended during the manoeuvres, the mass of the descent vehicle must have been 5-5.5 tonnes.

On 21 June the Cosmos was referred to as a "new 'Star' module" by Radio Moscow, leading to Western observers calling the modules "Star": with hindsight, it is possible that this was a mis-translation of a "new 'cosmic' [ie, space] module", and this class of Cosmos is usually called the Heavy Cosmos.

After docking with Salyut 6, a series of manoeuvres was completed during June and July which led to rumours that a new manned visit would be made to Salyut 6 to perform tests on the complex. Soviet commentators were equivocal about such plans, and no flights transpired. A further series of manoeuvres between 20-22 October raised the orbit of the complex to an altitude which matched that used in the summer of 1979 when Progress 7 boosted Salyut prior to the deployment of the KRT-10 radio-telescope.

From this point on, the orbit of the Cosmos 1267/Salyut 6 complex was allowed to decay naturally until the end of July 1982. On 28 July the Salyut propulsion system performed a manoeuvre which lowered the perigee by about 100km and the following day the Cosmos 1267 propulsion system ignited to de-orbit the complex over the Pacific Ocean, ensuring that any large pieces of debris would fall harmless into the sea far away from human activity.

In Retrospect

By the time that Salyut 6 was eventually de-orbited, Salyut 7 had been launched and its first resident crew had been in orbit for more than two months.

Prior to the launch of Salyut 6, the Soviet space station programme had shown a great deal of promise, but the flight record left much to be desired. There had been ten manned launches to Salyut 1, Salyut 3, Salyut 4 and Salyut 5. Of these one had failed to reach orbit (Soyuz 18-1), two had failed to dock with their Salyuts (Soyuz 15, Soyuz 23), one had docked but the crew had been unable to transfer to their Salyut (Soyuz 10) and one crew had perished during their return to Earth (Soyuz 11). This left the Soviets with a 50 per cent suc-

MANOEUVRES OF COSMOS 1267 (WITH AND WITHOUT SALYUT 6)			
PRE MANOEUVRE ORBIT		POST MANOEUVRE ORBIT	
Epoch (1981)	Altitude (km)	Epoch (1981)	Altitude (km)
Initial orbit		25.39 Apr	193-260
25.87 Apr	192-258	26.31 Apr	191-269
27.24 Apr	188-261	27.98 Apr	254-273
28.29 Apr	248-271	28.97 Apr	251-270
30.22 Apr	248-269	1.15 May	254-268
2.65 May	250-266	3.08 May	257-278
3.08 May	257-278	3.71 May	259-274
8.25 May	251-266	8.94 May	252-294
8.94 May	252-294	9.25 May	267-278
16.23 May	259-269	16.81 May	264-286
Capsule recovered: 24 May			
		25.22 May	258-278
27.22 May	254-274	27.72 May	306-317
6.22 Jun	299-311	6.59 Jun	304-315
16.16 Jun	298-309	17.22 Jun	340-348
17.22 Jun	340-348	17.35 Jun	340-359
18.25 Jun	341-358	18.63 Jun	342-361
19.20 Jun	340-363		
Cosmos 1267 docked with Salyut 6: 19 June			
		19.40 Jun	333-363
28.92 Jun	332-359	29.11 Jun	336-368
29.93 Jun	335-368	30.06 Jun	338-388
29.22 Jul	334-380	29.60 Jul	346-381
29.60 Jul	346-381	30.56 Jul	376-404
20.35 Oct	356-376	21.18 Oct	355-398
21.56 Oct	355-397	22.20 Oct	393-413
22.20 Oct	393-413	22.27 Oct	396-411
(1982)			
28.25 Jul	318-324	28.62 Jul	223-321
29.31 Jul	222-321	29 Jul	De-orbited

Notes: Salyut 6 and Cosmos 1267 were de-orbited as a combined spacecraft: the first de-orbited manoeuvre on July 28 was using the Salyut 6 propulsion system (its only use following the two craft docking), while the final de-orbit was completed by the Cosmos 1267 propulsion system.

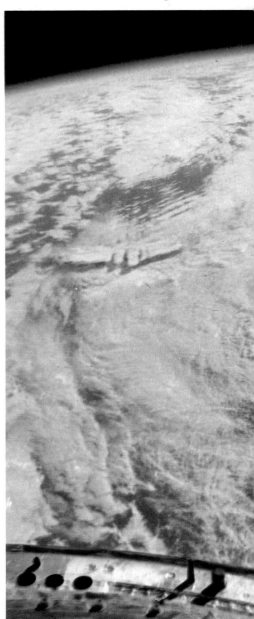

Right: *A close-up view of the rendezvous antenna at the rear of Salyut 6, as seen by an approaching visiting crew aboard their Soyuz craft. The curving Earth covered in cloud systems provides a spectacular background to the rim of Salyut. The station was eventually de-orbited in July 1981.*

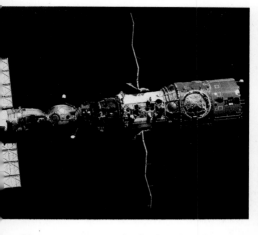

Above: *Soyuz-T 4 docked with Salyut 6, as seen from the Soyuz 39 spacecraft. This was the first photograph of Salyut docked with the newly designed ferry to be published by the Soviets.*

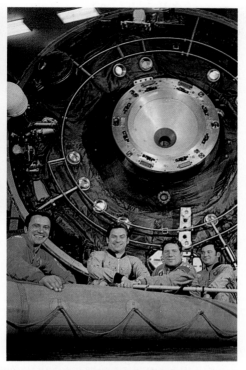

Right: *The four resident crew commanders for Salyut 6 pose at the rear of the station. Left to right they are Popov, Kovalyonok, Lyakhov and Romanenko.*

cess rate, if we deem Soyuz 21 as a successful mission even though it was terminated earlier than planned. This record was hardly one of which to be proud.

In the mid-1970s it appears that the whole of the Soviet manned and unmanned space programme was reviewed and improvements soon began to show through. The flight of Salyut 6 reversed the Soviet record of space station missions. During 1977-1981 there were 16 Soyuz spacecraft launched towards Salyut 6 and of these only one failed to dock (Soyuz 33) and one docked but the crew could not transfer (Soyuz 25); additionally, there were 4 launches of Soyuz-T craft, 12 launches of Progress craft and the Cosmos 1267 mission – all of which successfully docked with Salyut 6. For Salyut 6 the success rate was 94 per cent.

Prior to its launch the Soviets had planned for three main expeditions, lasting 3, 4-5 and 6 months respectively. The first expedition was expected to receive one international crew, the second two international crews and the third two international crews, with the option of the final three original international missions being flown to Salyut 6 if a fourth residency of the same length as the third could be supported (of course, Vietnam joining the Intercosmos programme added one more international flight to this schedule). Despite the set-backs with Soyuz 25 and Soyuz 33, these goals were accomplished and more. The new manned Soyuz-T ferry was also thoroughly tested prior to its becoming the manned spacecraft designated to support Salyut 7.

Salyut 6 had brought to the Soviet space programme a new level of success and prominence in the world. It marked the future direction of the Soviet manned space programme, pointing towards the setting up of permanently manned orbital stations, with missions lasting longer and longer as the medical researchers discovered how the human body adapted to long missions – and more importantly how it could readapt to conditions back on Earth.

FURTHER READING

There have been many books and articles written about the missions to Salyut 6, and each Intercosmos country produced its own book(s) when its cosmonaut was launched. For the daily operations of the Salyut missions, the "Mission Reports" by Gordon Hooper and later Neville Kidger in *Spaceflight* and the *Journal* of the British interplanetary Society are recommended. The following Soviet books relate specifically to Salyut 6 events:

Braikov, A.V. et al, *Nivigationnoe Obespechenye Polyeta Orbitalnogo Compleksa "Salyut 6-Soyuz-Progress"*, Nauka Press, Moscow, 1985. (This book is a mine of information concerning the day-to-day manoeuvres of the Salyut 6 complex to the end of 1979.)

Feoktistov, K.P., *Naychniya Orbitalnay Compleksa*, Zhanye, issue 3, Moscow, 1980.

Gazenko, O.G. et al, *Salyut 6-Soyuz-Progress: Rabota Na Orbite*, Mashinostroyeniye Press, Moscow, 1983.

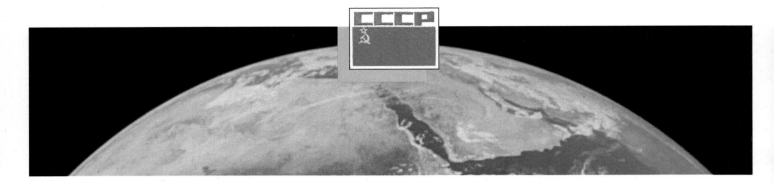

Chapter 12: **The Mission of Salyut 7**

With the successes of the Salyut 6 during 1977-1981, the launch of the new Salyut 7 was eagerly awaited; some western observers hoped that just as Salyut 6 had provided a quantum leap over the previous Salyut capabilities, Salyut 7 would provide a similar leap over Salyut 6. These hopes were to be proved groundless, because Salyut 7 was almost identical to Salyut 6: its equipment was improved, the experiments were changed but its overall capabilities did not go beyond Salyut 6.

The Soviet Union planned to use Salyut 7 to extend manned durations beyond the six months experienced on Salyut 6; before the station was abandoned in 1986, a mission lasting for eight months had been successfully completed, while a mission of 9-10 months had been scheduled but had to be re-arranged following the major in-flight failure of Salyut's systems in 1985 and then curtailed when one of the cosmonauts fell seriously ill. On the positive side,

during the Salyut 7 missions two further tests of the Heavy Cosmos modules were undertaken: the first in 1983 and the second in 1985.

After it was abandoned following a short manned visit in 1986, Salyut 7 was put into a high storage orbit for a possible manned visit in the early 1990s.

Salyut 7 Description

From its external appearance, Salyut 7 might have been the back-up vehicle for the Salyut 6 mission, although some improvements were evident. The exterior of the station had more hand-holds to assist in spacewalking and the three sets of solar panels included attachment points for sets of supplementary panels to be added if necessary. The perimeter of the front docking port was enlarged which allowed safer docking with the Heavy Cosmos modules.

The interior of Salyut 7 was dominated by the scientific equipment, the layout being almost identical with Salyut 6. Improvements to the Salyut 6 design included the installation of electric stoves for the heating of food, a refrigerator, constant hot water and newly-designed seats at the main command station in the smaller work compartment. These were more like bicycle seats than terrestrial chairs.

Two of the portholes were modified to allow ultra-violet radiation to penetrate

the station. Although this had the minor side-effect of allowing the cosmonauts to get a suntan, the main reason was that the radiation would help in killing any infections which the crew might fall victim to. The portholes all had transparent covers which could be closed when the portholes were not in use; this was to reduce the damage caused by meteoritic matter. A new, large porthole was added to the forward transfer compartment specifically for astronomical observations.

Since Salyut 7 was to support longer manned residencies than had Salyut 6, its medical, biological and exercise facilities were improved, as it was now realised how important a high level of exercise activity was to keep muscles from withering away and in preparing for the return to Earth.

Once more, the MKF-6M and KATE-140 experiments were carried, but replacing the BST-1M in the large work compartment cone was an X-ray detection system. This included the XT-4M telescope which had been prepared by scientists at the P. N. Lebedev Institute of Physics. There was also the XS-02M X-ray spectrometer, built by the P. K. Sternberg Institute: this had a sensitive surface of 3,000cm^2 compared with 450cm^2 of the Filin detector carried on Salyut 4. Further details of the Salyut 7 experiments will be given in the mission descriptions.

Despite western predictions, the Soviets never planned to operate Salyut 7 as a per-

SALYUT ALTITUDES FOR MANNED SOYUZ-T LAUNCHES		
TWO-MANNED MISSIONS	**STATION ALTITUDE (km)**	**AVERAGE ALTITUDE (km)**
Soyuz-T 2	333-349	341
Soyuz-T 4	338-350	344
Soyuz-T 5	342-346	344
Soyuz-T 9	324-336	330
Soyuz-T 10-1	334-352	343
Soyuz-T 13	337-352	344
Soyuz-T 15	335-343	339
THREE-MANNED MISSIONS	**STATION ALTITUDE (km)**	**AVERAGE ALTITUDE (km)**
Soyuz-T 3	285-298	291
Soyuz-T 6	283-306	294
Soyuz-T 7	290-299	294
Soyuz-T 8	287-300	293
Soyuz-T 10	289-296	292
Soyuz-T 11	287-298	292
Soyuz-T 12	334-354	344
Soyuz-T 14	337-353	353

Notes: This table shows the difference in orbital altitude when two- and three-manned Soyuz-T missions were launched. The first three missions were flown to Salyut 6, while the remainder flew to Salyut 7 (Soyuz-T 15 had first flown to the Mir space station core). For the two-manned missions the mean altitudes were 330-344km, while for the three-manned missions excluding Soyuz-T 12 and Soyuz-T 14 the mean altitudes were 291-294km. It is generally thought that the weight involved in carrying the third cosmonaut and his supplies meant that Soyuz-T could not reach the altitude of the "two-day repeater" orbit normally operated by the stations, but Soyuz-T 12 and Soyuz-T 14 were both three-manned missions launched when Salyut 7 was in the higher orbits.

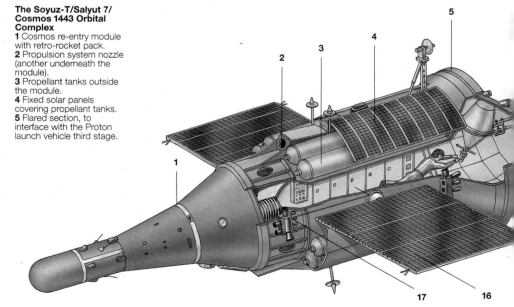

The Soyuz-T/Salyut 7/ Cosmos 1443 Orbital Complex
1 Cosmos re-entry module with retro-rocket pack.
2 Propulsion system nozzle (another underneath the module).
3 Propellant tanks outside the module.
4 Fixed solar panels covering propellant tanks.
5 Flared section, to interface with the Proton launch vehicle third stage.

manently manned station, although on two occasions they planned to hand the station over in fully operating mode from one crew to another (the first attempt failed when the launch vehicle exploded on the pad). Operations on Salyut 7 would closely follow the philosophy of Salyut 6, in that there would be extended main "resident" missions supported by shorter visiting missions. Unlike Salyut 6, a long series of guest cosmonaut missions was not planned. In 1982 a French cosmonaut was launched to Salyut and in 1984 an Indian cosmonaut completed a now-standard eight-day mission.

Salyut 7 Crews

During most of the Salyut 7 operations it was easy to predict how many cosmonauts were to be launched to the station, although the Soyuz-T spacecraft would be launched with either two or three cosmonauts on board. The table shows that (with the exceptions of Soyuz-T 12 and Soyuz-T 14) when a three-manned crew was to be launched Salyut was operating in an orbit some 35-45km lower than when a two-manned crew was scheduled. Therefore, as a landing window was approaching – implying the impending launch of a 1-2 week visiting mission – the altitude of Salyut 7 revealed the size of the crew to be launched.

The reason behind the difference in crew size comes when the propellant requirements for the different Salyut orbits are calculated. When manoeuvring to the lower Salyut orbit a minimum of about 110kg of propellant is required, while to reach the higher orbit the figure comes to a minimum of 175kg. The difference is approximately the mass of a cosmonaut. Therefore, in order to maintain an approximately constant Soyuz-T launch mass, the extra weight of a third cosmonaut was compensated for by a reduction in the propellant load, thus reducing the altitude which the Soyuz-T could reach.

As time went on, a further insight into the cosmonaut training programme was gained when the three-man crews were considered. It seems probable that the mission commander and flight engineer train together as a team, and if a three-man crew is required the third man is added to an existing two-man team. This was shown particularly by the pairing of Kizim and Solovyov, who appeared as the back-ups for the Franco-Soviet mission with Baudry "plugged on". Once the flight was over, the two men were transferred to the resident crew training group, training together as back-ups for the aborted Soyuz-T 10-1 mission. That mission having failed, they worked together on simulations for Soyuz-T 9 spacewalk and then had a third man added (the doctor Atkov) to form the Soyuz-T 10 crew. The pairing remained active into 1986, when Kizim and Solovyov were launched in Soyuz-T 15.

The Launch of Salyut 7

As the Salyut 6 mission was drawing to a close, it was noted by the East German

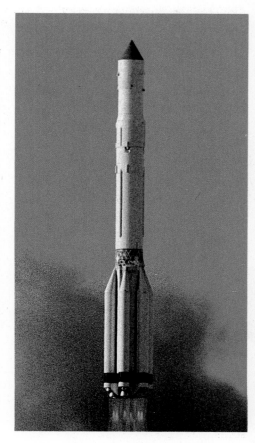

Above: *The four stage variant of Proton. For Salyut missions the station replaces the fourth stage and payload.*

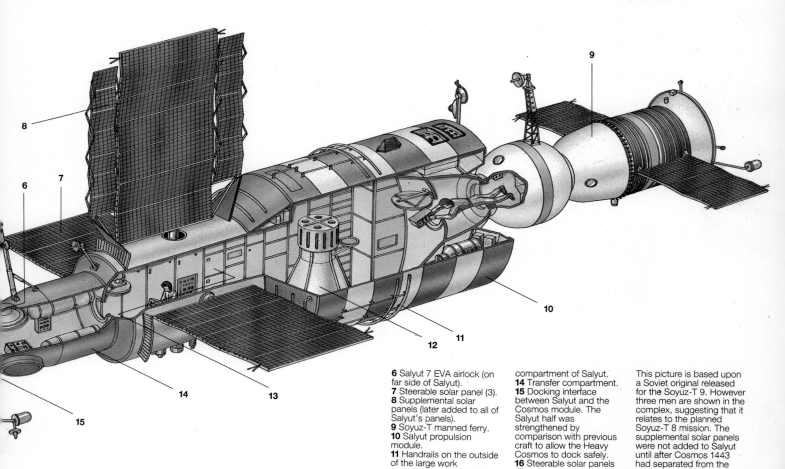

6 Salyut 7 EVA airlock (on far side of Salyut).
7 Steerable solar panel (3).
8 Supplemental solar panels (later added to all of Salyut's panels).
9 Soyuz-T manned ferry.
10 Salyut propulsion module.
11 Handrails on the outside of the large work compartment – added to aid EVA work.
12 Shroud covering the X-ray detection equipment.
13 Small work

compartment of Salyut.
14 Transfer compartment.
15 Docking interface between Salyut and the Cosmos module. The Salyut half was strengthened by comparison with previous craft to allow the Heavy Cosmos to dock safely.
16 Steerable solar panels (2).
17 Main Heavy Cosmos work module, based upon the smaller diameter Salyut work compartment.

This picture is based upon a Soviet original released for the Soyuz-T 9. However three men are shown in the complex, suggesting that it relates to the planned Soyuz-T 8 mission. The supplemental solar panels were not added to Salyut until after Cosmos 1443 had separated from the station, although this work probably was scheduled for the Soyuz-T 8 visit while Cosmos was still docked with Salyut.

magazine *"Neues Deutschland"* in March 1981 that while a Romanian cosmonaut was preparing for his launch, the Salyut 7 orbital station was being prepared for launch at Tyuratam. It was further reported that the French cosmonaut would be flown in 1982 and an Indian flight was due in 1983 (actually, this took place in 1984).

The launch of Salyut 7 came on 19 April 1982 – exactly eleven years after the first Salyut had been launched and twenty-one years after Gagarin's flight – at 19.45. Based upon the experience of previous launches, a manned visit was expected to begin within about two weeks, but in fact more than three weeks elapsed before the launch took place. It has been suggested in some quarters that this delay was because the Soviet spaceflight controllers were devoting their energies to the observation of the British Task Force heading towards the Falklands following the Argentinian invasion of the colony at the beginning of April. However, a careful study of the unmanned military flights of the time does not appear to bear out this suggestion.

While it had only taken eight days for Salyut 6 to attain its operational "two-day repeating" orbit, it was twice as long before Salyut 7 reached its operational orbit.

The Soyuz-T 5 Residency

By the time that Salyut 7 was launched, the American space programme had returned to manned flights, with the first flight in the Space Shuttle programme having begun on 12 April 1981. Although the Shuttle programme was being delayed because of technical and payload problems, it seems probable that the Soviets were trying to up-stage any "firsts" which the Americans might plan for their programme. Such "firsts" would only be attempted if the Salyut mission schedules allowed them, and there was no risk to the mission.

On 13 May, Soyuz-T 5, the first manned mission to Salyut 7 was launched: the mission commander was Lt-Col. Anatoly Berezovoi, on his first space mission, and the flight engineer was V. V. Lebedev who had trained for the Soyuz 35 residency on Salyut 6, but who had been replaced following a knee injury. During the flight and before initiating the rendezvous with Salyut, Berezovoi and Lebedev checked the control systems of their spacecraft, including the Igla radio guidance system, and re-set the docking probe; presumably at launch the probe is in its "retracted" position, and it required extending prior to docking. The docking with Salyut took place on 14 May at 11.36. Once on board the station, Berezovoi and Lebedev began to unpack the scientific equipment and de-mothball the station's systems.

The first experiment to upstage an American "first" came on 17 May, when the cosmonauts ejected a small amateur radio satellite called Iskra 2 from one of the airlocks on Salyut. The satellite – with a mass of 28kg – had been built by the student office of the Moscow Sergo Ordzhonikidze Aviation Institute, and it carried a radio repeater for experiments in amateur radio communications, a computer

memory, a command radio channel and a radio-telemetric system for the transmission of information from the satellite.

The launch of Iskra 2 (Iskra 1 had been launched with the 31st Meteor-1 satellite in July 1981, but apparently failed to deploy in orbit) allowed the Soviets to claim the first launch of a satellite from a manned spacecraft, prior to the November 1982 flight of the fifth Space Shuttle mission which carried two commercial satellites for deployment. Unfortunately, this claim had a major flaw, in that the first satellite deployments from manned vehicles had actually been the two particles and fields sub-satellites left in lunar orbit by the American Apollo 15 and Apollo 16 missions (July-August 1971 and April 1972 respectively).

The cosmonauts prepared their Delta on-board computer complex for work: the theory was that Delta would take over many of the routine attitude correction and general navigation work from the cosmonauts, allowing them more time to devote to their scientific research programme. On 19 May the Oazis "orbital garden" was activated and it was planned to use this to study the growth of plants on orbital flight. Similar experiments on Salyut 6 had initially proved disappointing, but further studies were planned for Salyut 7.

The cosmonaut routine on Salyut 7 owed much to Salyut 6, and it was therefore no surprise when Progress 13 was launched on 23 May. The spacecraft completed an automatic docking at the rear of Salyut on 25 May at 07.57. The cargo on this mission included: 660kg of propellant for Salyut; 290 litres of water for the Rodnik supply system; film for the various camera systems carried on Salyut; clothing and other hygienic supplies; 895kg of scientific equipment which included 242kg of equipment in support of the up-and-coming French mission; an EFO-1 electrophotometer and two new furnaces, Kristall and Magma-F.

Above: *The first two commanders aboard Salyut 7: Dzhanibekov (left) from the Franco-Soviet Soyuz-T 6 crew and Berezovoi (right) from the resident Soyuz-T 5 crew.*

On 21 May it was announced that the Salyut 7 complex had been put into a gravity stabilised mode, something which was originally discovered by Romanenko and Grechko during their residency of Salyut 6. However, on this occasion there was a difference; previously the cosmonauts had manoeuvred the Salyut into the stable mode, but this time it was done automatically, using the Delta system.

On 30 May an overview of the Salyut 7 equipment was given, describing it as falling into three main areas: medical, astrophysical and technical. The medical equipment comprised two parts: Ekhografiya (Echography) for research into the dynamics of the cardiovascular systems and Poza (Posture) for research into the cosmonauts' balancing mechanism during the space mission. The astrophysical experiments included Pironik. These experiments are designed for the study of various astronomical objects, as well as having applications in the study of the Earth and its atmosphere. The third group of experiments would concentrate on materials science, continuing the research begun with Splav and Kristall on Salyut 6.

Progress 13 undocked from Salyut on 4 June at 06.31 and was de-orbited two days later. A Salyut landing window was approaching and an international crew was due to be launched to Salyut.

In September 1980 two French cosmonaut-trainees (the French call their spacefliers "spationauts") arrived in the Soviet Union: Jean-Loup Chretien and Patrick Baudry. After undergoing the standard Soyuz training programme they were assigned to crews, ready to begin specific training for their visit to Salyut 7. In September 1981 Chretien joined Malyshev and

Ivanchenkov in the prime crew for the mission, while Baudry joined Kizim and V. Solovyov, a rookie cosmonaut. Unfortunately, there was apparently a serious personality clash between Malyshev and Chretien which resulted in Malyshev being taken off the mission and replaced by Dzhanibekov who had been working in the Salyut 7 resident crew training group with three other cosmonauts (V. G. Titov and G. M. Strekalov was one pairing and Dzhanibekov had been training with A. P. Alexandrov, who should not be confused with the Bulgarian back-up cosmonaut for Soyuz 33); in turn, Dzhanibekov's place was taken by V. A. Lyakhov.

The French Flight

On 24 June the first international mission to Salyut 7 was launched, Soyuz-T 6, with Col. V. A. Dzhanibekov, A. S. Ivanchenkov and Jean-Loup Chretien. The automatic rendezvous system on Soyuz-T failed, and Dzhanibekov was forced to complete docking manually, at 17.46 the day after launch. The transfer to Salyut came on 25 June at about 21.00 and for the first time there were five cosmonauts working on a Salyut station. There was a full programme of scientific work for the cosmonauts to complete in their seven days together. The French experiments included:

DC-1 (or Minerve) which was used to measure the speed at which blood passes through the vessels in the brain.

Braslet which checked the acuity and depth of vision, and was in the Ekhografiya and Poza series.

Biobloc-3 to study the effects of the heavy nuclei from cosmic radiation on living organisms.

Cytos-2 which studied the effects of various antibiotics on various types of bacteria in orbital flight.

Neptun which studied the acuity of vision in orbit.

Mars which studied the thresholds of colour sensitivity in orbit.

Anketa which was noted as a medical experiment, but details were immediately specified.

Piramig which was an infra-red astronomy experiment.

PCN which studied the night sky from orbit.

Kalibrovka which was part of the materials processing programme, studying the temperature variations while an electric furnace was being operated.

Liquation which studied the influence of capillary forces on the formation of an aluminium and indium alloy. Kalibrovka and Liquation used the Kristall furnace.

Elma-2 which was a continuation of an engineering experiment undertaken in 1979, and was designed to produce new alloys by smelting and cooling in a weightless condition.

In addition to this full programme of research, the Franco-Soviet crew also took part in other experiments which involved the more routine work undertaken by the resident crew. On 2 July at 11.04 Soyuz-T 6 vacated the rear port of Salyut, and was

Above: *The roll-out of a Soyuz-T launch vehicle. This normally takes place 1-2 days before the planned launch, with only propellant loading being required at the pad before the cosmonauts board.*

Below: *The first international crew to visit Salyut 7 included the first French "spationaut". Left to right are Chretien (France), Dzhanibekov and Ivanchenkov. Chretien will fly again on Soyuz-TM 7.*

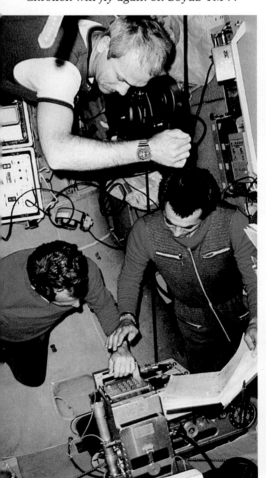

Above: *The Soyuz-T 6 descent module underneath its parachute, heading for a perfect landing. Of the guest cosmonaut visits to Salyuts, the French was perhaps the most significant for the Soviets.*

Below: *After the successful landing the three Soyuz-T 6 cosmonauts sit in the recovery couches as Dzhanibekov reports the successful conclusion of the mission to the controller.*

successfully recovered just over three hours later.

While the international mission was in progress, a surprise announcement was made by General G. Beregovoi, the Soyuz 3 cosmonaut, who revealed that a group of women cosmonauts were in training for the first time in nearly twenty years. This was surprising, since in recent years a number of statements had been made by Soviet space managers and former cosmonauts saying in effect that a woman's role in space was virtually non-existent and women should remain in terrestrial kitchens – something which would have upset any Soviet-based Women's Liberation movement! However, Beregovoi stated that there were four women trainee cosmonauts, and that two were being readied for flight: one was a pilot-engineer and the other a flight engineer. The selection of women was seen as an answer to the American plans to fly women routinely as mission specialists on Shuttle missions; indeed, the first American woman, Sally Ride, was due to fly into space in the middle of 1983. The new group of women cosmonauts had begun training for their missions in 1980 and it was learned later that a further group of women cosmonaut-trainees had been selected in 1982 – possibly six women.

Such considerations aside, the Soyuz-T 5 mission continued. Once the rear port of Salyut was free, preparations for the new Progress mission began. The launch of Progress 14 came on 10 July and the routine docking came two days later at 11.41. On board the freighter were 175 litres of water, 700kg of propellant for Salyut and 1,100kg of lesser items which included food, spare parts, camera film, clothing and scientific apparatus.

The First EVA

On 30th July the first spacewalk from Salyut 7 was conducted. At 02.39 the hatch on the side of the forward transfer compartment opened, and Lebedev climbed out, followed by Berezovoi. It was Lebedev's job to complete the main part of the work, with Berezovoi ready to help him. Lebedev was anchored outside Salyut, near an area where a series of exposure experiments had been positioned prior to launch. First, Lebedev replaced some canisters of the Medusa experiment which was designed to study the effects of weightlessness and direct cosmic radiation on amino acids and biopolymers. He then removed part of the MMK micrometeorite detector.

A major part of the spacewalk was connected with the testing of methods and equipment for the later assembly of structures in space. One piece of equipment was Pamyat which studied how best to make joints between pipes and girders. Resurs was a device which showed how different materials behave under a heavy metal load, while Istok checked the reliability of bolting various materials together. The complete spacewalk lasted for 153 minutes, after which the cosmonauts returned to the confines of Salyut for most of the remainder of the mission.

On 4 August it was reported that an arabidopsis plant growing in the Fiton facility had produced a pod, and this was thought to be the first time that a plant had developed from a seed to produce its own seeds in orbit. On 10 August at 22.11 Progress 14 undocked from Salyut and was de-orbited on 13 August.

A new landing window was approaching for Salyut, and it was expected that one of the women cosmonaut-trainees would be making a flight. Fulfilling these expectations, on 19 August Soyuz-T 7 was

Above: *Lebedev pictured outside Salyut 7 by Berezovoi during the first spacewalk of the station's operation. Experiments involving the assembly of components were connected by the cosmonauts.*

launched with Col. Popov on this third flight, flight engineer A. A. Serebrov and research-cosmonaut Svetlana Savitskaya who became the second woman to fly in space. The docking with Salyut came the following day at 18.32. Naturally, the media emphasis was on the inclusion of Savitskaya in the crew, the fact of her being christened "Miss Sensation" in 1970 (when she was in London for a world aerobatics competition) being recalled by the Soviet media. Interestingly, little comment from Valentina Tereshkova about the flight was reported.

A full scientific programme was undertaken by the five cosmonaut crew on Salyut. Use was made of the equipment supplied by the French mission, since the emphasis would be on the biological sciences, studying a woman's adaptation to the space environment. An experiment conducted for the first time on a Salyut station was Tavria, which studied the electrophoretic method of separating mixtures of biologically active substances. Holography was used for the first time to record the separation of the components.

The flight of the Soyuz-T 7 crew ended after a standard eight-day mission. They returned to Earth in the older Soyuz-T 5 spacecraft, undocking from Salyut on 27 August and returning to Earth at 15.04. Two days later Berezovoi and Lebedev left Salyut for a second time, but this time they were inside Soyuz-T 7: undocking came at 16.47 and re-docking at the front port was achieved during the same orbit. There was not an immediate Progress flight to fill the rear port of Salyut, as the cosmonauts had plenty of work to complete on their own.

The launch of Progress 15 finally came in 18 September, the successful docking being achieved on 20 September at 06.12. As usual, the spacecraft was fully laden with supplies for the cosmonauts: about 150 litres of water, 700kg of propellant for Salyut, spare parts, scientific equipment and film. Progress remained docked with Salyut until 11.46 on 14 October and was de-orbited two days later.

While Progress was docked with Salyut 7, the normal Salyut routine of experiments continued. On 28 September the cosmonauts were tending their "orbital garden", Oazis with Vazon containers, which proudly boasted peas, oats and onions. The effects of a non-uniform magnetic field was being studied on the development of flax seeds with the Magnitogravstat equipment. In 1 October Berezovoi and Lebedev conducted experiments with the Yelena gamma ray telescope and the Astra-1 spectrometric equipment.

Since the departure of Progress 15 occurred as a landing window was opening, it was thought that perhaps a third visiting mission would be made to Salyut 7. Progress had raised the orbit of Salyut to an altitude which suggested that a two-man crew would be launched, but no manned flight appeared. In fact, there was no particular requirement for such a mission since Soyuz-T 7 had only been in orbit for two months.

The regular medical check-ups continued on Salyut: the recording of physiological parameters was completed using the Aelita-1 and Rheograph equipment. A series of astrophysical observations was undertaken with the Pyramig equipment, objects in the constellations of Andromeda, Cassiopeia, Cetus and Pegasus being carried out in late October.

The Launch of Iskra

Progress 16 was launched towards Salyut on 31 October, with the successful docking coming on 2 November at 13.22. The normal payload was carried by Progress, although this time it also carried a new small sub-satellite. Iskra 3 was launched from a Salyut airlock hatch, like its predecessor, on 18 November. It is possible that each Iskra might have carried a small detachable "kick" motor which would ensure that the Iskra orbit was sufficiently different from that of Salyut to prevent a collision during the short Iskra lifetimes. When each launch took place, a second object was always tracked close to Iskra, which could have been a discarded "kick" stage; alternatively, it may have been another garbage container which was jettisoned at the same time as Iskra.

Meanwhile, on 5 November Berezovoi and Lebedev exceeded the official duration record set by Lyakhov and Ryumin on the Soyuz 32 mission (the longer Soyuz 35 flight was not recognised as a new record because the old one had not been exceeded by the statutory 10 per cent).

Berezovoi and Lebedev experimented with the Korund computer-operated smelting kiln, which had a mass of 136kg. This kiln could work with larger samples than the Kristall or Magma-F apparatus.

The use of Korund was expected to be a "great step forward towards the establishment of space factories".

The nominal landing window for Soyuz-T 5 was Christmas week, and this would have permitted a flight of seven-and-a-half months; it was with some surprise that news of the imminent return of Soyuz-T 7 was announced on 9 December. It was noted that recently the cosmonauts had been taking air samples for return to Earth and analysis, and this raises the question of possibly minor contamination of the atmosphere inside Salyut.

Soyuz-T 7 undocked from Salyut on 10 December, and Berezovoi and Lebedev were safely recovered at 19.03. The cosmonauts landed in a blizzard, and must have spent some anxious hours before their return to Tyuratam.

It was unusual for a crew to leave a Progress docked with a Salyut: the reason that this had been done with Salyut 6 was its troublesome propulsion system, but there

had been no such failure on Salyut 7. Progress 16 undocked on 13 December at 15.32, and was de-orbited the following day. One can speculate that originally the Soyuz-T 5 cosmonauts were to return to Earth following the Progress being de-orbited, but some (still unknown) problem required their vacating Salyut early.

It was revealed five years later that Berezovoi and Lebedev, who were both "nit-pickers", had verbally clashed early in the mission, and had hardly spoken to one

Above: *The Soyuz-T 7 crew at the Zvezdny Gorodok training centre prior to launch. Left to right, they are Popov, Savitskaya and Serebrov. Savitskaya became the second woman to complete a space mission.*

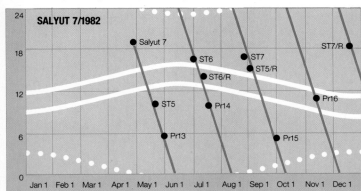

Launch and Landing Windows

The launch and landing graph for Salyut 7 during 1982 shows how the Soyuz-T 6 and Soyuz-T 7 visiting missions were timed for the landing windows. In fact, the Soyuz-T 7 crew were launched a day or two earlier than the nominal conditions predict for their eight day mission returning in Soyuz-T 5. The Soyuz-T 5 resident crew came down outside a landing window, suggesting either an early recall or a mission extended beyond the boundaries of the November window.

MANOEUVRES IN SUPPORT OF THE SOYUZ-T5 RESIDENCY

PRE MANOEUVRE ORBIT		POST MANOEUVRE ORBIT	
Epoch (1982)	Altitude (km)	Epoch (1982)	Altitude (km)
Initial Orbit		19.94 Apr	213-261
21.30 Apr	209-253	22.29 Apr	208-262
22.48 Apr	206-257	23.03 Apr	207-326
24.34 Apr	205-322	25.15 Apr	297-364
25.15 Apr	297-364	25.28 Apr	309-353
4.26 May	305-347	5.84 May	346-347
Soyuz-T 5: docked 14 May			
		14.98 May	342-346
Progress 13: docked 25 May			
		25.76 May	338-343
2.36 Jun	335-341	2.62 Jun	302-334
2.62 Jun	302-334	2.81 Jun	291-321
Progress 13: undocked 4 June			
		4.88 Jun	291-319
Soyuz-T 6: docked 25 June			
		26.11 Jun	283-306
1.57 Jul	280-299	1.63 Jul	283-320
Soyuz-T 6: undocked 2 July			
		2.82 Jul	283-319
3.20 Jul	283-319	3.64 Jul	307-329
Progress 14: docked 12 July			
		13.79 Jul	306-325
Progress 14: undocked 10 August			
		11.54 Aug	290-306
17.44 Aug	286-299	17.94 Aug	294-298
17.94 Aug	294-298	18.25 Aug	292-301
Soyuz-T 7: docked 20 August			
		21.26 Aug	290-299
26.22 Aug	287-297	26.72 Aug	293-315
Soyuz-T 5: undocked 27 August			
		28.98 Aug	291-316
Soyuz-T 7: redocked 29 August			
		30.05 Aug	292-314
31.25 Aug	291-313	31.75 Aug	314-334
Progress 15: docked 20 September			
		20.37 Sep	302-327
28.24 Sep	296-320	28.37 Sep	308-367
29.19 Sep	308-366	29.32 Sep	364-374
Progress 15: undocked 14 October			
		15.26 Oct	360-369
Progress 16: docked 2 November			
		2.67 Nov	351-366
28.23 Nov	347-353	28.36 Nov	354-367
8.04 Dec	350-367	8.16 Dec	351-354
Soyuz-T 7: undocked 10 December			
		10.83 Dec	350-355
Progress 16: undocked 13 December			
		13.82 Dec	348-354

SALYUT 7/1982

(graph: launch and landing windows, Jan 1 – Dec 1 1982, altitude axis 0–24, with data points Salyut 7, ST5, ST6, ST6/R, ST7, ST5/R, ST7/R, Pr13, Pr14, Pr15, Pr16)

LAUNCHES IN SUPPORT OF THE SOYUZ-T 5 RESIDENCY

LAUNCH DATE AND TIME (G.M.T.)	RE-ENTRY DATE AND TIME (G.M.T.)	SPACECRAFT	MASS (kg)	CREW
1982 19 Apr 19.45	In Orbit	Salyut 7	19,920	
13 May 09.58	27 Aug 15.04	Soyuz-T 5	6,850?	A. N. Berezovoi, V. V. Lebedev
23 May 05.57	6 Jun 00.05*	Progress 13	7,000?	
24 Jun 16.29	2 Jul 14.21	Soyuz-T 6	6,850?	V. A. Dzhanibekov, A. S. Ivanchenkov, J-L. Chretien
10 Jul 19.58	13 Aug 01.29*	Progress 14	7,000?	
19 Aug 17.12	10 Dec 19.03	Soyuz-T 7	6,850?	L. I. Popov, A. A. Serebrov, S. Y. Savitskaya
18 Sep 04.59	16 Oct 17.08*	Progress 15	7,000?	
31 Oct 11.20	14 Dec 17.17	Progress 16	7,000?	

Notes: The layout of this table – and also that which follows – is identical with the corresponding tables for Salyut 6 residencies. Berezovoi and Lebedev returned to Earth in Soyuz-T 7 with a mission duration of 211d 9h 5m; Popov, Serebrov and Savitskaya returned to Earth in Soyuz-T 5 with a mission duration of 7d 21h 54m.

another for four months. This must have set back the selection of crews on psychological grounds somewhat!

Soyuz-T 8 and Soyuz-T 9

Following the return of the Soyuz-T 5 cosmonauts, it was expected that the next step in the Soviet manned programme would be the extension of the duration record to 8-9 months. A resident crew of the new cosmonaut V. G. Titov and the experienced G. M. Strekalov was expected to man the station in 1983.

The orbit of Salyut 7 was allowed to decay without any manoeuvres until after the first man-related launch of 1983. On 2 March 1983 the unmanned Cosmos 1443 module was launched into orbit, and the launch announcement said that:

"The aim of the launching is a try-out of on-board systems, equipment and structural elements of the satellite in various modes of flight, including a joint flight with the Salyut 7 station.

Cosmos 1443 is similar in design to the artificial Earth satellite Cosmos 1267 which was tested in 1981-1982 in autonomous regime and in the course of a joint flight with the Salyut 6 orbital station."

Clearly, one could expect that Cosmos 1443 would become the first Heavy Cosmos to host a manned mission. The docking with Salyut came on 10 March at 09.20, the Heavy Cosmos using the front port of Salyut.

The two spacecraft complex continued to orbit with no manoeuvres effected until 3-5 April, this being followed by a further manoeuvre on 11 April. The altitude following the last manoeuvre suggested that a three-manned Soyuz-T could be expected.

The launch of Soyuz-T 8 came on 20 April and the three man crew comprised Lt-Col Titov (who was *no* relation to the Vostok 2 cosmonaut), Strekalov and Serebrov (undertaking a rapid return to space

following his Soyuz-T 7 mission, and thus becoming the first man to be launched on successive spacecraft). Only Titov had no previous spacecraft experience. The now-standard initial approach manoeuvres were made towards the Salyut 7 complex, but when the cosmonauts tried to use the main radar system to track Salyut for the final approach no data were returned. The radar system was not working. The cosmonauts tried a manual rendezvous and docking with Salyut 7, but they approached at too high a speed and had to pass the complex. They manoeuvred Soyuz-T 8, hoping for a second manual docking attempt, but then realised that they were running seriously short of propellant. If the second docking failed, they would have already begun to use the propellant which was reserved for the de-orbit manoeuvre. The mission was therefore cancelled, and the cosmonauts returned to Earth safely after only two days in orbit.

Soyuz-T 8 had been launched during a landing opportunity for Salyut, implying that possibly a mission of 4, 6 or even 8 months had been planned. Comments were made at the time of launch that this was not expected to be a record-breaking mission, and therefore a four month mission seems to be likely.

The exact reason for the failure was not officially stated by the Soviet Union, but one reliable source indicates that when the payload shroud separated during the ascent to orbit, it somehow snared the rendezvous antenna and tore this off the spacecraft. From just after launch it seems that the mission was doomed not to dock with Salyut.

Between 28-29 April Salyut 7 – still docked with Cosmos 1443 – was

manoeuvred into a "two-day repeater" orbit, and two months later the orbit was adjusted slightly, back to the repeating altitude. This indicated that another manned launch was due soon and that it would be a two-manned mission.

On 27 June Soyuz-T 9 was duly launched, carrying the back-up "core" crew from the Soyuz-T 8 assignment: Col V. A. Lyakhov and the Soviet cosmonaut A. P. Alexandrov on his first flight. For some obscure reason the third crew member from the intended Soyuz-T 8 mission repeat had been dropped (possibly Serebrov was flying on Soyuz-T 8 for a specific experiment which had included perishables which could no longer be used?). It should be born in mind that if Soyuz-T 8 had been successful, the launch of the

Above: *A Soviet painting, showing the Salyut 7/Cosmos 1443 complex being approached from the rear by a visiting Soyuz-T manned ferry.*

Below: *The unsuccessful first crew, launched towards Salyut 7/Cosmos 1443. Left to right, they are Serebrov, Titov and Strekalov.*

MANOEUVRES OF THE COSMOS 1443 MODULE			
PRE MANOEUVRE ORBIT		**POST MANOEUVRE ORBIT**	
Epoch (1983)	Altitude (km)	Epoch (1983)	Altitude (km)
Initial orbit		2.45 Mar	195-252
3.26 Mar	192-242	3.50 Mar	193-262
4.25 Mar	191-259	5.48 Mar	193-396
5.48 Mar	193-396	5.61 Mar	259-349
5.61 Mar	259-349	5.67 Mar	280-314
9.25 Mar	278-313	9.37 Mar	260-313
9.37 Mar	260-313		
Docked with Salyut 7: 10 March			
		11.32 Mar	325-327
14.51 Aug	314-330		
Undocked from Salyut 7: 14 August			
		14.70 Aug	314-326
18.17 Aug	313-326	18.62 Aug	326-348
Descent craft recovered: 23 August			
		23.56 Aug	326-347
16.74 Sep	322-338	17.18 Sep	288-337
19.01 Sep	287-336	19 Sep	De-orbited

Notes: Details of the manoeuvres of Cosmos 1443 while it was docked with Salyut 7 are given in the table relating to the Soyuz-T 8 and Soyuz-T 9 missions.

Soyuz-T 9 cosmonauts probably would not have occurred until early 1984, and the cosmonauts were not therefore fully prepared for a mission to Salyut 7 in the summer of 1983. Soyuz-T 9 approached the rear docking port of Salyut 7 and successfully docked at 10.46 on 28 June. It was clearly said on that day that the mission would not be lasting as long as that of Soyuz-T 5. The main part of the Soyuz-T 9 mission seemed to be the testing of the Salyut 7/Cosmos 1443 complex.

Cosmos 1443 was carrying approximately 3 tonnes of cargo, for the Soyuz-T cosmonauts, some 3.5 times that which a Progress could carry. Most of the equipment carried on the Cosmos, in fact, could have been carried by successive Progress missions. Additionally, it was confirmed that the module was carrying a descent craft which could return some 500kg of equipment to Earth.

On 7 July it was announced that Lyakhov and Alexandrov had adapted to the spaceflight environment and that they had completely emptied the descent module of Cosmos 1443 of its cargo. The cosmonauts had loaded the tape recorders, the MFK-6M and KATE-140 were ready for operations, the equipment having been successfully tested. It was announced that the cosmonauts were hoping to go further than previous crews in the growing of plants in orbit, and it was hoped that their work would bring the time nearer when fresh vegetables would be available from an orbiting "garden".

Earth Observation

A week later it was stated that the cosmonauts had completed the first in a series of studies of the Earth's surface, work which was directed towards the economic exploitation of space information. Amongst the equipment used for the observations were the Bulgarian Spektr-15 and the East German MKS-M instruments. The Yelena gamma-ray telescope was being prepared for observations.

The flight of Lyakhov and Alexandrov quickly settled into the routine for Salyut residency missions. On 4 August the mission controllers said that the cosmonauts were beginning to load the descent module of Cosmos 1443 with material. Among the items being planned for the return from orbit were:

"... all the cine and photo materials. We also plan to bring back to Earth several instruments which have been operating in space for a long period of time and for this reason are of interest to specialists. They include an air regenerator which has completed its period of service. A defunct memory unit of the Delta autonomous navigation system will also be returned to Earth."

While preparing the descent craft of Cosmos 1443 for its return to Earth, the cosmonauts had still to complete the unloading of the main Cosmos spacecraft. The freight had been tightly secured in it, and some small trolleys had been included to assist in the unloading of the module. By the time that Cosmos 1443 was ready to end its flight with Salyut, 350kg of material had

been loaded into the descent module for the return of Earth.

Since Cosmos 1443 was so large — and therefore allowed so much more room for the cosmonauts to work and rest — it was generally expected that the main spacecraft would remain docked with Salyut, while the descent module was recovered. If the Cosmos had been fitted with the standard Soyuz-T/Salyut docking units at each end (probe to dock with Salyut and drogue at the end connecting with the re-entry vehicle) then it was thought that the re-entry vehicle would return to Earth and Soyuz-T 9 could re-dock at the free Cosmos 1443 port, allowing the rear Salyut port to be used for visiting Progress missions. However, these predictions proved to be incorrect, because as far as we know, the Heavy Cosmos only carries the single docking unit (for Salyut). The tunnel connecting the main module and the descent craft does not incorporate such a system.

The undocking of the complete Cosmos 1443 vehicle came on 14 August at 14.04; the unmanned vehicle would be tested further prior to the end of its mission. Two days later, Soyuz-T 9 undocked from the rear port of Salyut 7 at 14.25, and pulled 250m away from the station, allowing it to complete a 180° rotation in space. Within 20 minutes of the undocking, Soyuz-T 9 with its cosmonauts had redocked at the front of Salyut. Once more, the Salyut mission would settle down to a normal series of Soyuz-T and Progress visits. This was emphasised on 17 August when Progress 17 was launched to dock with Salyut two days later at 13.47.

Progress 17 was carrying the normal set of supplies for Salyut. There was a major problem which was not fully acknowledged at the time: during the transfer of propellant from Progress, one of the propellant lines in Salyut sprang a leak. Rumours of the problem appeared in the western aerospace press, and when

Soyuz-T 10 was launched in 1984 a major part of the mission EVA would be dedicated to attempts to repair the propellant line.

Activity in the programme continued with the return to Earth on 23 August of the Cosmos 1443 descent module at 11.02. The main spacecraft module remained in orbit, conducting system tests. After Progress 17 had been unloaded and had then been filled with garbage, it was jettisoned on 17 September at 11.44 and it was de-orbited later that day. A further clearing of the skies came on 19 September when the main Cosmos 1443 module was de-orbited to destruction over the Pacific Ocean.

On-The-Pad Abort

With the rear docking port of Salyut 7 free and a Salyut landing window approaching, a visiting mission was expected. The altitude of the orbital complex indicated that a two-manned mission should be flown, although it had been earlier expected that the back-up 3 person crew for Soyuz-T 7, with the woman cosmonaut-trainee Irina Pronina, would fly.

No launch towards Salyut apparently came, and then at the end of September rumours of a launch failure surfaced: the rumours related to the expected two-men, one-woman crew, and suggested that the cosmonauts had to complete an on-the-pad abort on 26 September when their launch vehicle exploded. The true story was not long in coming out from Soviet officials: they were questioned at a major congress the following month, and revealed a number of mission details.

The crew had comprised Titov and Strekalov, who were being quickly recycled from the Soyuz-T 8 mission in April (the

Below: *The rear of Salyut 7 as Soyuz-T 9 approaches for docking. The two sets of solar panels (one set on Salyut, another on Cosmos 1443) are just discernible.*

reason for this became clear in November). The intended Soyuz-T 10 – here called Soyuz-T 10-1 following Glushko's designation system for the Soyuz 18 launch failure in April 1975 – had been undergoing its terminal countdown when a fire began in the base of the SL-4 launch vehicle some 90 seconds before the planned launch. It was rumoured that the fire burnt through the cables which allowed the ground controllers to activate the rocket tower atop the Soyuz shroud and it took a few seconds for the cosmonauts to realise that the automatic abort system would not work; it was further rumoured that they had to abort the mission themselves. The payload shroud was severed across the plane of the Soyuz-T instrument module and descent module interface, carrying the orbital and descent modules to a peak altitude of 950m inside the upper part of the payload shroud. At this point, the descent module separated and emergency parachutes lowered the descent module to the ground some 2.5km from the launch pad. Meanwhile the launch vehicle had exploded, with its debris burning on the pad for some 20 hours. Titov later admitted that he had suffered peak overloads of 17g (17 times the acceleration due to gravity) during the abort.

A Revised Schedule

The intended mission for Soyuz-T 10-1 can only be deduced with hindsight, and discussion of it will therefore not be found until after the Soyuz-T 9 mission description has been concluded.

Once more, a crew launched towards Lyakhov in orbit had failed to complete the rendezvous (he had been on Salyut 6 when

Below: Cosmonauts training in the hydrotank at Zvezdny Gorodok. Before any spacewalk operations are conducted, they are carefully simulated using this facility with spacecraft mock-ups.

Above: *Lyakov (top) and Alexandrov beginning their journey to the top of their Soyuz-T 9 launch vehicle. They were the first crew to operate with a Heavy Cosmos satellite.*

Soyuz 33 failed to dock). The cosmonaut work programme continued while the ground controllers decided what to do. Soyuz-T 9 with Lyakhov and Alexandrov should have returned to Earth during the first half of October, but this schedule was now changed. The cosmonauts were to continue in orbit for the time being.

It was realised that Soyuz-T 9 would soon exceed the longest period after which a manned spacecraft had been safely recovered, and the less-knowledgeable sections of the British Press began to circulate "scare stories" that the launch failure of Soyuz-T 10-1 had left Lyakhov and Alexandrov stranded in orbit. Soviet controllers were supposedly trying to discover how they could rescue their cosmonauts

who only had a time-expired spacecraft docked with Salyut. Perhaps the most ill-informed stories suggested that the Americans had been contacted and secretly one of the impending Space Shuttle launches was being re-scheduled to allow a rendezvous and rescue mission with Salyut. In fact, all that these stories did was to show the world how little British journalists knew about the Soviet space programme.

In truth, if the Soviet authorities had been worried, then following the launch failure, the Soyuz-T 9 cosmonauts would have immediately returned to Earth within the so-called time-limit of their spacecraft or, failing that, a new unmanned Soyuz-T could have been launched for a spacecraft switch similar to that undertaken in 1979 when the unmanned Soyuz 34 was launched to replace the time-expired Soyuz 32. All of this was academic, however, because the Soyuz-T had been designed with a lifetime limit of about 6 months; there was no emergency.

The Soviet answer to the scaremongering was to continue the Soyuz-T 9 mission on Salyut 7 in as routine a manner as possible, with the cosmonauts undertaking their programme of scientific, technical, biological and medical experiments: in early October they were adjusting a colorimeter Tsvet-1. Although this mission would not be setting a new duration record, the cosmonauts were still continuing a full programme of physical exercise. The launch of Progress 18 came on 20 October and a successful docking at the back of Salyut was accomplished two days later at 11.34. Amongst other supplies, the craft carried a further 500kg of propellant.

On 1 November a new spacewalk was undertaken by Lyakhov and Alexandrov. The Cosmos 1443 module had delivered a pair of additional solar cell arrays to the orbital station, and it was planned that these would be installed to the upper solar array on the station. On the ground cosmonauts Kizim and Solovyov, who had been the back ups for the aborted mission, had been training in the water tank facility at Zvezdny Gorodok to prove that it would be possible for Lyakhov and Alexandrov to undertake the assembly of the solar cells, a task for which they had not been trained. This training complete, Lyakhov and Alexandrov began their spacewalk at 04.47. During the spacewalk the cosmonauts carefully attached one additional solar cell assembly to the upper main solar array. This work done, they returned to Salyut after being outside for 170 minutes.

However, only half of the work had been completed and a second spacewalk was undertaken two days later, beginning 03.47. This time, the cosmonauts stayed outside for 175 minutes and they successfully erected the second additional solar array onto the upper main array.

With this excitement over, Lyakhov and Alexandrov returned to the routine of Salyut operations. On 13 November Progress 18 undocked at 03.08 and was de-orbited three days later (an unusually long time). With the rear port vacant, if it was necessary the Soviets could have launched an unmanned Soyuz-T to bring the cosmo-

Above: *The crew of Soyuz-T 10. Left to right, they are Atkov (a doctor), Kizim and Solovyov. They spent eight months aboard Salyut 7.*

Left: *Leonid Kizim photographed outside Salyut 7 during one of the Soyuz-T 10 residency spacewalks. During this mission, and the later Soyuz-T 15 visit, he and Vladimir Solovyov became the world's most experienced spacewalkers.*

nauts back to Earth, but this contingency was not required. Soyuz-T 9 with the cosmonauts aboard undocked from Salyut on 23 November at 16.40 and later that day made a successful landing inside the Soviet Union. It was well outside the normal landing window, for the nominal Salyut landing opportunity was not until the end of the year. The elements of the British press which had been so vocal in spreading the "disaster in space" stories less than two months earlier could hardly be bothered to report the successful conclusion of the mission.

The Soyuz-T 10-1 Mission

Activity on the Soyuz-T9 mission following the launch abort can be explained

only when the mission for which the intended Soyuz-T 10 has been deduced. Strangely enough, this can be done from open Soviet sources. In December 1983 the Soyuz-T 9 cosmonauts held a post-flight press conference, and in reply to questions from the western press it was confirmed that they were intending to hand Salyut 7 over to the Soyuz-T 10 crew. Additionally, the work accomplished during the two Soyuz-T 9 spacewalks had been planned as part of the work programme for the new Soyuz-T crew to undertake prior to the Soyuz-T 9 return to Earth during the first ten days of October.

In connection with the joint flight of Soyuz-T 9 and the Salyut 7/Cosmos 1443 complex, the magazine *Soviet Union*

Launch and Landing Windows
Cosmos 1443 was launched towards Salyut 7 just as a nominal landing window was closing. The Soyuz-T 8 launch came in the middle of a landing window suggesting a duration of an even number of months. The launch abort of Soyuz-T 10-1 came about two weeks before a landing window opened, and if it had been successful, the Soyuz-T 9 crew would have returned in the second half of October. The failure meant that the Soyuz-T 9 mission was extended.

SALYUT 7/1983

[Chart showing launch and landing windows across 1983, with data points labelled: C1443, ST8, ST8/R, ST10-1, ST9/R, Pr17, C1443/R, Pr18, ST9]

LAUNCHES IN SUPPORT OF THE SOYUZ-T 8 AND SOYUZ-T 9 MISSIONS

LAUNCH DATE AND TIME (G.M.T.)	RE-ENTRY DATE AND TIME (G.M.T.)	SPACECRAFT	MASS (kg)	CREW
1983 2 Mar 09.37	19 Sep 00.34*	Cosmos 1443	20,000 ?	
20 Apr 13.11	22 Apr 13.29	Soyuz-T 8	6,855	V. G. Titov, G. M. Strekalov, A. A. Serebrov
27 Jun 09.12	23 Nov 19.58	Soyuz-T 9	6,850 ?	V. A. Lyakhov, A. P. Alexandrov
17 Aug 12.08	17 Sep 23.45*	Progress 17	7,000 ?	
26 Sep 19.38	Launch Abort	Soyuz-T 10-1	6,850 ?	V. G. Titov, G. M. Strekalov
20 Oct 09.59	16 Nov 04.18	Progress 18	7,000 ?	

Notes: Cosmos 1443 carried a re-entry vehicle which returned to Earth on 23 August at 11.02. The launch time of the Soyuz-T 10-1 mission is estimated: the launch vehicle exploded on the pad at Tyuratam prior to the planned launch. This is the only launch pad abort experienced by the Soviet Union so far in their manned programme.

MANOEUVRES IN SUPPORT OF THE SOYUZ-T 8 AND SOYUZ-T 9 MISSIONS

PRE MANOEUVRE ORBIT		POST MANOEUVRE ORBIT	
Epoch (1983)	Altitude (km)	Epoch (1983)	Altitude (km)
Cosmos 1443: docked 10 March			
		11.20 Mar	326-326
3.52 Apr	318-320	5.95 Apr	296-319
11.24 Apr	294-316	11.87 Apr	293-305
Soyuz-T 8: docking failure 21 April			
		21.85 Apr	287-300
28.18 Apr	282-295	28.62 Apr	291-333
29.18 Apr	290-333	29.62 Apr	329-341
23.19 Jun	315-328	24.20 Jun	326-337
Soyuz-T 9: docked 28 June			
		29.00 Jun	324-336
Cosmos 1443: undocked 14 August			
		14.64 Aug	314-330
14.64 Aug	314-330	14.83 Aug	315-346
Soyuz-T 9: redocked 16 August			
		16.73 Aug	315-346
18.18 Aug	314-346	19.06 Aug	313-326
Progress 17: docked 19 August			
		20.25 Aug	313-326
20.25 Aug	313-326	20.70 Aug	319-341
26.39 Aug	321-338	26.58 Aug	332-358
16.20 Sep	332-355	17.09 Sep	335-354
Progress 17: undocked 17 September			
		17.53 Sep	334-354
Soyuz-T 10-1: launch abort 26 September			
		27.24 Sep	334-352
Progress 18: docked 22 October			
		23.16 Oct	329-347
4.26 Nov	327-345	4.45 Nov	326-342
Progress 18: undocked 13 November			
		13.25 Nov	324-340
14.58 Nov	324-340	14.64 Nov	322-337
19.26 Nov	322-334	20.01 Nov	322-337
Soyuz-T 9: undocked 23 November			
		23.81 Nov	321-337

(issue 3, 1984) carried two illustrations which showed a complex which was never actually flown in orbit. The cover picture showed the three spacecraft complex as an exterior representation: Salyut has its extra solar arrays deployed, although they were not actually deployed until *after* Cosmos 1443 had undocked. A cut-away of the complex inside the magazine shows three men in the Soyuz-T/Salyut 7/Cosmos 1443 complex, but Soyuz-T 9 only carried two cosmonauts; once more, Salyut was shown as having the added solar panels.

The discrepancies between the depictions and what actually happened can be resolved if it is realised that they show *not* the Soyuz-T 9 mission but the Soyuz-T 8 mission. The spacewalk activity to assemble the extra solar arrays would have been completed by the Soyuz-T 8 cosmonauts Titov and Strekalov (spacewalks are conducted by the mission commander and flight engineer), while Serebrov remained inside Salyut (he was the research engineer). Therefore, Titov and Strekalov had been specifically trained for the assembly of the solar cells, and when Lyakhov and Alexandrov were launched on the replacement mission they had not been fully trained for this work.

The Soviets therefore decided to insert a new mission into the planned Soyuz-T 9 flight programme. Titov and Strekalov would be flown on a two-manned Soyuz-T mission as Soyuz-T 9 was drawing to a close and while there were four men on Salyut 7 they would perform their solar cell assembly work outside Salyut – probably during a single spacewalk since they had been specifically trained for the work. The Soyuz-T 9 crew would return to Earth in October and Titov and Strekalov would

probably have returned following a single landing window cycle at the very end of December or early in January 1984.

However, the abort meant that once more Salyut's new solar cell arrays could not be erected by the intended crew. Lyakhov already had spacewalking experience (from Soyuz 32) and Alexandrov has probably done some basic spacewalking training; it was decided that following extra ground testing and training in orbit, Lyakhov and Alexandrov would try to deploy the arrays. The Soyuz-T 9 mission was therefore extended by about five weeks to allow for the extra preparations and at the beginning of November the solar cells were successfully deployed during two spacewalks.

However plausible this scenario might appear – it fits with all of the flight evidence, picture evidence and what the Soyuz-T 9 cosmonauts said at their press conference – it has not been confirmed by the Soviet authorities to the end of 1987. Therefore, it cannot be taken as an exact representation of what was planned.

The Soyuz-T 10 Residency

After the successful completion of the Soyuz-T 9 mission (although that mission had its problems) in 1984, the Soviets were to return to more normal operations with Salyut 7, a long duration resident crew hosting manned visiting missions and unmanned Progress cargo ferries.

Following the Soyuz-T 9 return, Salyut 7's orbit was allowed to decay until 11-13 January 1984 when manoeuvres deliberately lowered the orbit; a further reduction of the orbit took place during 1 February and the resulting orbit indicated that a three-man mission to Salyut was likely.

The successful launch of Soyuz-T 10 came on 9 February, carrying Col Leonid Kizim, flight engineer V. A. Solovyov and cosmonaut-researcher O. Y. Atkov. Of these men, only Kizim had previously flown in space, although he had been paired with Solovyov for more than two years as a Salyut "core" crew for the Franco-Soviet mission in 1982. Atkov was a qualified doctor, specialising in heart diseases. His inclusion in the crew strengthened western expectations that this would be another long mission.

Soyuz-T 10 docked at the front of Salyut at 14.43 on 9 February and the cosmonauts quickly began the reactivation of the station's facilities. The next step in the mission was the launch of Progress 19 on 21 February with a rear docking on Salyut two days later at 08.21. The freighter was carrying "supplies of propellant for the station's combined propulsion unit, various equipment, instruments and materials for scientific research and for life-support systems as well as mail".

The cosmonauts began photography of Comet Crommelin on 24 February, whilst work began on the unloading of Progress. Work with Progress was completed on 31 March and undocking came at 08.40 with the de-orbiting taking place the following day.

The flight programme for Soyuz-T 10 included the second and final international expedition to Salyut 7. The Soviet Union had signed an agreement with the

Below: *The next Soyuz-T launch with a crew which included the back-ups for the aborted mission was more satisfactory. The Soyuz booster made many successful flights between the abort and this launch.*

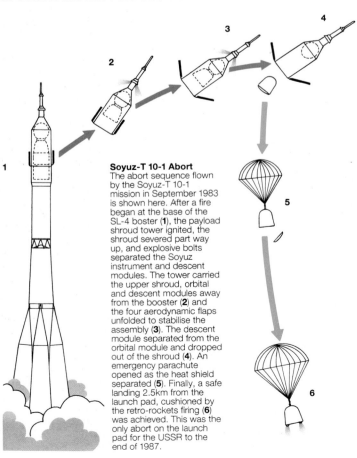

Soyuz-T 10-1 Abort
The abort sequence flown by the Soyuz-T 10-1 mission in September 1983 is shown here. After a fire began at the base of the SL-4 boster (1), the payload shroud tower ignited, the shroud severed part way up, and explosive bolts separated the Soyuz instrument and descent modules. The tower carried the upper shroud, orbital and descent modules away from the booster (2) and the four aerodynamic flaps unfolded to stabilise the assembly (3). The descent module separated from the orbital module and dropped out of the shroud (4). An emergency parachute opened as the heat shield separated (5). Finally, a safe landing 2.5km from the launch pad, cushioned by the retro-rockets firing (6) was achieved. This was the only abort on the launch pad for the USSR to the end of 1987.

Above: *The Soviet-Indian crew, launched aboard Soyuz-T 11. Left to right, Malyshev, Sharma and Strekalov. Strekalov was quickly re-cycled after the Soyuz-T 10-1 launch failure, replacing Rukavishnikov.*

Left: *Kizim outside Salyut 7 during the extensive spacewalks which were required to attempt repairs to the Salyut propulsion system. Five spacewalks were completed in less than a month.*

Indians to fly an Indian cosmonaut, and in 1982 two Indians, R. Sharma and R. Malhotra, arrived in Zvezdny Gorodok to begin their training programme. The following year the prime crew for the mission was announced as Malyshev, Rukavishnikov and Sharma, with Berezovoi, Grechko and Malhotra being the scheduled back-up crew. Due to medical problems, Rukavishnikov was removed from the prime crew and was replaced by Strekalov — being quickly recycled following the Soyuz-T 10-1 launch failure in September 1983. Incidentally, this showed the Soviet reluctance to use an "official" back-up to replace an incapacitated cosmonaut: a substitute would be brought into the flight crew, leaving the back-up crew intact.

On 3 April Soyuz-T 11 was successfully launched with Col Malyshev, Strekalov and Sharma, and a successful docking was achieved at the rear port the following day at 14.31. Once they had transferred to Salyut, for the first time the Soviet Union had six men working on a Salyut, and since the Americans were then flying the Shuttle mission 41C, a total of 11 men were in orbit at once — another record. The Indians had planned a number of scientific experiments for the flight under the names Optokinesis, Profilaktika, Membrana, Anketa and Yoga. The latter experiment was an attempt to see how Indian yoga helped in relaxing while in orbit. A remote sensing experiment named Terra was used to allow the Indian to observe a major forest fire in Burma. Most of the Terra experiment was scheduled to allow the photography of some 40 per cent of the Indian sub-continent from orbit.

Experiments in Orbit

On 8 April the experiments Poll, which evaluated the psychological condition of the crewmen during various stages of the flight, and Anketa, which called for the cosmonauts to answer a number of questions which made it possible to evaluate their vestibular conditions, were undertaken. A further biological experiment was Ballisto-3.

A further manned flight to Salyut 7 was not scheduled for some four months, and therefore Soyuz-T 11 would be traded for the older Soyuz-T 10. The international crew transferred to Soyuz-T 10 on 11 April, undocking at 07.33, and the spacecraft was safely recovered later the same day. Two days later Soyuz-T 11 with the resident three cosmonauts undocked from Salyut at 10.27 and after Salyut had performed its standard rotation through 180°, Soyuz-T 11 re-docked at the front port of Salyut. There then followed an intense period of work which involved the flights of three Progress missions in as many months, during which there would be five spacewalks.

Progress 20 was launched on 15 April and docked at the back of Salyut two days later at 09.22. The cosmonauts transferred the equipment into Salyut, the supplies including food, "elements of the life support system" and hygienic equipment. On 20 April they installed a new block of chemical batteries into the Salyut system.

The first spacewalk began at 04.31 on 23 April; it was intended that this would be a preliminary exercise only, with Kizim and Solovyov moving a folded ladder, containers with tools and "necessary" materials to the area planned for later work. The ladder was then unfolded and installed outside the station and equipment containers were also fixed to the exterior of the station. Once this work had been completed, the cosmonauts returned inside Salyut having completed a spacewalk lasting for 4h 15m.

On 26 April a second spacewalk took place. Beginning at 02.40, this was to last for 4h 56m. Once outside Salyut, the cosmonauts took about 20 minutes to transfer to the rear of the station, where they began to work in the Salyut engine bay. Using special tools the cosmonauts opened a protective cover near the switched-off part of the "reserve conduit" of the integrated propulsion system and installed a valve. The conduit was blown through to check that it was airtight. This complete, Kizim and Solovyov returned to Salyut's interior.

A third spacewalk was completed on 29 April. It began at 01.35, and Kizim and Solovyov returned to the area where they had been working three days earlier. They installed an additional conduit and checked its airtightness. This proven, the protective cover was replaced, the equipment stored outside the station and the cosmonauts went back into Salyut. This spacewalk had lasted for 2h 45m.

Continuing this heavy programme, on 3 May at 23.15 the hatch on the side of Salyut 7 opened again and Kizim and Solovyov began their fourth walk in space. They removed the thermal covering over part of the propulsion system which had been installed on the previous spacewalk and installed a second additional conduit and checked its airtightness. The thermal cover was replaced, the tools replaced and the cosmonauts returned to Salyut. They had been outside for 2h 45m.

Progress 20 had been supporting these spacewalks, having carried the equipment for the operations and also supplying the air supply required to replace the small quantity which was lost each time the EVA hatch was opened at the beginning of every spacewalk. The undocking of Progress 20 came on 6 May at 17.46 and was de-orbited the following day.

Immediately, another freighter was launched, Progress 21 beginning its flight to Salyut on 7 May. Docking came on 10 May at 00.10. It is possible that the solar cell add-on arrays to be used on the next spacewalk were carried by Progress 21, although they could have been on board Salyut since Cosmos 1443 had been unloaded the previous year. The cosmonauts returned to a normal Salyut work and experiment schedule for a few days before a fifth spacewalk was scheduled.

The last of the series of five spacewalks began on 18 May at 17.52 when Kizim and Solovyov went back outside Salyut. This time they took outside the two new supplemental solar panels and these were attached to the main solar cell on the same side of Salyut as the EVA hatch (the particular solar 'wing' was not announced, but photographs from a later spacewalk show that this main panel had the added units as well as the vertical panel which was known to have been expanded the previous November). It is interesting to note that Kizim and Solovyov completed the installation of the *two* new solar panels in a single spacewalk lasting for 3h 5m, compared with the performance of a relatively untrained (for this specific task) Soyuz-T 9 crew who needed two spacewalks lasting a total of 5h 45m. When, during this spacewalk, the first added panel had been installed, Atkov – inside Salyut – rotated the main panel, thus allowing Kizim and Solovyov to complete the second installation from the same work place.

At launch, Salyut 7 had carried three solar cells with a total area of about 60m², and each of the four new sections which had been installed added a further 4.6m².

After all this activity no more spacewalks were planned in the short term; Progress 21 undocked on 26 May at 09.41 and was de-orbited at an unspecified time later the same day.

A piece of personal news was passed to Kizim on 24 May, his wife having given birth to a daughter that day. He would have to be patient, however, because it would be more than four months before he would meet the new arrival.

The cosmonauts returned to routine operations with Progress 22 being launched on 28 May; after the standard approach procedures, a successful docking was achieved two days later at 15.47. Progress was carrying fresh propellant for the Salyut propulsion system, mail for the cosmonauts as well as new scientific equipment, ciné and photographic materials and food. More specifically, it was stated that 40kg of medical equipment had been delivered, as well as 70kg of scientific apparatus of which 45kg was ciné equipment. Following the unloading of Progress it was said that a series of geophysical observations would be undertaken, the programme lasting for about six weeks. Progress 22 remained docked with Salyut until 15 July, undocking at 13.36 and being de-orbited later that day. Once more, the rear port of Salyut was free, and a Salyut landing window was approaching. Therefore, a fresh visiting crew was expected, and the high altitude of the station suggested that it would be a two-manned Soyuz-T mission.

Soyuz-T 12

The altitude of Salyut up to this time has proved an infallible indicator of Soyuz-T crew sizes, but the next launch disproved the "rule". On 17 July Soyuz-T 12 placed two men and a woman in orbit. In command was Col V. A. Dzhanibekov, with the flight engineer Svetlana Savitskaya and cosmonaut-researcher I. P. Volk. This was the first time a woman had made a second spaceflight (thus beating America's Sally Ride to this claim). The background of Igor Volk was impressive. He was an ex-perienced test pilot, having flown the Tu-144 supersonic transport plane; when Savitskaya had taken an examination at test pilot's school he had been one of her examiners. He held the qualification "Test Pilot, First Class" and the honorary title of "Merited Test Pilot of the USSR". Thus, he seems to have been very over-qualified for the third seat on Soyuz-T 12, and it was

Above: *The return of the Soviet-Indian crew who used the Soyuz-T 10 spacecraft. Left to right, in festive garlands, are Sharma, Malyshev and Strekalov.*

Below: *The second flight of Savitskaya was the Soyuz-T 12 visit. Behind her is Igor Volk who later worked on shuttle orbiter landing tests.*

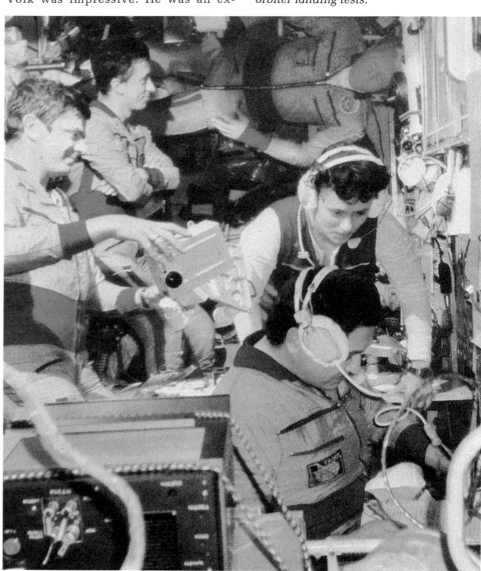

speculated in the West at the time that he might possibly be a trainee space shuttle pilot being proven on a spaceflight before a shuttle assignment. This speculation was later to be borne out when it was learned in 1987 that he had been undertaking the landing tests of the Soviet shuttle orbiter.

Early in the flight it was stated that this would be a visiting crew to Salyut, rather than a new resident crew. Docking with Salyut came on 18 July at 19.17. A full programme of work was planned for the two crews, including the Franco-Soviet Cytos-3 biological experiment. However, another "first" was planned.

On 25 July the EVA hatch on Salyut again opened, but this time Dzhanibekov and Savitskaya stepped outside: this was the first spacewalk to be undertaken by a woman. Starting at 14.55, the spacewalk lasted for 3h 55m. The cosmonauts had installed in the transfer compartment and prepared for use a portable electron beam device (mass 30kg), a control panel, a transformer and board with metal samples. Using the general-purpose hand tool, Savitskaya carried out a succession of operations, involving the cutting, welding and soldering of metal plates; meanwhile, Dzhanibekov carried out a running commentary on the work and also conducted the photography for the operation. The cosmonauts then changed places, and Dzhanibekov tested the welding apparatus. This done, the equipment and samples were transferred back into Salyut. Before ending the spacewalk, the cosmonauts took the opportunity of retrieving some samples which had been left outside Salyut to test the effects on them of long duration exposure to space conditions.

The Soyuz-T 12 crew had some further work to do with the Soyuz-T 10 cosmonauts, but the nature of this would not become clear until after the visiting mission ended. The undocking and recovery of Soyuz-T 12 came on 29 July, after an unusually long visiting mission.

With the return of the Soyuz-T 12 cosmonauts, it was expected that the Soyuz-T 10 cosmonauts would continue a routine Salyut programme until their return to Earth in early August – with the routine only being disturbed by visiting Progress freighters. However, this was not what the Soviet planners had in mind.

Salyut's Problems

It had become clear to the ground controller that the work completed on the Salyut propulsion system – trying to overcome the damage caused by the propellant leak during the operations with Progress 17 the previous year – had not been totally successful and some further work for which the cosmonauts had not trained was called for. Dzhanibekov was trained for the work outside Salyut, but the resident crew decided that they wanted to complete the work themselves. As a result, tutorials were held in orbit, with Dzhanibekov teaching Kizim and Solovyov what they had to do on a sixth spacewalk. During the Soyuz-T 12 press conference some details of the Salyut 7 problems were given for the first time by the Soviets; to quote the Deputy Director of the Soyuz-T 12 mission:

"...the unified power unit to which the relevant pipeline ... belongs is a highly reliable system on the Salyut 7 station. It carries out many functions, has several sections and tanks, pipelines and units. As was reported at a news conference in this hall, last September one of the reserve propellant lines of this power unit became de-pressurised. This did not affect the flight programme of the main crew which was carried out fully. Later two visiting crews' expedition programmes were carried out ... and while these flights were taking place there was intensive work in progress on Earth to prepare for the repair of this reserve propellant line. A special procedure was done in hydro-weightlessness, including the training of the flight control personnel and the crews, and eventually the decision was made to instruct the crew of the main expedition in space to begin this work.

During the first four spacewalks the crew brought out the tools, set up the folding ladder on the equipment compartment, set up two cross-pieces to by-pass the faulty propellant line and on the fourth spacewalk precisely located the spot of de-pressurisation. This operation was carried out at a time when the crew naturally was already in orbit and we understood that it was possible to eliminate the de-pressurisation fault by pinching off the pipeline in a certain place. As the crew was already in space, it was decided to train Vladimir Dzhanibekov for this work, to give him the opportunity to train in hydro-weightlessness, to deliver a piece of this pipeline to the station as well as a

Above: *On Soyuz-T 12 Savitskaya became not only the first woman to be launched for a second time, but also the first to complete a spacewalk.*

video film on his work on Earth and the new procedure which was developed on the basis of the training. It was initially proposed that Dzhanibekov should do this work but the crew of the main expedition, having carried out four spacewalks to work on the unified power unit, asked permission of the flight control centre to let them complete this job."

On 8 August at 08.46 the side hatch on Salyut opened for a final time during the Soyuz-T 10 mission, and Kizim and Solovyov climbed outside the orbital station. Having moved back to the exterior of the propulsion system, they removed part of the heat insulation covering the back of the service compartment and with the aid of their tools they closed off a piece of the propellant line. The insulation was replaced and secured. On their way back into Salyut the cosmonauts dismantled part of a solar cell array which would be returned to Earth for subsequent analysis by specialists. This spacewalk lasted for a total of about five hours.

The launch of Progress 23 came on 14 August, with a docking taking place two days later at 08.11. The usual supplies were transferred to Salyut by the crew, and after a short mission, Progress was undocked on 26 August at 16.13 and de-orbited two days later.

From this point on, the Soyuz-T 10 cosmonauts remained alone in orbit. On 29 August they took part in the international Black Sea and Gyunesh-84 experiments, which involved not only the Soviet Union, but also Bulgaria, Hungary, the GDR, Mongolia, Poland and Czechoslovakia. This called for co-ordinated photography of the Earth from orbit and from aircraft flying at various altitudes.

On the night of 6-7 September (Moscow Time) the Soyuz-T 10 cosmonauts beat the previous duration record which had been set by the Soyuz-T 5 cosmonauts two years earlier. As the landing opportunity drew near, the cosmonauts increased their exer-

MANOEUVRES IN SUPPORT OF THE SOYUZ-T 10 RESIDENCY			
PRE MANOEUVRE ORBIT		POST MANOEUVRE ORBIT	
Epoch (1984)	Altitude (km)	Epoch (1984)	Altitude (km)
11.24 Jan	312-325	13.26 Jan	298-323
1.27 Feb	293-317	1.83 Feb	292-303
Soyuz-T 10: docked 9 February			
		10.12 Feb	289-296
Progress 19: docked 23 February			
		23.35 Feb	281-286
25.40 Feb	282-283	25.46 Feb	280-309
26.28 Feb	279-308	26.53 Feb	306-311
30.20 Mar	291-292	30.83 Mar	289-303
Progress 19: undocked 31 March			
		31.64 Mar	290-301
Soyuz-T 11: docked 4 April			
		4.78 Apr	287-298
Soyuz-T 10: undocked 11 April			
		13.18 Apr	283-293
Soyuz-T 11: redocked 13 April			
		14.25 Apr	282-293
Progress 20: docked 17 April			
		18.25 Apr	278-290
18.25 Apr	278-290	18.57 Apr	285-326
Progress 20: undocked 6 May			
		7.23 May	277-318
Progress 21: docked 10 May			
		10.24 May	276-317
20.35 May	272-312	23.23 May	296-347
25.25 May	295-347	25.76 May	334-359
Progress 21: undocked 26 May			
		26.58 May	335-358
Progress 22: docked 30 May			
		30.96 May	334-358
10.17 Jul	328-354	11.00 Jul	307-354
14.22 Jul	307-353	14.92 Jul	334-355
Progress 22: undocked 15 July			
		15.94 Jul	334-355
Soyuz-T 12: docked 18 July			
		19.23 Jul	334-354
28.31 Jul	333-353	28.88 Jul	342-372
Soyuz-T 12: undocked 29 July			
		30.28 Jul	342-372
15.25 Aug	341-369		
Progress 23: docked 16 August			
		17.95 Aug	351-375
25.25 Aug	351-376	25.37 Aug	373-375
Progress 23: undocked 26 August			
		28.25 Aug	374-374
25.28 Sep	370-374	27.96 Sep	370-375
Soyuz-T 11: undocked 2 October			
		3.20 Oct	370-375

cise programme and on 1 October it was announced that they would be returning to Earth the following day. Undocking from Salyut came at about 08.40 on 2 October and the landing came later the same day. As the cosmonauts began their adaptation to Earthly conditions, the duration of their mission was directly equated with the time required for a one-way trip to Mars, although it was by no means certain that the Soviets had an actual manned Mars expedition planned for the twentieth century.

1985 Plans for Salyut 7

With the success of the Soyuz-T 10 long duration mission behind them, the Soviets were planning an even longer flight for 1985: three cosmonauts were to spend 9-10 months in orbit with perhaps one or two visiting missions (involving all-Soviet crews). The two-man "core" back-up cosmonauts for Soyuz-T 10 were assigned for the long Soyuz-T 13 mission: commander V. V. Vasyutin, a rookie, and flight engineer Savinykh. However, Atkov's back-up – a doctor called Poliakov – was not assigned to the mission once it was decided that someone who was medically trained to Atkov's degree was too much of a specialist for the routine Salyut operations. A doctor had a place in space, but it would probably be in the third seat of an all-Soviet visiting crew. Poliakov's place was taken by another rookie cosmonaut, A. A. Volkov.

Two possible visiting missions for 1985 can possibly be identified. During 1984 it was reported that the experienced cosmonaut B. V. Volynov was training for a short mission to Salyut, and this could have been one of the visiting missions – probably the first. The next mission was indicated by Rukavishnikov who said that an

all-woman cosmonaut crew was in training for a Salyut visit. Since Savitskaya had only flown to Salyut in 1984, she would probably not have flown the Soyuz-T 14 visiting mission in the first half of 1985. She had to be on the prime crew, since every crew must include at least one experienced cosmonaut; the further implication of this is that the back-ups for this mission must have included at least one man, since Savitskaya was the only experienced woman flier still in training.

Bearing in mind what actually happened, we can speculate that the latter part of the Soyuz-T 13 mission would have involved the use of a new variety of Heavy Cosmos module, which would have remained docked with Salyut after the station was abandoned when the Soyuz-T 13 cosmonauts returned to Earth. Additionally, while the Heavy Cosmos was attached to Salyut, Vasyutin and Savinykh would have performed either one or two spacewalks to test the erection and stowing of large structures in space.

Assuming that this information is correct, then the launch of Soyuz-T 13 for a 9-month mission was expected to take place in late February 1985 with the mission ending during the November-December landing window. The orbit of Salyut was slowly decaying and in February it was in its normal "two-day repeater" slot ready for another crew, but none was launched.

The Rescue of Salyut 7

All the above plans had to be scrapped in February 1985 when the Soviet mission controllers lost all contact with Salyut 7. Something serious had gone wrong with the station, but without telemetry they could not ascertain exactly what was the fault. The Soviets must have tried all that

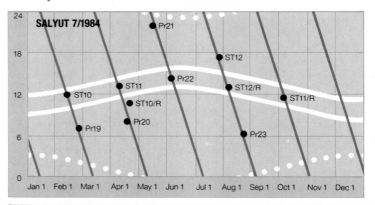

Launch and Landing Window The launch of Soyuz-T 10 during a landing window indicated that a mission lasting for an even number of months was planned, since the landing windows come approximately every two months. The Soyuz-T 11 and Soyuz-T 12 launches allowed landing during the nominal windows. The three Progress missions during April-June were a result of the heavy EVA schedules during April and May. After Soyuz-T 11, problems meant that the 1985 missions had to be revised.

LAUNCHES IN SUPPORT OF THE SOYUZ-T 10 RESIDENCY				
LAUNCH DATE AND TIME (G.M.T.)	RE-ENTRY DATE AND TIME (G.M.T.)	SPACECRAFT	MASS (kg)	CREW
1984 8 Feb 12.07	11 Apr 10.50	Soyuz-T 10	6,850 ?	L. D. Kizim, V. A. Solovyov, O. Y. Atkov
21 Feb 06.46	1 Apr 18.18*	Progress 19	7,000 ?	
3 Apr 13.09	2 Oct 10.57	Soyuz-T 11	6,850 ?	V. A. Malyshev, G. M. Strekalov, R. Sharma
15 Apr 08.13	7 May	Progress 20	7,000 ?	
7 May 22.47	26 May	Progress 21	7,000 ?	
28 May 14.13	15 July	Progress 22	7,000 ?	
17 Jul 17.41	29 Jul 12.55	Soyuz-T 12	6,850 ?	V. A. Dzhanibekov, S. Y. Savitskaya, I. P. Volk
14 Aug 06.28	28 Aug 01.28	Progress 23	7,000 ?	

Notes: Kizim, Solovyov and Atkov returned to Earth in Soyuz-T 11 after a flight of 236d 22h 50m; Malyshev, Strekalov and Sharma returned in Soyuz-T 10 after 7d 21h 41m. The de-orbit times for some of the Progress missions were not announced, and therefore the descents are given by date only.

which the cosmonauts would participate.

The careful rendezvous with Salyut took a full two days, and after approaching Salyut, the cosmonauts circled it, checking its exterior condition. The station was slowly tumbling in orbit, its solar panels were only randomly aligned and therefore the cosmonauts knew that they would be docking with a dead, frozen station. The rendezvous with Salyut had been a completely manual operation, with Soyuz-T 13 carrying (for the first time in the programme) a laser range-finder to assist the operation. The delicate manual docking at

Above: *A view of the Mission Control at Kaliningrad, a suburb of Moscow. This picture was taken during the launch of Soyuz-T 5. The cosmonauts are shown in the top right-hand screen, the launch groundtrack on the centre screen, and the spacecraft positions over the Earth on the lower left screen.*

Left: *Salyut 7, a dead orbital station, photographed by the approaching Soyuz-T 13 rescue crew. This docking was the first performed by the Soviets with an unco-operative satellite.*

they could to regain contact with, and therefore control of, the errant orbital station, but by the end of the month it looked as if nothing could be salvaged. On 1 March TASS announced that "In view of the fact that the planned programme of work aboard the Salyut 7 orbital station has been fulfilled completely, at present the station is mothballed and continues its flight in an automatic regime." This was taken as indicating that Salyut had been abandoned, although the statement did not say this in so many words.

The Soviet space planners would not give up easily. The Soviets had widely criticised the Americans for being unable to control the re-entry of the large Skylab station in 1979 (ignoring the uncontrolled re-entries of the failed Soviet stations Salyut 2 and Cosmos 557), and therefore the uncontrolled re-entry of Salyut 7 would be an embarrassment to them to say the least. In theory, it might have been possible to dock an unmanned Progress with Salyut, but since the station would not be responding to the Progress transmissions and was slowly tumbling, the operation would be a very difficult one.

During March the Salyut training groups were re-assigned, with the intended Soyuz-T 13 cosmonauts being split. The flight engineers Savinykh and his back-up

A. P. Alexandrov were assigned to the revised Soyuz-T 13 mission with new commanders: Dzhanibekov, backed-up by Popov. The remaining four cosmonauts from the original Soyuz-T 13 crewing — Vasyutin and A. A. Volkov, backed-up by Viktorenko and Saley — were re-assigned to Soyuz-T 14, with new flight engineers: Grechko, backed-up by Strekalov (reversing their roles from Soyuz-T 11). The four new Soyuz-T 13 cosmonauts began to train for any eventuality in orbit, including a manual rendezvous and docking with an unco-operative satellite, and repairs to the station (if they could enter it) in a freezing environment.

Running Repairs

With no prior warning to western observers, on 6 June the Soviets launched Soyuz-T 13 containing Col Dzhanibekov and Victor Savinykh. Prior to this assignment, Dzhanibekov had trained extensively for the Salyut 7 repairs scheduled for the Soyuz-T 12 mission, while Savinykh was a fully Salyut-trained engineer who was scheduled to fly to Salyut that year anyway. The initial launch announcement simply said that the flight programme provided ". . . for carrying out joint work with the orbital research station Salyut 7 . . ." with no hint of the drama in

the front port of Salyut came on 8 June at 08.50. After thorough checks of the docking seal and sampling of Salyut's atmosphere, the cosmonauts entered the station wearing breathing apparatus and layers of warm clothing (including woolly hats).

On entering Salyut the cosmonauts described it as being "stuffy and cold", which must have been an understatement. The first job of the cosmonauts was to stabilise the station and re-align the solar panels with the Sun so that the station's batteries could begin recharging. This was completed by 11 June. The cosmonauts continued to de-mothball the station. As they continued their work, they sent a "shopping list" to the ground controllers of items which they needed to be included on the next Progress supply mission. On 12 June the cosmonauts began the activation of the life-support systems, with regenerators, absorbers of harmful impurities, gas analysers and a heater being connected. Up to this point the cosmonauts had only been able to work inside Salyut for short periods, after which they would retreat to the warmer environment of their Soyuz-T spacecraft. The same day, the cosmonauts began to check the Salyut communications systems; previously they had relied on those of the Soyuz-T 13 for all communications.

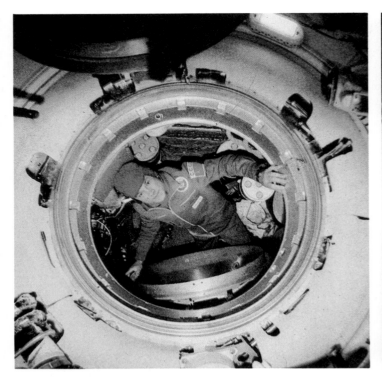

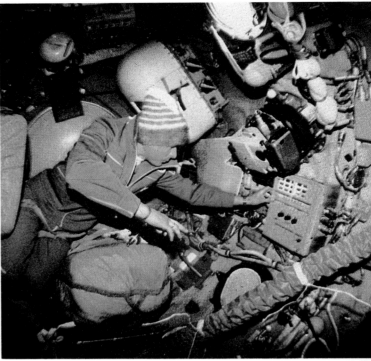

Above: *Dzhanibekov (left) and Savinykh (right) inside Salyut 7 during the rescue of the station. When they entered, the life support systems were not working, and the temperature was below freezing. Note their protective warm clothing.*

On 13 June, the cosmonauts brought the television system back on-line. After a week's hard work in orbit, however, there was still plenty to do before Salyut 7 was safe for another long mission. The next day the cosmonauts began operations with the Salyut control systems and they were delighted that the station responded to commands once more. By 17 June the Soviet controllers announced that the cosmonauts were carrying out the concluding operations for the full reactivation of Salyut. The manoeuvring system was operating, the station's atmosphere had been thoroughly checked and the crew had replaced a number of items which had exceeded their operating life. That day the cosmonauts further checked the orientation system and did some more minor repair work.

The first set of supplies to be launched to the Soyuz-T 13 rescue mission appeared on 21 June with the launch of Progress 24. With Salyut operating, Progress was able to make a normal two-day rendezvous and approach to the station, docking at the rear port on 23 June at 02.34. Meanwhile, the cosmonauts found time to take part in the Kursk-85 Earth survey experiment – a follow-on to the successful work completed the previous year by the Soyuz-T 10 cosmonauts. The cosmonauts were busy with unloading Progress 24, repairing and replacing faulty Salyut equipment. The freighter was carrying propellant for Salyut which would be used exclusively by the attitude control system. Following the propellant leak in 1983, the Salyut engines had not been used to correct the orbit of the station since the manoeuvres prior to the Soyuz-T 10 launch, and indeed would not be used again for orbital changes. The repair attempt in 1984 had ensured that Salyut could be refuelled and that the attitude control system was still operating, but no more.

On board Progress was equipment which allowed the cosmonauts to replace faulty units in the Salyut temperature control system. On 2 July the cosmonauts began operations with the MKF-6M and KATE-140 camera systems for the first time since the major systems failure. Additionally, the cosmonauts undertook a full programme of Earth observations and exercise experiments. It was announced on 15 July, that the cosmonauts had finished work with Progress 24, and the following day at 12.28 it was undocked, with de-orbit coming later the same day. The final Progress mission to Salyut 7 had been completed.

Cosmos 1669

The orbital plane of Salyut 7 was drifting towards one of the two-monthly landing opportunities, and it was expected that a new crew might be launched, to take over now that Dzhanibekov and Savinykh had brought the station back to operational status. It was therefore something of a surprise when on 19 July the launch of Cosmos 1669 was announced. The orbit was

MANOEUVRES OF THE COSMOS 1686 MODULE

PRE MANOEUVRE ORBIT		POST MANOEUVRE ORBIT	
Epoch (1985)	Altitude (km)	Epoch (1985)	Altitude (km)
Initial orbit		27.42 Sep	172-302
28.90 Sep	172-302	29.90 Sep	284-319
1.91 Oct	281-315	2.35 Oct	290-336
2.35 Oct	290-336		
Docked with Salyut 7: 2 October			
		3.24 Oct	335-352

Notes: This table lists the manoeuvres of Cosmos 1686 to the time that it docked with Salyut 7; its subsequent manoeuvres are detailed in the table relating to the Soyuz-T 13, Soyuz-T 14 and Soyuz-T 15 missions.

similar to a Progress and the orbital plane was correct for a joint flight with Salyut 7. The launch announcement made no comment about possible joint work, the first indication coming on 21 July when the docking at 15.05 with the rear port of Salyut was announced.

At first it was thought that this might be a relative of the Heavy Cosmos modules, but Cosmos 1669 had been launched by the SL-4 vehicle. When this was realised, it was speculated that Cosmos 1669 might be carrying solar panels or that it might be a refined Progress with a descent module. The spacecraft, however, seems to have carried out a standard Progress mission. In fact, some dispatches from Soviet journalists had talked of a Progress freighter being expected, and one report was even more specific: Dzhanibekov and Savinykh "are getting ready for a meeting with the Progress 25 cargo ship that was launched at the Baikonur cosmodrome yesterday." When talking of the docking, the Radio Moscow World Service described Cosmos 1669 as a "support satellite", "similar to spacecraft of the Progress series". One may speculate that perhaps the flight of the intended Progress 25 had proved less than trouble-free initially, and that the "Cosmos" cover name was used in case the flight failed.

Interestingly enough, few comments were initially made about the unloading of Cosmos 1669 and the cargo which it was presumably carrying. While it was docked, on 2 August at 17.15 Dzhanibekov and Savinykh began a spacewalk, during which they attached two supplementary solar cell arrays to the third solar array which had not previously been enlarged. These had been carried into orbit during the Progress 24 mission. This work complete, the cosmonauts returned to the EVA hatch on the side of Salyut, and before ending their spacewalk they retrieved a micrometeorite detector and cartridges containing samples of biopolymers and various "design materials" for later return to Earth. The spacewalk lasted for about five hours.

With the spacewalk complete, the cosmonauts worked on the unloading of the Cosmos 1669 freighter. It included the Mariya unit which was to study the generation of high energy particles in the Earth's radiation belts; it was described as being more effective than the Yelena gamma radiation detector because its results were available in the space of only a few minutes. Another experiment called Medusa, which was designed to study the synthesis of the components of nucleic acid in outer space conditions, had earlier been left outside Salyut during the Soyuz-T 13 spacewalk.

The routine of Dzhanibekov and Savinykh now conformed to that of a normal Salyut residency, with the usual array of scientific, technological, biological and medical experiments being conducted, as well as a heavy exercise programme. They undertook Earth observation work connected with Intercosmos Gyunesh-85 experiments. Cosmos 1669 completed the filling of the Salyut 7 propellant tanks – thereby proving that the Cosmos carried no descent craft (the descent craft would have taken the place of the propellant load) – and its mission complete, it undocked from the station on 28 August at 21.50 and two days later it was de-orbited over the Pacific Ocean. Apparently, a standard Progress mission had been completed.

The two cosmonauts on Salyut were subjected to a thorough medical examination using Salyut's equipment on 30 August. From the announcements made about the daily work of the cosmonauts, it would appear that the cosmonauts were concentrating on Earth observation experiments and bio-medical work, both disciplines that required equipment which utilised a lower power supply than the materials processing systems. However, there was still work done in the materials processing programme.

As another Salyut 7 landing opportunity was approaching it came as no surprise when Soyuz-T 14 was launched on 17 September: the three cosmonauts were Lt-Col V. V. Vasyutin, G. M. Grechko and Lt-Col A. A. Volkov. Both Vasyutin and Volkov were from the original Soyuz-T 13 long duration crew grouping and were rookie cosmonauts. The following day the new Soyuz-T docked at the back of Salyut at 14.14, and soon Salyut had five men working once more on board the complex. The Soviets announced that there would be a partial crew rotation: Dzhanibekov and Grechko would return to Earth in Soyuz-T 13 while the original Soyuz-T 13 crew would be operating as a team on Salyut with Soyuz-T 14 docked.

A normal programme of joint work was undertaken by the cosmonauts, while Dzhanibekov began an intensive programme of exercises, preparing for his return to Earth after the relatively "short" stay of 112 days. Soyuz-T 13 undocked from the front of Salyut on 25 September at 03.58, but an immediate landing was not planned. Following the undocking the cosmonauts tested differing methods of approaching Salyut, although a re-docking

was not to be attempted. After an independent flight lasting for about a day, Soyuz-T 13 returned to Earth safely.

With the original long duration crew reunited in orbit as a team, the question was raised as to the length of the Soyuz-T 14 mission. Since a nine month mission had probably been scheduled for the cosmonauts, it was thought that this would be attempted by Savinykh while the other cosmonauts remained in orbit for only six months. This would imply a landing in March 1986.

A Heavy Cosmos

Only a day after the Soyuz-T 13 recovery, a new spacecraft was launched towards Salyut 7. Called Cosmos 1686, this was a new Heavy Cosmos module – a supposition confirmed by Soyuz-T 14 remaining at the rear port of Salyut: the previous Heavy Cosmos modules had docked at the front of Salyut, while the Cosmos 1669 Progress craft had docked at the back of Salyut. Cosmos 1686 took five days to reach Salyut, docking on 2 October at 10.16. With this in place, it appeared that the Soyuz-T cosmonauts would operate this configuration until their mission ended. The Soviets did not release any drawings of Cosmos 1686, but said that "the amount of cargo carried . . . is about the same as that carried by a Progress, excluding fuel: it includes food, gas regenerators, new scientific apparatus and individual assemblies and parts". The module could be used as "a research laboratory, a space tug or as a hothouse". Elsewhere, it was said that Cosmos 1686 was carrying five tonnes of cargo, including more than a tonne of scientific equipment.

The three cosmonauts began work on

the unloading of Cosmos 1686. Unlike Cosmos 1443 which had been operated during 1983, Cosmos 1686 did not carry a descent module. A battery of astronomical telescopes was carried by the module, probably in place of the descent module. However, the full potential of the module would not be realised on this mission.

On 13 November, the Kettering Group started to pick up encoded transmissions between Salyut and the ground controllers, and this was thought to be related to some classified observations which might be taking place (Volkov was a military cosmonaut, and there was a school of thought which suggested that Cosmos 1686 might have a reconnaissance capability). It is now known, however, that Vasyutin had fallen ill and the cosmonauts were asking for medical instructions and advice on a "secure" radio channel. Originally, Vasyutin had been suffering from a minor complaint, but as he was unable to fulfil his full programme of work he became depressed and withdrawn. Luckily, a Salyut landing window was

Above: *The rescue mission over, the Soyuz-T 13 descent module has returned to Earth with Grechko (launched on Soyuz-T 14) and Dzhanibekov.*

Below: *Repaired and revitalised, the Salyut 7/Soyuz-T 14 complex is photographed by the departing Soyuz-T 13 crew on 25 September 1985.*

approaching and on 21 November the three cosmonauts undocked Soyuz-T 14 from Salyut and returned to Earth: this was the first time that a mission was curtailed because of medical problems.

In 1988 A. A. Volkov stated privately that if Vasyutin had not fallen ill, the Soyuz-T 14 mission would have continued until 15 March 1986: this would have meant that he and Vasyutin would have been in orbit for 179 days, while Savinykh would have been in orbit for 282 days – a new duration record. Surprisingly, there was no attempt made to repeat the nine month mission on either Salyut 7 or the new Mir orbital complex and next long duration flight would last for eleven months, exceeding the previous record by three months.

The Soyuz-T 15 Visit

When the Soyuz-T 14 cosmonauts left Salyut 7 they mothballed the complex as best they could, probably not knowing whether a further visiting mission would be scheduled to complete their work. Almost certainly, the Soviets had not planned to re-man Salyut after the scheduled nine-month mission. The Salyut complex continued to decay from orbit until 6 February 1986 when a very small manoeuvre might have taken place (it is on the borderline of the Two-Line Orbital Elements error limits): perigee may have been reduced by 2km, but this is very uncertain.

The launch of a new space station core Mir (see next chapter), took place in February 1986, and for a while it was thought that the new station might dock with Salyut 7; certainly, their orbital planes were close and their operating altitudes were such that Mir could have completed a rendezvous and docking with Salyut, had that been intended. In reality, the Salyut 7/Cosmos 1686 complex continued to orbit with no further orbital corrections.

On 13 March Soyuz-T 15 was launched towards Mir and it docked with the new station two days later. On board the spacecraft was the experienced Salyut 7 crew of Col L. Kizim and V. A. Solovyov. They continued a programme of work on Mir, supplied by Progress ferry craft, and no hint of any plans for Salyut 7 emerged until 3 May when it was announced that:

"The first stage of the station's [= Mir] manned flight is nearing completion. In accordance with the programme for further work in space, Leonid Kizim and Vladimir Solovyov are to fly over to the Salyut 7/Cosmos 1686 orbital complex, which has been operating in automatic mode since 21 November 1985. Separation of the Soyuz-T 15 craft from the Mir station is planned for 5 May."

When Soyuz-T 15 undocked from Mir on 5 May the station had an orbit of 309-345km, while the orbit of the Salyut 7 complex was 358-360km. In order to be able to counter the distance *around* the orbits of Mir and Salyut, some manoeuvres by Soyuz-T had to be completed:

    5.75 May   311-343km
    5.94 May   307-342km

The successful docking of Soyuz-T 15 at

the rear of the Salyut 7 orbital complex took place on 6 May at 16.58. Once on board the familiar station – where they had spent eight months some two years earlier – Kizim and Solovyov began work on de-mothballing the systems and completing minor work as necessary. Because each of the Salyut docking ports was in use, there were no plans for a Progress mission to Salyut, and therefore the cosmonauts had taken all their supplies from the Mir complex. It appears that the cosmonauts were quick to settle down to the standard routine of experiments, observations, medical tests and exercises.

Building in Space

The main objective of the Soyuz-T 15 visit to Salyut 7 was to complete the experiments which had been intended for the Soyuz-T 14 cosmonauts. The major experiment began on 28 May when the cosmonauts opened the Salyut EVA hatch at 05.43 to begin a 3h 50m spacewalk. The main task of this spacewalk was to master methods of assembling large structures in space. The cosmonauts used a lattice and pin-connected structure which had been carried to Salyut in a folded position (on Cosmos 1686, presumably). The cosmonauts unfolded the frame and then returned it to its folded position. They then mounted the new instrument near a porthole on the work compartment, this being intended for experiments related to the future transmission of telemetry using optical frequencies. Once more, before returning to Salyut's interior the cosmonauts retrieved samples of the micrometeorite detector and other specimens from outside Salyut for later return to Earth.

The structure which the cosmonauts had erected and then stored again during

their spacewalk had been 15m high, with the cross-section of a rhombus with 40cm sides. It had been manufactured by the Paton welding institute in Kiev.

With an openness that surprised observers of the Soviet space programme, it was announced on 30 May that a further spacewalk was scheduled for the following day. At 04.57 on 31 May the cosmonauts again stepped outside Salyut to begin a

Above: *Cosmonauts Solovyov (left) and Kizim arrive at the Tyuratam-Baikonur launch site, in preparation for their Soyuz-T 15 mission. After visiting Mir, they paid a six-week visit to Salyut 7.*

Below: *The Soyuz-T 13 and 14 crews on Salyut 7 – Vasyutin, Grechko, Savinykh, Volkov and Dzhanibekov. Savinykh was scheduled for a nine month mission, but Vasyutin's illness curtailed it.*

long spacewalk. The structure tested on the earlier spacewalk was unfurled again and the cosmonauts were to use it to construct a tower which was said to resemble an "oil derrick". Instruments were mounted on the top of the structure, including the Fon waveband instrument which would monitor the vibrations of the frame. Live television coverage showed the frame being extended to a height of 12m. This completed the first phase of the work.

The two cosmonauts again folded up the frame, with the instrument package still mounted on top. Then they began to weld the frame: the welding unit was the one which had been tested by Dzhanibekov and Savitskaya during their Soyuz-T 12 visit nearly two years earlier. They mounted a micro-unit for testing any deformations on the outside of Salyut to test different materials for strength; the unit was designed to strain samples made of an aluminium and magnesium alloy.

This work completed, the cosmonauts returned inside Salyut for the last time; they had been outside Salyut for 4h 40m. In 1988, Volkov stated that the work performed during these EVAs was scheduled for completion by Vasyutin and himself during the Soyuz-T 14 mission. The T 15 work programme would now include the more routine Earth observations and other experiments. On 23 June it was announced that the cosmonauts were completing their work on board Salyut. They mothballed the station, at the same time removing some scientific equipment and transferring it to Soyuz-T 15. Undocking from Salyut would take place on 25 June. During 24 June the cosmonauts transferred cassettes, ciné film, spectrograms, biological equipment and spectrometers to Soyuz-T. The orbit of Salyut was 356-359km, and at 14.58 on 25 June Soyuz-T 15 undocked from Salyut. It was tracked in a 325-338km orbit (epoch 26.43 June) as it manoeuvred back to the Mir orbital station, which by this time was in a 332-366km orbit.

Manned operations with the Salyut 7/Cosmos 1686 complex were now complete for the foreseeable future, although the Soviets did not rule out a possible mission to the station after some years to see how it had stood up to so many years in orbit. One possibility is that it may be visited and even retrieved during a flight of the Soviet space shuttle system.

Meanwhile, the Soviets wanted to move Salyut out of the orbital regime in which the new Mir complex would be operating. During 17-23 August the Cosmos 1686 propulsion system was used to manoeuvre the complex into a storage orbit some 475km high. The complex was then abandoned for the foreseeable future.

Conclusions

While Salyut 7 had not been the quantum leap over Salyut 6 that some western observers had been expecting, its operations were able to advance Soviet spaceflight experience greatly. With the Salyut 6 and Salyut 7 stations, the Soviet Union had made manned flights to Earth orbit routine events, to the point where a new launch would receive little coverage in the West unless something went wrong.

The Soviets had managed to extend the manned duration record to eight months, and must have been disappointed that after the hard work of the Soyuz-T 13 rescue mission, the nine months mission had to be terminated because of ill-health of the mission commander.

On a more positive note, the Soviets had now gained a great deal of experience in the repair and general overhaul of orbital stations, as well as flight testing two variants of the Heavy Cosmos module (Cosmos 1443 with a descent craft and Cosmos 1686 with a scientific payload) with men on board. The next logical step in the space station programme would be the assembly of a modular complex in orbit around the Earth which could be permanently manned. This project is the subject of the following chapter.

FURTHER READING

There have been no western books published which have been devoted to Salyut 7 operations. As with Salyut 6, the daily operations have been covered by Neville Kidger in *Spaceflight*.

The following three books (in Russian) deal with the first resident mission to Salyut 7, the Indo-Soviet mission and the Franco-Soviet mission respectively.

Korolev, M. Y., *211 Sutok Na Bortu "Salyut 7"*, Mashinostroeniya Press, Moscow, 1983

Malyshev, Y. et al, *CCCP-India Na Kosmysheskych Orbita,* Mashinostroeniya Press, Moscow, 1984.

Rebrov, M. et al *CCCP-France Na Kosmysheskych Orbita,* Mashinostroeniya Press, Moscow, 1982.

The following book (in French) is a preflight account of the Soviet programme with particular reference to Franco-Soviet co-operation and Soyuz-T 6.

Chretien, J.-L. et al, *Spatiale Premiere – Le Premier Français Dans L'Espace,* Plon Publishers, Paris, 1982.

SALYUT 7/1985

Launch and Landing Windows
The Salyut 7 missions planned for 1985 were rescheduled after the station's failure. Soyuz-T 13 was launched on a rescue flight and after the work was completed, Soyuz-T 14 was launched for a long flight. One of the Soyuz-T 13 cosmonauts stayed with two Soyuz-T 14 men and should have remained in orbit until March 1986, but the mission had to be curtailed. Cosmos 1669 was a Progress-type vehicle while Cosmos 1686 was a Heavy Cosmos variant.

LAUNCHES IN SUPPORT OF THE SOYUZ-T 13 AND SOYUZ-T 14 MISSIONS

LAUNCH DATE AND TIME (G.M.T.)	RE-ENTRY DATE AND TIME (G.M.T.)	SPACECRAFT	MASS (kg)	CREW
1985 6 Jun 06.40	26 Sep 09.52	Soyuz-T 13	6,850 ?	V. A. Dzhanibekov, V. P. Savinykh
21 Jun 00.40	15 Jul	Progress 24	7,000 ?	
16 Jul 13.05	30 Aug	Cosmos 1669	7,000 ?	
17 Sep 12.39	21 Nov 10.31	Soyuz-T 14	6,850 ?	V. V. Vasyutin, G. M. Grechko, A. A. Volkov
27 Sep 08.41	In orbit	Cosmos 1686	20,000 ?	

Notes: Dzhanibekov and Grechko returned to Earth in Soyuz-T 13: their durations were 112d 3h 12m for Dzhanibekov and 8d 21h 13m for Grechko. Vasyutin, Savinykh and Volkov returned in Soyuz-T 14: the flight times for Vasyutin and Volkov were 64d 21h 52m and that for Savinykh was 168d 3h 51m.

MANOEUVRES IN SUPPORT OF THE SOYUZ-T 13, SOYUZ-T 14 AND SOYUZ-T 15 MISSIONS

PRE MANOEUVRE ORBIT		POST MANOEUVRE ORBIT	
Epoch (1985)	Altitude (km)	Epoch (1985)	Altitude (km)
Soyuz-T 13: docked 8 June			
		8.79 Jun	356-359
Progress 24: docked 23 June			
		24.06 Jun	355-358
Progress 24: undocked 15 July			
		15.76 Jul	354-358
Cosmos 1669: docked 21 July			
		21.74 Jul	354-358
5.81 Aug	352-355	6.83 Aug	353-358
28.70 Aug	352-353	28.83 Aug	339-353
Cosmos 1669: undocked 28 August			
		28.96 Aug	339-353
Soyuz-T 14: docked 18 September			
		19.85 Sep	337-353
Soyuz-T 13: undocked 25 September			
		25.69 Sep	337-352
Cosmos 1686: docked 2 October			
		3.24 Oct	335-352
3.87 Oct	336-353	4.32 Oct	338-358
4.32 Oct	338-358	4.44 Oct	358-359
Soyuz-T 14: undocked 21 November			
		21.92 Nov	354-355
(1986)		(1986)	
Soyuz-T 15: docked 6 May			
		7.46 May	335-343
Soyuz-T 15: undocked 25 June			
		26.88 Jun	333-338
17.97 Aug	331-333	19.81 Aug	333-385
19.81 Aug	333-385	20.00 Aug	332-468
20.00 Aug	332-468	21.10 Aug	466-468
21.10 Aug	466-468	22.79 Aug	470-475
22.79 Aug	470-475	22.86 Aug	475-475

Notes: Soyuz-T 15 originally docked with the Mir core station, and following its visit to Salyut 7 it returned to Mir. At the end of 1987 Salyut 7 was still docked to Cosmos 1686, the orbit of the complex slowly decaying.

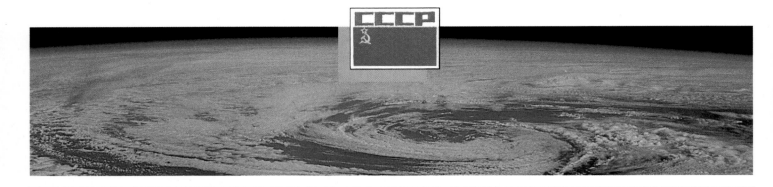

Chapter 13: Mir: The First Modular Space Station

F ollowing the successful operations with the second generation Salyut 6 and Salyut 7 orbital stations, Soviet cosmonauts and spokesmen considered the next stage in the manned space station programme; according to some reports, Salyut 8 was planned to be a modular space station, possibly with the Heavy Cosmos technology being used.

The history of modular space station concepts in the Soviet space programme is a long one, with realistic artist's impressions being published as early as the 1970s. However, the modular Salyut 7 follow-on station did not conform exactly to such concepts. Mir, as the core station was called, suffered from delays in the preparation of associated modules, and therefore operations with the station were slow in evolving.

The Mir programme coincided with the ascent to power of Mikhail Gorbachev and the resulting policy of "Glasnost". Glasnost – or openness – has meant that more and more information has been released about up-and-coming all-Soviet missions as time has gone on. The impending

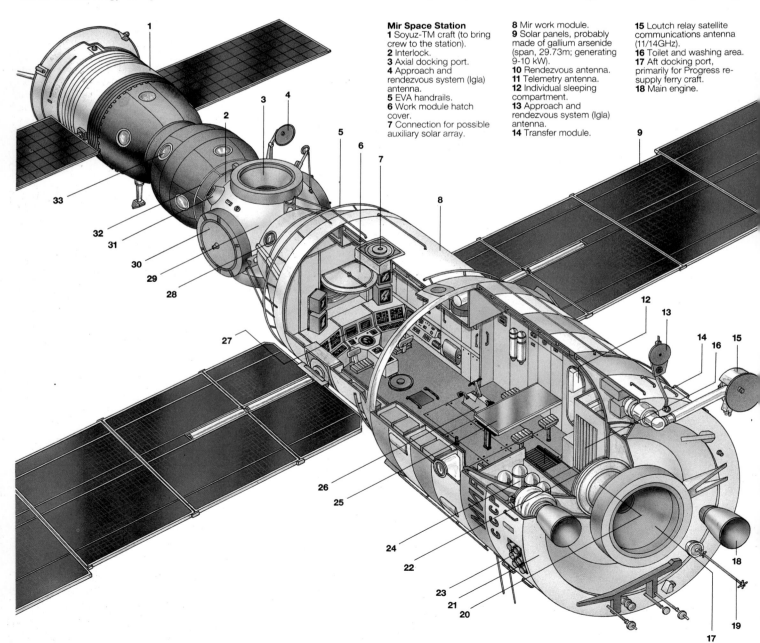

Mir Space Station
1 Soyuz-TM craft (to bring crew to the station).
2 Interlock.
3 Axial docking port.
4 Approach and rendezvous system (Igla) antenna.
5 EVA handrails.
6 Work module hatch cover.
7 Connection for possible auxiliary solar array.

8 Mir work module.
9 Solar panels, probably made of gallium arsenide (span, 29.73m; generating 9-10 kW).
10 Rendezvous antenna.
11 Telemetry antenna.
12 Individual sleeping compartment.
13 Approach and rendezvous system (Igla) antenna.
14 Transfer module.

15 Loutch relay satellite communications antenna (11/14GHz).
16 Toilet and washing area.
17 Aft docking port, primarily for Progress re-supply ferry craft.
18 Main engine.

launch of the first manned visit to Mir was announced the day before launch, with the crew being named. For the main 1987 residency the prime crew were named more than a week before the launch, while both the prime and back-up crews were named more than a week before the launch of the second all-Soviet crew in 1987. More live television coverage was given of space-flight events during this period, and the overall scope and timetable for the Mir programme has been publicly announced.

Modular Space Stations

Almost as soon as Salyut 1 had been unveiled in 1971, paintings by Leonov and Sokolov were published showing two such Salyuts docked nose-to-nose with a multiple docking unit; two Soyuz ferries docked radially are shown with this complex. From the design of Salyut, it was clear that with few exterior changes, the space station could be adapted as a basic module for a larger space station structure.

As the Salyut programme progressed, further depictions of modular space stations appeared. Another painting, dating

from the time of the launch of Salyut 4, showed three Salyut modules docked at 120° intervals around a central core unit. A final depiction using the basic Salyut design indicated that two second generation Salyuts could be docked nose-to-nose, with manned ferries, or Progress freighters, docked at the rear port of each Salyut.

One of the most interesting paintings showed a Salyut-shaped module without solar panels, with multiple docking ports available. When symmetry is considered, the central module would apparently have four radial docking ports as well as a front longitudinal port; in addition, there might also have been a rear docking port. Modules were shown docked with the core station, each module being of the same general size as the core station itself. Would this be the basis of the next Salyut?

In 1985 the cosmonaut Rukavishnikov talked about the future of the Soviet manned space programme, and he described Salyut 8 as being a modular station, with the modules being replaceable. At the time, the life of Salyut 7 was believed to be coming to a close, and the

Left: *One of a number of Soviet paintings which have been published showing how the final assembly of the Mir complex will appear during 1990-1992.*

launch of Salyut 8 was expected in 1986. In fact, when the launch of the new station came it was named Mir, which has been variously translated as "peace" (this is the normal translation supplied by the Soviet Union), "world" or "commune". Political considerations aside, it seems probable that "commune" might be the most appropriate translation.

The Mir Core Station

The Mir orbital station core was based upon the shell of the second generation Salyut, and the Soviets described it as a third generation station. Its shape was that of two cylinders and a connecting frustrum, with a spherical module at the front end.

The overall length of the Mir core was said to be 13.13m and the maximum diameter 4.15m. The Soviets described the station as having four sections: hermetic transfer compartment; hermetic working compartment; hermetic transfer adapter; non-hermetic service compartment.

The transfer compartment was spherical in shape, with a diameter of 2.2m. One can speculate that this might be a modification of the Soyuz orbital module. Including the frustrum which connected the sphere to the work compartment, the length of the transfer compartment was 2.84m. There were five docking units on the transfer compartment, each with a diameter of 0.8m. One docking unit was at the front of the station, and the other four radiated from the station at 90° intervals. The design of the five ports were different. Three of the ports incorporated the usual conical drogue docking system, while the other two were simply flat docking units. Outside the transfer compartment were sited the rendezvous antennas, on-board lights and the docking targets. In addition there were manipulator sockets for the Liappa module transfer system.

Initially, spacecraft docked with the Mir core using the front longitudinal ports, but a system has been devised which will allow the modules to be transferred to the radial ports. The Mir transfer compartment is provided with two sockets, and modules which dock at the front of Mir will carry short "arms". After the initial docking and checkout, an arm will be extended from the module to one of the sockets, and after the arm has been secured the module will undock from the front port and be swung around to one of the other ports.

The work compartment comprised the two main cylinders which form the general shape of the Mir core. The first part of the work compartment was a cylinder 2.9m in diameter, with the transfer compartment at one end. A connecting frustrum into the rear work area was 4.15m in diameter; the total length of the work compartment was 7.67m. The work compartment was basically devoid of major scientific equipment, and was intended as the living and central control area. The station's manual control systems were located here, and the wider

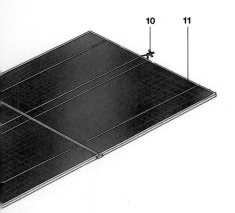

19 Docking target.
20 Docking hatch.
21 Transfer module.
22 Treadmill exerciser.
23 Attitude control thrusters (32).
24 Propellant tanks.
25 Work and dining table.
26 Exercise bicycle (velo-ergometer).
27 Station control consoles, provide access to the eight control computers.
28 Observation window.
29 Multiple docking adapter, one aft docking port and four axial ports.
30 Axial docking port.
31 Socket for attachment of remote manipulator arm on docking modules.
32 EVA handrails.
33 Observation window.

The Modular Space Station (right)
1 Experiment module.
2 Progress re-supply craft.
3 Mir.
4 Experiment module.
5 Two Kvant-like experiment modules.
6 Soyuz-TM craft.
7 Cosmos-type experiment module.
The diagram shows how modules can be attached to the core Mir station to allow the Soviet Union to build an impressively large and capable laboratory in space. Fundamental to the design is the multiple

docking adapter, which permits four modules to be docked axially to Mir, while a fifth port can accommodate a Soyuz. Each experiment module first docks on the aft port; a remote manipulator arm then attaches to a socket on the Mir hub, and after checkout, the arm then swings round the socket through 90° and re-docks the module on one of the vacant side docking ports.

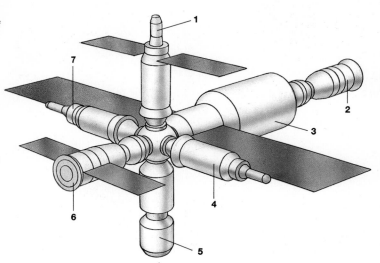

part of the compartment housed two small bays which the resident cosmonauts could use for sleeping. Outside the work compartment were the station's temperature control system radiators, as well as two large solar panels.

The service propulsion compartment was a stubby cylinder mounted at the back of the large-diameter area of the work compartment. It was 2.26m in length and 4.15m in diameter. It contained the propulsion unit (including both the main engines and the orientation engines), rendezvous antennas, lights, docking targets and a radio communications antenna.

Within the service propulsion compartment was a transfer adapter, 1.67m long and 2m in diameter. This allowed spacecraft to dock at the rear of the core station.

The orbital control system on Mir comprised both attitude control thrusters and the main propulsion system. The main system, used for actual orbit altitude corrections comprised two engines, each with a thrust of 300kg. The attitude control system included 32 thrusters, each having a thrust of 14kg. Four propellant tanks were carried, and – having learned lessons from Salyut 6 and Salyut 7 – the membrane in each tank was now metallic and therefore not susceptible to disintegration. Additionally, there were balloons for the pressurant gas.

Two large solar panels were carried outside Mir, these arrays having a total span of 29.73m and a surface area of 38m² each.

At the rear of the station was a large antenna which was used for space-space-ground communications (i.e. from Mir to a Cosmos satellite to ground control). When Mir was launched, it used the Cosmos 1700 satellite which was located over 95°E as a relay; the satellite system was announced as being Loutch. However, the nearest announced Loutch location was 90°E, and the Cosmos 1700 location had been announced for the Statsionar-14 or the SDRN-B satellites. The reason for this anomaly has not been clarified – perhaps the Loutch location had simply been shifted?

A new rendezvous and docking system was carried on Mir – Kurs. The older Igla system required the mutual search and manoeuvring of the two spacecraft (the orbital station manoeuvring in attitude, but not in altitude), something which would be difficult to control with an asymmetric Mir complex sporting radially-docked modules. The new Kurs system allows the approaching spacecraft to track the Mir complex but only the approaching spacecraft manoeuvres in attitude; the complex maintains its attitude and altitude. The rear port of Mir carries both the Kurs and Igla systems, allowing the approach of Progress freighters which will carry Igla or the approach of modules or Soyuz-TM craft using Kurs.

The launch mass of the Mir core station was stated by Glushko to be 20,900kg.

New Spacecraft for Mir

Since Mir was described as the command core for a modular space station, a series of modules was expected to be launched to rendezvous with the station. However, development of the modules progressed at a slow rate and the launches were therefore delayed.

The flight of Soyuz-T 15 was announced as the final flight of the Soyuz-T spacecraft, and observers of the Soviet programme wondered what the replacement craft would be – especially since there had been no tests within the unmanned Cosmos programme that might have been indicative of a new spacecraft under testing. The answer came in May 1986 when the first Soyuz-TM spacecraft was launched: the "M" indicated "modified".

Outwardly, the Soyuz-TM appeared to be almost identical with the Soyuz-T, and it was configured for the launch of either two or three cosmonauts. The launch mass of the first Soyuz-TM (it was unnumbered by the Soviet Union) was 7,070kg, some 220kg more than the standard Soyuz-T mass (6,850kg). The descent module of the Soyuz-TM had a mass of 3,000kg, and was equipped with new primary and reserve parachutes.

The spacecraft carried modernised power supply systems and an improved altimeter for the solid-propellant retro-rocket system carried by the descent module. The cosmonaut seats were described as "unified", and space had been set aside for equipment. While the Soyuz-T spacecraft could return 50kg of cargo and experimental results, the Soyuz-TM could manage 150kg; additionally, 200kg more payload could be carried into orbit. The propulsion system of the Soyuz-TM was an improved version of that used on Soyuz-T, once more having metallic tank separators (as with the Mir propulsion system).

The dimensions of the Soyuz-TM were the same as those for Soyuz-T: length (with the docking probe withdrawn), 6.98m, maximum diameter, 2.72m and the span of the solar cells, 10.6m. The launch vehicle was described as improved, allowing a heavier payload to be orbited.

It appears that the Progress spacecraft have been modified for the Mir missions. Earlier Progress vehicles had masses of about 7,000kg, but those launched towards Mir had masses of 7,150-7,200kg, according to cumulative payload masses announced by Glushko. The amount of cargo which could be carried was

Below: *A spectacular view of Mir in orbit, with the Kvant module docked at the rear of Mir, and Soyuz-TM 3 docked at the rear of Kvant. The picture was taken from the returning Soyuz-TM 2.*

increased to about 2.5 tonnes (rather than 2.3 tonnes), although how this was split between propellant and other cargoes was not stated.

It was expected that the Heavy Cosmos modules would be used as the basis of the modules to be launched to Mir, although the first module to be added to the station was actually of a new design. In 1986, before a module was launched to Mir, photographs of the Zvezdny Gorodok training centre showed that a stubby module had been added to the rear of the Mir mock-up. In shape and size, this seemed to be derived from the Proton 4 satellite which had been launched in 1968, the Proton satellite having the same general size as the main work module on the Salyut 1 and Salyut 4 stations. The new module was launched in 1987 and named Kvant (Quantum).

The Kvant spacecraft was in two sections: a main experimental module and a service module which would be separated in orbit. At launch the mass of Kvant was 20.6 tonnes. The main experimental module was 5.8m long with a maximum diameter of 4.15m (matching that of Mir). The module when docked with Mir would add a further 40m³ volume to the work area of the complex; its mass was 11 tonnes. The service module was not described in detail, but some Soviet sketches suggested that it might be built around a shortened variant of the Heavy Cosmos main module, having solar panels and a domed end. Kvant carried a probe docking system at one end (where it docked with the rear of Mir) and a drogue docking unit at the other end (to allow spacecraft with probes to continue docking at the rear of the complex). Presumably, the service module carried a probe docking system, to attach with the main Kvant module.

The service module apparently carried the same external propellant tanks and

engine system as the Heavy Cosmos modules. Based upon the performance of the first Kvant, the propellant load must be at least 800kg, implying a dry mass of the service module of about 8.8 tonnes.

Some pictures showing the development of the Mir complex over the years were released during 1986-1987, and these showed that normally Soyuz-TM and Progress craft would be docked at the longitudinal ports of the core station. The rear port sometimes had a Kvant module attached, sometimes not. The modules on the radial ports appeared to be derived from the Heavy Cosmos, although the solar cells were redesigned to be similar to the larger Mir style. The possible use of the Earth return spacecraft on the Heavy Cosmos variants could not be clearly deduced from the pictures. Taking the known Soyuz-TM, Progress, Mir core, Heavy Cosmos and Kvant masses, the total mass of the Mir complex could probably reach 125 tonnes.

The Launch of Mir

Early in the morning (Moscow Time) of 20 February 1986 the Soviet Union launched the Mir core station using a standard three-stage Proton SL-13 vehicle.

The launch marked a totally new openness in the Soviet space programme, as it was shown on television, providing the West with its first clear view of this launch vehicle variant (the first clear Proton photographs to be released were those of the two SL-12 variants which had launched the two VEGA – Venus-Halley – space probes in December 1984). Unusually in the light of such openness of information, the initial orbit of the station was apparently not announced.

On the day of launch, Konstantin Feoktistov, the former cosmonaut who was a senior spacecraft designer, appeared on Moscow Television giving a press briefing about the new station which featured fully detailed cutaway representations of the station. The improved facilities on the station were stressed, as well as the introduction of the six docking ports. Mir was actually described as:

". . . representing a base module for assembling a multipurpose permanently operating manned complex with specialised orbital modules for scientific and national economic purposes."

When it was launched, the orbital plane of Mir was close to that of the Salyut 7/Cosmos 1686 orbital complex, and there was initial speculation that Mir might dock with that complex – at least as an interim modular space station assembly exercise. However, this was not to be, although the first mission to Mir would also visit the Salyut complex.

Four days after the launch of Mir, it was stated that following a series of manoeuvres its orbit was then 51.6°, 324-352km; this was typical of a civilian Salyut orbital station.

Soyuz-T 15: First Operations

The new openness in the Soviet manned space programme continued on 12 March, when it was announced that the launch vehicle for a new manned mission had been taken to the pad, and the launch was scheduled for the following day at 15.33 Moscow Time (12.33 GMT). The crew of the Soyuz-T 15 spacecraft would be Col L. D. Kizim and V. A. Solovyov, who were still paired after first coming together when they were the back-up crew for the Franco-Soviet mission in 1982. They were chosen for the Mir mission in November 1985.

The Soyuz-T 15 flight had two main objectives. The crew had trained primarily for Salyut and had not specialised with the Mir systems. Kizim and Solovyov were to activate Mir and check out its essential systems, and after conducting this work, they would use their Soyuz-T ferry to manoeuvre to and dock with the Salyut 7 complex, completing the programme which had only partially been carried out by the Soyuz-T 14 crew (who had returned to Earth prematurely in November 1985 following the commander's illness). This done, they would then return to Mir for some final experiments and tests.

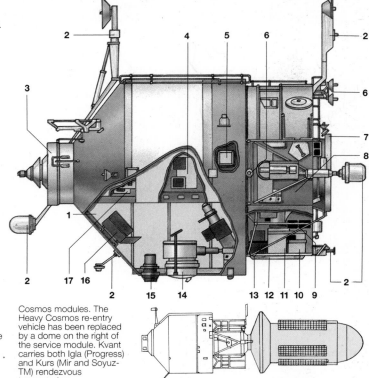

The Kvant Astrophysics Module
1 Laboratory compartment.
2 Antenna for the Igla rendezvous system.
3 Docking probe.
4 Mir crew supplies and equipment inside Kvant.
5 Svetlana electrophoresis installation.
6 Antennas for the Kurs rendezvous system.
7 Drogue docking unit (to allow further spacecraft to dock with Mir/Kvant).
8 Research instrument compartment.
9 X-ray telescope.
10 Siren-2 spectrometer.
11 Glazar ultra-violet telescope.
12 TTM X-ray telescope.
13 Pulsar X-1 X-ray telescope.
14 Optical sight.
15 Astro-orientation equipment.
16 Central control panel.
17 Life support system equipment.

Kvant plus Service Module
The Kvant module (left) is seen attached to its service module (right). The service module has not been fully described by the Soviets, but it is clearly a stripped-down variant of the Heavy Cosmos modules. The Heavy Cosmos re-entry vehicle has been replaced by a dome on the right of the service module. Kvant carries both Igla (Progress) and Kurs (Mir and Soyuz-TM) rendezvous transponders.

The Soviets admitted later in the year that there were problems in getting the intended Mir modules ready for flight, and therefore one can assume that they would not have wanted to commit a fully-trained Mir crew to a mission which would only involve the basic station. An experienced Salyut 7 team of men was therefore flown as an interim crew.

The launch of Soyuz-T 15 came at the announced time of 12.33 on 13 March, the event being shown "live" on Moscow television. This was only the fourth time that a Soviet launch had been shown "live", the previous such missions having been the Soviet half of ASTP (Soyuz 19), the Franco-Soviet Soyuz-T 6, and the Indo-Soviet Soyuz-T 11 missions. With the Soyuz-T launch, it was stated that future modules intended for docking with Mir would be for "nature (Earth resources, presumably), technological, biological and astrophysical studies".

Soyuz-T 15 took two days to reach Mir, docking on 15 March at 13.38. The approach using the Igla system brought Soyuz-T to within 200m of the rear port of Mir, and the cosmonauts then flew the craft around to the front of the station, docking with the front longitudinal port. The connecting hatches opened and the two cosmonauts floated into their new home.

Preparing for Work

During the first few days, the cosmonauts were hard at work de-mothballing the station. It was not realised in the West that the Mir flight would use the standard Salyut/Soyuz/Progress profile, especially while the scheduled modules that would be added to the station were delayed. The first indication came on 19 March when Progress 25 was launched. The docking came following a standard two-day approach, on 21 March at 11.16. It was announced that the freighter was carrying propellant, food, water, "apparatus and equipment necessary for the long work of the Mir station". Unloading of the freighter began the day following docking, with about two tonnes of supplies to be off-loaded; specifically, 200 litres of water was initially identified as being carried.

On 29 March the cosmonauts used – apparently for the first time – the satellite Cosmos 1700 for communications with the mission controllers. The transmissions from Mir were relayed to the satellite nearly 36,000km above the equator, and in turn they were relayed by the satellite to Moscow control. Three days later the cosmonauts completed a Resonance experiment, testing the rigidity of the new Mir complex with spacecraft docked at each end. During early April the cosmonauts were engaged in the assembly of instruments on board Mir, as well as conducting a routine exercise programme and various Earth observation experiments. By this time, the cosmonauts had settled into a regular work routine, having two days rest following five days work.

On 20 April it was noted that Progress 25 had been used to manoeuvre the Mir complex on a number of occasions and was left with the minimum amount of propellant

required for de-orbit manoeuvres. On that day at 19.24 Progress undocked from Mir and at an unannounced time the same day the freighter was de-orbited.

The cosmonauts were not to be without a freighter for long; Progress 26 was launched 23 April, with docking coming on 26 April at 21.26 – unusually, after a three-day rendezvous. Meanwhile, the cosmonauts had been using the Gamma equipment to check Solovyov's cardiovascular system and tested the gas analyser which checked the carbon dioxide level in the station's atmosphere. Once more, more than 2 tonnes of supplies equipment were carried, described as "foodstuffs, drinking water, extra equipment and apparatus, sanitary requisites, medical apparatus" and propellant. In fact, 200kg of food was carried and 300 litres of water.

Work with Progress would not last for long, because it was announced on 3 May that two days later the cosmonauts would be leaving Mir for a visit to the Salyut

7/Cosmos 1686 complex. They had begun to mothball Mir, and it was not clear whether or not they would be returning to the station. In connection with the Salyut visit, Progress 26 had been loaded with 500kg of materials which would be transferred to Salyut. Of course, with Cosmos 1686 occupying the front port of Salyut and the Soyuz-T the rear port, no Progress could supply the cosmonauts when they reached Salyut; all their supplies had to be taken with them. The undocking from Mir came on 5 May at 12.12 and the older complex was reached the following day. Details of the work undertaken on Salyut 7 were given in the previous chapter.

While it was unmanned, work continued with the Mir complex. On 22 May

Below: *Leonid Kizim aboard the Mir station during the Soyuz-T 15 visit. The station was launched with little scientific instrumentation and this was later supplied by Progress missions.*

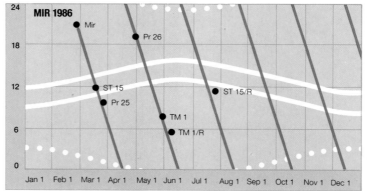

Launch and Landing Windows
Mir was launched close to the time required for a mission to Salyut 7, thus allowing a crew transfer between the stations. Early hopes in the West that Mir would actually dock with Salyut were erroneous. Soyuz-T 15 completed a visit to Mir and then the crew transferred to Salyut 7, to complete experiments scheduled for Soyuz-T 14. This done, they returned to Mir for three weeks with equipment from Salyut 7. Soyuz-TM 1 was an unmanned test flight of the ferry.

LAUNCHES IN SUPPORT OF THE SOYUZ-T 15 RESIDENCY				
LAUNCH DATE AND TIME (G.M.T.)	**RE-ENTRY DATE AND TIME** (G.M.T.)	**SPACECRAFT**	**MASS** (kg)	**CREW**
1986 19 Feb 21.19	In Orbit	Mir	20,900	–
13 May 12.33	16 Jul 12.34	Soyuz-T 15	6,850?	L. D. Kizim, V. A. Solovyov
19 Mar 10.08	20 Apr	Progress 25	7,150?	–
23 Apr 19.40	23 Jun 18.40*	Progress 26	7,150?	–
21 May 08.22	30 May 06.49	Soyuz-TM	7,070	

Notes: This table includes all of the launches during 1986. The masses of the later Progress missions (following 1984) are increased from the average of 7,020kg of the earlier missions according to cumulative payload masses published by Glushko. Glushko has also quoted the actual masses of the Mir core station and the first Soyuz-TM.

the launch of the unmanned Soyuz-TM was announced (the Soviets did not number the spacecraft), and it was described as a modernised version of the Soyuz-T vehicle. The new ferry docked at the front longitudinal port of Mir on 23 May at 10.12. The Soyuz-TM carried improved computer systems, presumably modified because of (apparently) the regular failing of the Soyuz-T system during the final approach to Salyuts (resulting in manual control being used for the final docking manoeuvre).

Soyuz-TM remained docked with the Mir complex until 29 May at 09.23 and it landed at an unspecified time the following day (the landing time quoted in the table is estimated, and should be correct to ±1 minute). The flight was pronounced a success, and it was announced the next flight would be as the operational manned ferry.

Work continued with the Mir/Progress 26 complex, and in June the propellant tanks in Mir were automatically topped up with supplies brought by the freighter. On 22 June at 18.25 Progress 26 undocked from Mir and the freighter was de-orbited the following day.

On 25 June Soyuz-T 15 with Kizim and Solovyov undocked from Salyut 7 to begin the journey back to the Mir station; docking with their earlier base came on 26 June at 19.46. The cosmonauts had brought back some 20 instruments from Salyut with a mass of 350kg. Once more, the cosmonauts settled down to their routine aboard an orbital station, using the Svetoblok unit for biological experiments and exercising on the veloergometer and the closed-circuit running track. On 2 July the cosmonauts finished installing a new experimental Strela computer-based information system which they tested thoroughly. However, their mission was coming to an end. They completed their experiments and mothballed the Mir station.

The new openness in news coverage of the Soviet space programme meant that the undocking of the Soyuz-T 15 from Mir and the landing were to be seen "live" in the Soviet Union. On 16 July at 09.07 the manned ferry undocked from the station, and the television reporter counted off the landing events as they took place: orbital module separation at 10.23, braking retro-fire for three-and-a-half minutes, separation of the descent and instrument modules at 12.05, entry into the atmosphere at 12.11 and finally the landing at 12.34. The lighting conditions at the landing site meant that Soyuz-T 15 had to complete a landing while passing southbound over the Soviet Union – normally the landings have taken place when passing northbound (Soyuz 1 was the only previous southbound landing).

Soyuz-TM 2: 11 Months in Orbit

Mir was left to itself for the rest of the year, without even being manoeuvred following the return of Soyuz-T 15. The Soviets announced that the delays in the Mir module production meant that no more manned flights were scheduled for 1986.

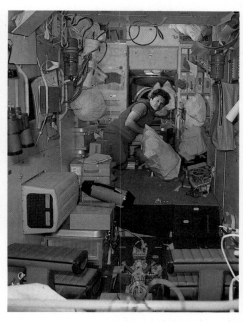

Above and right: *Leonid Kizim at work aboard the Mir core station. The above picture is looking towards the multiple docking adapter, while the picture at right looks towards the aft docking port.*

MANOEUVRES IN SUPPORT OF THE SOYUZ-T 15 RESIDENCY

PRE MANOEUVRE ORBIT		POST MANOEUVRE ORBIT		SALYUT 7 ORBIT	
Epoch (1986)	Altitude (km)	Epoch (1986)	Altitude (km)	Epoch (1986)	Altitude (km)
Initial Orbit		19.95 Feb	171-302	19.91 Feb	347-349
21.12 Feb	172-294	21.93 Feb	170-335		
21.93 Feb	170-335	22.05 Feb	210-335	22.07 Feb	346-349
22.05 Feb	210-335	22.18 Feb	270-340		
22.19 Feb	270-348	22.31 Feb	324-340		
6.89 Mar	322-337	7.08 Mar	333-342	7.08 Mar	345-348
Soyuz-T 15: docked 15 March					
		16.01 Mar	330-341		
Progress 25: docked 21 March					
		22.28 Mar	333-338		
24.94 Mar	332-339	25.64 Mar	338-358	26.05 Mar	340-345
7.97 Apr	337-358	8.54 Apr	339-360	8.93 Apr	338-344
16.93 Apr	338-360	17.24 Apr	336-343	17.94 Apr	337-344
Progress 25: undocked 20 April					
		22.69 Apr	335-343		
Progress 26: docked 26 April					
		27.89 Apr	335-345		
3.91 May	335-345	4.67 May	309-345	4.86 May	335-343
Soyuz-T 15: undocked 5 May					
		5.94 May	310-345	5.93 May	335-343
8.84 May	309-345	10.61 May	331-344		
Soyuz-TM 1: docked 23 May					
		24.75 May	331-342		
24.75 May	331-342	25.44 May	335-343		
25.44 May	335-343	25.63 May	340-355		
25.63 May	340-355	25.69 May	341-359	26.84 May	334-341
28.93 May	340-359	29.31 May	334-340		
Soyuz-TM 1: undocked 29 May					
		29.94 May	334-341		
17.88 Jun	333-339	18.96 Jun	332-335	17.82 Jun	333-339
Progress 26: undocked 22 June					
		22.88 Jun	332-335		
24.78 Jun	332-337	25.80 Jun	332-366		
Soyuz-T 15: redocked 26 June					
		26.88 Jun	332-366	26.88 Jun	333-338
Soyuz-T 15: undocked 16 July					
		16.94 Jul	329-365	16.95 Jul	332-338

Notes: This table is similar in format to those presented for the earlier Salyut missions: however, because Soyuz-T 15 visited both Mir and Salyut 7, comparative Salyut orbits are given in this table with further Salyut data given in the equivalent table in the previous chapter.

As 1986 drew to a close, the Soviet authorities hinted that a new manned mission would be launched to Mir at the beginning of 1987, and that this would represent the start of the permanent manning of the station. It was expected that the mission would last for about ten months. A launch in January would result in a recovery during late October after a ten month flight. However, the Soviet Union, like most of Europe, was suffering from heavy snow falls in early 1987, and apparently the manned launch was delayed slightly.

In early 1987 the Cosmos 1700 data relay communications satellite drifted off-station, and the Soviets indicated that there had been some difficulties in communications via a geosynchronous orbit satellite. On 30 January Cosmos 1817 was launched, and this was identified as a failed geosynchronous orbit communications satellite; it is unclear whether this was to replace Cosmos 1700 or part of the normal Statsionar communications satellite programme. Mir would operate without a geosynchronous data relay link for most of 1987.

In order not to delay the Mir programme too much, as the weather conditions improved slightly, Progress 27 was launched on 16 January. Possibly it was being launched on time. Two days later at 07.27 Progress docked at the rear port of Mir.

On 28 January it was announced that a new crew was being readied for launch to Mir; the crew was named as the veteran commander Romanenko and rookie flight engineer A. I. Leveykin. It was only a year later that it became clear from a press release issued by Novosti in London that this was not the original prime crew for the mission. Commenting on the manned launch in December 1987 it was said that:

"Vladimir Titov was in training for his third mission one year ago and . . . Serebrov was to be his partner. However, Yuri Romanenko and Alexandr Leveykin were destined to make that flight."

It appears that Serebrov had failed the crew medical three weeks before the mission, and therefore the back-up crew was required to fly.

The impending launch of the new manned space mission was announced the day before the event, and Moscow Television featured live coverage of the cosmonauts preparing for the launch, their final pre-launch press conference and the launch itself. Soyuz-TM 2 was launched on 5 February, and a two-day journey to Mir ensued. The Soviets indicated that the longer approach profile was more economical on propellant than the previous 25-hour docking process. On 7 February at 22.28 Soyuz-TM 2 docked at the front longitudinal port of Mir; the permanent Soviet presence in space had begun.

After entering Mir the cosmonauts' first job was to begin de-mothballing the station. They transferred the temperature control to a new "operational mode", switched on the water regeneration system, checked the performance of equipment and started medical experiments. On 10 February the unloading of Progress 27 began. Three days later the de-mothballing of Mir was complete. The cosmonauts had a rest day on 14 February and it was noted that Leveykin was needing his rest badly because his adaptation to weightlessness had been "rather painful".

The following day, preparations began for the transfer of propellant from Progress 27 to Mir, while 16 February saw the cosmonauts undergoing a major medical check-up using the Gamma-1 system. The unloading of Progress was completed on 19 February, and the freighter's mission drew to an end. Four days later at 11.29 Progress undocked from Mir and on 25 February it was de-orbited.

Meanwhile, on 24 February the cosmonauts checked the suits which would be used for spacewalks, ensuring that the pressurisation and communications systems were in order. A further three days later, the cosmonauts started their study of the processes of heat and mass transfer in liquids while in weightlessness, using the Pion-M apparatus.

Above: *Romanenko and Leveykin (rear) prepare for the launch of Soyuz-TM 2 to Mir. Leveykin would be in orbit for six months, while Romanenko spent 326 days in orbit — a new record.*

Left: *The night-time launch of Soyuz-TM 2 signalled the beginning of permanent occupation of space by the Soviets; starting with this mission, Mir is scheduled to be permanently manned.*

MANOEUVRES BY THE KVANT SERVICE MODULE			
PRE MANOEUVRE ORBIT		**POST MANOEUVRE ORBIT**	
Epoch (1987)	Altitude (km)	Epoch (1987)	Altitude (km)
Initial orbit		31.06 Mar	168-278
31.06 Mar	168-278	31.12 Mar	172-300
31.99 Mar	169-296	1.17 Apr	172-314
1.54 Apr	170-313	2.23 Apr	297-345
2.29 Apr	298-344		
Docking failure with Mir: 5 April			
		5.38 Apr	345-364
8.88 Apr	345-364		
Docked with Mir: 9 April			
		10.22 Apr	344-364
Service module undocked from Mir/Kvant: 12 April			
		12.83 Apr	344-364
12.83 Apr	344-364	13.21 Apr	341-363
19.62 Apr	340-361	20.14 Apr	383-406

Notes: Kvant and its service module were launched together. After docking with Mir, the service module separated and manoeuvred to a higher orbit.

On 3 March Progress 28 was launched towards Mir, and docking came two days later at 11.43. Progress seems to have been the first freighter to carry a heavy load of experimental equipment. Its cargo included the Korund unit, which was described as a semi-industrial unit for growing crystals in weightlessness, and a wide-lens photographic camera, KATE-140, which would be used for geophysical research. A further experiment named Kolosok was designed to find out how aerosols and hydrosols behave in weightlessness.

After Progress had been unloaded and propellant transferred to Mir, the freighter was undocked on 26 March at 05.07; two days later it was de-orbited. The rear port clear for the time being, the cosmonauts continued their programme of medical exercises and experiments. The new Korund and Pion-M materials processing units had been undergoing successful tests. However, preparations were underway for another visitor.

Kvant is Launched

Shortly after midnight (GMT) on 31 March, a Proton SL-13 booster was launched from Tyuratam carrying the first scientific module for Mir. Called Kvant, it was the module which had been seen in photographs taken at Zvezdny Gorodok during 1986. The launch mass of the two-part spacecraft (main experiment module and service module) was 20.6 tonnes. The main module had a mass of 11 tonnes, and it carried 1.5 tonnes of scientific instruments and more than 2.5 tonnes of equipment for Mir. Kvant was carrying a series of international astrophysical experiments. Outside the main module and surrounding the transfer tunnel at its rear was a truss system, on which were mounted the major experiments. There was an international X-ray observatory called Rentgen, with a mass of 800kg; this included the Pulsar X-1 hard X-ray telescope/spectrometer, a Fosvich high-energy scintillation telescope/spectrometer, a telescope with an aperture mask and a Siren-2 proportional gas scintillation spectrometer. This set of equipment had been developed in co-operation with the United Kingdom, The Netherlands, West Germany and the European Space Agency. The outer scientific bay also carried the Glazar telescope and the Svetlana automated electrophoresis unit for experiments in biotechnology.

The initial orbit of Kvant was announced as 51.6°, 177-320km, and a docking with Mir was planned for 5 April. After manoeuvring in orbit, early on 5 April Kvant approached the rear port of Mir. Apparently, all was going well up to a point 200m from the manned complex, but the Kvant thrusters failed to slow down the module and it flew past Mir. The Soviet controllers had to decide whether they could attempt a second docking and salvage what was a major international mission, or whether the flight had to be abandoned. While they were pondering the problem, the cosmonauts on Mir reverted to their regular routine. The decision was made to try a second docking with Kvant, and the cosmonauts readied Mir for this on

9 April, the orbit of Mir being adjusted slightly to assist in the docking. On 9 April at 00.36 a soft docking was made between Kvant and Mir, although a hard docking could not be achieved. This mission was not yet saved. It was later stated that the probe on the Kvant docking unit had penetrated 365mm, but had then got stuck.

The only way to discover what exactly had gone wrong was for Romanenko and Leveykin to perform an unscheduled spacewalk, and try to sort out the problem. On 11 April at 19.41 one of the hatches on the front transfer compartment opened and the cosmonauts climbed out of the station. It was commented later that they had difficulties in doing this, because the docking hatches were only just large enough to allow the cosmonauts in full spacesuits to pass through. A later module was scheduled for Mir which would carry a special EVA hatch for easier access. Prior to the spacewalk, V. Titov and another cosmo-

naut (Serebrov or Manarov?) had duplicated the planned operations in the hydrotank facility in the training centre.

The cosmonauts climbed to the rear of the station and ground controllers extended the docking probe on Kvant to allow a detailed inspection of the system. The cosmonauts could see that a white sheet was fouling the docking system, and that this was preventing the completion of the hard docking; the sheet had been loaded on to Progress 28, but not secured and when Progress undocked the sheet was pulled out and remained attached to the Mir docking unit. The cosmonauts were able to pull the obstruction free, and this done the controllers were able to complete the hard docking between Kvant and Mir. The first module was successfully docked with Mir. After spending 3h 40m outside their orbiting home, Romanenko and Leveykin returned inside.

In order to free the rear port of the Kvant module, the service module was undocked on 12 April at 20.18 and a small phasing manoeuvre was completed to distance it from the manned complex. The Soviets had planned to de-orbit the module after it

Left: *The Yuri Gagarin tracking ship between periods "on station". The Soviets have a fleet of dedicated space-tracking ships, allowing the gaps in Soviet-based tracking facilities to be filled.*

Below: *Vladimir Solovyov (left) and Leonid Kizim on their return to Earth. They were the first crew to transfer from one orbital station to another.*

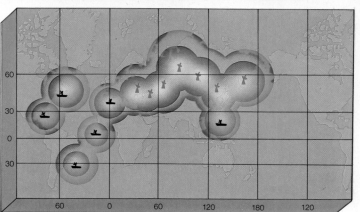

The Tracking Network
The Soviet tracking and communications network for spacecraft is based upon ground-based stations and ships at sea. Cosmonauts are in direct contact with controllers whenever they pass over the Soviet Union (the distorted circles give the zone visible from each station) and ships at sea communicate with controllers via satellites. The ship in the South Atlantic is positioned to monitor retro-fire when a spacecraft is returning to Earth. The Soviets are also planning to establish geosynchronous satellites for total coverage.

had delivered the main Kvant module to Mir, but the extra manoeuvres involved in the second docking attempt had left the service module short of propellant. Therefore, instead of trying the de-orbit manoeuvre, the service module was manoeuvred to a higher orbit during 19-20 April, taking it away from the orbital path of the Mir complex.

The cosmonauts first entered Kvant on 13 April and began the inspection and testing of its equipment. Fresh equipment carried inside Kvant was transferred to Mir for installation.

A new freighter, Progress 29, was launched towards the complex on 21 April, and two days later at 17.05 it docked at the rear port of Kvant. For the first time, the Soviets were operating a space laboratory consisting of four separate spacecraft. Almost immediately, the cosmonauts began work unloading their new visitor. As well as normal supplies, Progress was carrying 138.5kg of scientific equipment and 275kg of "replacement equipment" for the station's systems. The cosmonauts experienced problems in unloading Progress, because Kvant still housed a third solar panel which should have been erected before Progress 29 was launched; this operation had been delayed because of the problems in docking Kvant and Mir.

The transfer of propellant from Progress was a different operation this time, because it had to be pumped through pipes running through Kvant into the Mir tanks. Kvant had no main propulsion system of its own, and it obstructed the two main Mir engines. Therefore, Mir itself could only perform attitude control operations, and any orbital altitude manoeuvres had to be performed by either the Soyuz-TM or Progress craft.

After completing its operations, on 11 May at 03.10 Progress 29 was undocked from Kvant and de-orbited at an unspecified time the same day. The rear port of the complex did not remain empty for long, however. On 19 May Progress 30 was launched, and it docked on 21 May at 05.53. The cosmonauts quickly began the unloading operations, as well as continuing with the testing of Kvant equipment and their own experiments and exercises. A new system of stabilising the complex based upon gyrodenes, was being used; this was installed in the Kvant module, and elements of the system were designed to be easily replaced by the cosmonauts. The first operational work with the Kvant module was carried out 9 June when the supernova which had been discovered in the Greater Magellanic Cloud (a satellite galaxy of our Milky Way system) in February 1987 was put under scrutiny. Later a neutron star in the constellation of Cygnus was observed using the Kvant equipment. During 12 June the cosmonauts also prepared for their spacewalk which would involve the installation of new solar arrays outside Mir. At 16.55 on that day the cosmonauts climbed out of Mir to begin the assembly of the two-part solar cell system. Each part consisted of an extensible girder and two sections of solar cells. A unit had been installed on Mir prior to its launch in the position where the third solar panel had been on the earlier Salyuts; this clearly indicates that some assembly work was planned at some time.

After leaving the station, the cosmonauts removed the packages of solar cells and erected the first part of the system. The operation was televised to mission controllers, and the spacewalk lasted for 1h 53m. A second spacewalk lasting for 3h 15m began on 16 June to assemble the second section of the new solar array. When completed, the new assembly stood 10.5m high and had a surface area of 22m^2. This would greatly assist in remedying Mir's electrical power shortage.

Observing the Routine

Once more, the crew returned to their routine of exercise and experimental work. On 23 June they completed the final electrical connections between Mir's new solar arrays and the station's power supply "grid". They replaced some systems of the Strela computer in order to expand the computer's capabilities to monitor Mir's systems. An experiment named Biostoykost began, this using polymers to make test space equipment, and the Yantar installation was used which involved the depositing of thin metal coatings on materials. Additionally, the cosmonauts were busy conducting observations with the equipment carried by Kvant. On 9 July Romanenko and Leveykin set up the new

Below: *In June 1987 Leveykin and Romanenko performed a spacewalk, during which they erected a new set of solar panels on Mir, thus easing electrical power shortage problems.*

Kristallizator unit, as a further experimental apparatus for materials processing work. An orbiting greenhouse named Fiton had been set up in Mir, and the growth of onions and other plants was being monitored.

The undocking of Progress 30 came on 19 July at 00.20 and it was de-orbited later the same day. The first visiting mission to the Mir complex was about to begin. An agreement had been reached that the Soviet Union would launch a Syrian cosmonaut to an orbiting station, and in October 1985 two Syrian trainee-cosmonauts had arrived at Zvezdny Gorodok to begin training: Mohammad Al Faris and Munir Habib. The two crews were later announced for the flight: the prime crew for the mission would be Lt-Col A. S. Viktorenko (commander), A. P. Alexandrov (flight engineer) and M. A. Faris, while the back-ups were Lt-Col A. Y. Solovyov, V. P. Savinykh and M. Habib. Initially, it was planned that the three-man crew would remain intact, but this was not to be.

The launch of Soyuz-TM 3 came early in the morning of 22 July, and the now-standard two-day approach was made to the Mir complex. The docking at the rear of Kvant came on 24 July at 03.31. For the first time, Mir was hosting two cosmonaut crews. A full programme of joint experiments was planned for the Syrian visit: these included the Apamea experiment, using the Kristallizator to produce crystals of gallum antimonide, and the Kasyun experiment using the same equipment which would study the effects of microgravity on the structure of an eutectic alloy of aluminium and nickel.

The health of Leveykin had been giving cause for concern, with minor heart problems being noted. Although these did not endanger the flight, it was decided that for safety reasons he should be rotated out of the Mir residency and his place taken by Alexandrov. Alexandrov was re-trained for his role change for about a month prior to the launch of Soyuz-TM 3.

On 25 July Romanenko acted as cameraman while the Palmyra experiment was conducted; this used seven syringes and it was planned to combine two substances and observe the formation of crystals in the weightless environment. Euphrates was the name given to a series of Earth resources observations conducted while the complex was passing over Syria. Bosra was an experiment to study the upper atmosphere and the ionosphere. The Ruchy experiment involved cleaning by electrophoresis several batches of interferon and an anti-influenza preparation.

Medical experiments were conducted by the cosmonauts on themselves, the three launched on Soyuz-TM 3 undertaking the Ballisto experiment, which measured the functioning of the cardiovascular system in orbit.

Cosmonauts Viktorenko, Leveykin and Faris transferred to the Soyuz-TM 2 spacecraft on 29 July, the spacecraft undocking at 20.34. Landing within the Soviet Union came the following day at 01.04. Leveykin was the last man to be removed from the descent module, and he was described as being pale. Whatever his problems, they

Above: *Romanenko aboard Mir with the EVA suit preparing for a spacewalk. This suit was based upon those used aboard Salyut 6 and Salyut 7, but was of an improved design.*

did not prove serious, and he was soon fully recovered from his space mission.

The resident crew on Mir was now Romanenko and Alexandrov, and they had a duty to perform almost immediately. The cosmonauts transferred to Soyuz-TM 3 and undocked it from the rear of Kvant on 30 July at 23.28. The Mir complex was then rotated through 180° and at 23.48 Soyuz-TM 3 redocked at the front longitudinal port of Mir, leaving the rear port of Kvant free for visiting Progress craft.

More Supplies

Progress 31 was launched on 3 August and on 5 August at 22.28 it docked at the rear port of Kvant. Almost immediately, the cosmonauts began to unload the freighter. A full load of supplies, including water, propellant, food and mail was awaiting them. The cosmonauts continued work in the normal routine established for Salyut and Mir residencies. On 31 August an exercise was announced, to check whether the cosmonauts were able to cope with unforeseen problems on Mir, when they were instructed without prior notice to put on their pressure suits, transfer to the Soyuz-TM and prepare it for return to Earth.

Progress 31 was undocked from Mir on 21 September at 23.58 and early on 23 September it was de-orbited. Already, a new freighter was ready for launch, and later that day Progress 32 left the launch pad bringing regular supplies to the Mir complex. The docking came on 26 September at 01.08, and the new freighter was said to be carrying 850kg of propellant, 315kg of food, while the remainder of the 2 tonnes of supplies comprised oxygen, replacement equipment, scientific equipment, camera film, research equipment and mail. It was noted during the approach of Progress 32, that the solar cells erected by Romanenko and Leveykin were bending and shaking, and the ground controllers wondered whether the impact of a spacecraft docking might dislodge the assembly.

Despite these worries, the solar panel assembly continued to function without any problems.

On 30 September Romanenko exceeded the duration record which had been set by the Soyuz-T 10 cosmonauts in 1984, although, of course, to claim the new record he had to spend at least another 24 days in orbit (adding a further 10 per cent). In fact, he would do far more than this. As the mission progressed, Romanenko felt more and more tired, and the working day was shortened to allow him more rest time.

The unloading of Progress 32 continued, and on 6 October it was noted that a series of experiments had begun using the Svetoblok-T device, this studying the synthesis of polyacrylamide gel for obtaining biologically active compounds. Additionally, the crew began to use the new EFO-1 electronic photometer to study the dust layer which is about 100km above the Earth. Three days later a series of experiments were initiated using the Biryuza installation, these relating to the study of the dynamics of physio-chemical processes in the conditions of microgravity. In a comment relating to long duration missions, the science correspondent of Radio Moscow, Boris Belitskiy, noted that approximately 10kg of food, water and oxygen was consumed by each cosmonaut every day, in addition to the propellant which the station required. Clearly, regular flights of the Progress freighters would have to continue, and as larger crews were stationed in orbit this problem would increase.

A unique experiment was conducted with Progress 32 on 10 November. After it had been packed with garbage from the cosmonauts, it undocked from the Mir complex at 04.09 and retreated to a distance of 2.5km. It then re-approached the complex, docking again with Kvant at 05.47. The Soviets indicated that this test was conducted to prove a new control algorithm for spacecraft manoeuvres, which was designed to reduce propellant consumption during dockings. One may also speculate that the test was connected with Mir being used to service free-flying experimental modules, which would only need to be docked with the station occasionally for equipment to be replaced. The cosmonauts continued with Progress operations until 17 November. At 19.25 it undocked from Kvant and early the following morning it was de-orbited.

The final Progress mission of 1987 was launched on 20 November, with Progress 33 docking with Kvant on 23 November at 01.39. Compared with the three previous missions, the visit of Progress 33 would be a short one – only a month. On 26 November the cosmonauts began to assemble the Mariya astrophysical equipment which had been carried by Progress; it was used to monitor streams of highly-charged particles. A further piece of equipment carried by Progress was an all-new solar-powered mirror beam furnace which would be used to study the formation of crystals in the weightless environment.

The flight of Romanenko and Alexandrov was now drawing to a close, with the landing window opening in the last few days of 1987. On 6 December the cosmonauts began to increase their exercise regime, with two-and-a-half hours each day being devoted to physical exercises.

Soyuz-TM 4: A New Resident Crew

A new data relay satellite was launched on 26 November: Cosmos 1897 was placed into a geosynchronous orbit over 95°E, thus allowing continuous communications with the Mir complex once more. As Romanenko and Alexandrov continued with their work in orbit, on 9 December the deputy head of the Zvezdny Gorodok cosmonaut training centre, Aleksei Leonov, introduced the next Mir cosmonauts to journalists. Two three-man crews were announced: the prime crew comprised Col V. G. Titov, M. K. Manarov and A. S. Levchenko (a Merited Test Pilot of the Soviet Union), while the back-ups were Col A. A. Volkov, A. Y. Kaleri and A. V. Shchukin (Test Pilot, First Class). The two commanders had completed previous space missions, while the other four men were rookies.

The naming of the prime crew confounded observers of the Soviet space programme, because it had been widely reported that the next crew to fly to Mir would be Titov, Serebrov and Valeriy Poliakov, the latter having been the

back-up doctor to Atkov on the Soyuz-T 10 mission. Serebrov was not being flown and Poliakov was removed because the flight of Levchenko was required urgently. Along with Volk – who flew on Soyuz-T 12 – probably Shchukin and other unknown (at the end of 1987) cosmonauts, Levchenko was being trained for space shuttle operations, and the shuttle planners wanted to ensure that the shuttle cosmonauts would not have any problems in adapting to weightlessness. In a way, the shuttle cosmonauts were being medically "debugged" on short Soyuz-TM missions to ensure that they would not be subject to space sickness upon entering orbit. When Soyuz-TM 3 returned to Earth with Romanenko and Alexandrov, Levchenko would accompany them. Interestingly, when Romanenko and Alexandrov encountered Levchenko in orbit, it was the first time that they had met; this indicates that the shuttle trainee-cosmonauts are working away from Zvezdny Gorodok, and only go there as they are assigned to short Soyuz-TM missions.

Before the new launch, it was stated that the next Mir crew would be in orbit for about a year; prior to that, there had been reports that the mission would last for about 400 days. While Romanenko and Alexandrov continued to prepare for their

Above: *The three cosmonauts who returned aboard Soyuz-TM 3 in December 1987. Alexandrov (left) had been in orbit for 160 days, Romanenko (centre) for 326 days and Levchenko for 8 days.*

Right: *Soyuz-TM 3 on its side after more than five months in space. The six nozzles in the base are the solid-propellant retro-rockets, used to cushion the touchdown on solid ground.*

return to Earth, on 21 December at 11.18 Soyuz-TM 4 was launched into orbit carrying Titov, Manarov and Levchenko. A two-day approach was completed and Titov reported that they could see the Mir complex at a distance of 35km. The cosmonauts were soon only 1km from the complex and approaching at 4m/s. At 11.28 the Soyuz-TM was close to the complex and began to fly-around to dock at the rear of Kvant on 23 December at 12.51.

Having checked the integrity of the docking seal, at 14.20 the three new cosmonauts opened the connecting hatch and floated into the Kvant/Mir complex, to begin a period of joint work with the resident crew. Romanenko and Alexandrov took time to show Titov and Manarov the operations aboard the station.

The return to Earth of Romanenko, Alexandrov and Levchenko was scheduled for 29 December. The undocking of Soyuz-TM 3 from the front port of Mir came at 05.55. At 08.23 the retro-fire burn was completed and landing came at 09.16. This was the first time that a spacecraft was recovered carrying three cosmonauts, all of whom had been launched on separate spacecraft. Full attention was given to the return of Romanenko, who had established a new duration record of more than 326 days. There were serious worries about his health before the landing, but he was able to sit upright inside the rescue helicopter and he was walking when he left the helicopter. Confounding the experts, he completed a 100m run the day after landing. His clean bill of health opened the way to even longer missions, including a manned flight to Mars, possibly in the next decade.

Meanwhile, Titov and Manarov were settling into the routine of spaceflight. On 30 December at 09.10 Soyuz-TM 4 was undocked from the rear of Kvant and at

MIR MANOEUVRES IN SUPPORT OF THE SOYUZ-TM 2, SOYUZ-TM 3 AND SOYUZ-TM 4 MISSIONS							
PRE MANOEUVRE ORBIT		**POST MANOEUVRE ORBIT**		**PRE MANOEUVRE ORBIT**		**POST MANOEUVRE ORBIT**	
Epoch (1987)	Altitude (km)	Epoch (1987)	Altitude (km)	Epoch (1987)	Altitude (km)	Epoch (1987)	Altitude (km)
Progress 27: docked 18 January				Progress 30: undocked 19 July			
		20.89 Jan	315-344			20.84 Jul	312-360
24.87 Jan	315-344	26.63 Jan	329-363	Soyuz-TM 3: docked 24 July			
Soyuz-TM 2: docked 7 February						24.95 Jul	311-359
		8.91 Feb	328-363	Soyuz-TM 2: undocked 29 July			
9.86 Feb	328-363	10.94 Feb	342-368			30.08 Jul	309-360
19.90 Feb	339-368	21.87 Feb	346-370	Soyuz-TM 3: redocked 30 July			
Progress 27: undocked 23 February						31.92 Jul	310-359
		23.78 Feb	345-370	Progress 31: docked 5 August			
Progress 28: docked 5 March						6.04 Aug	310-360
		5.90 Mar	344-369	17.06 Sep	306-354	17.50 Sep	298-358
8.89 Mar	344-369	9.59 Mar	354-369	Progress 31: undocked 22 September			
Progress 28: undocked 26 March						22.56 Sep	295-357
		26.59 Mar	349-368	Progress 32: docked 26 September			
Kvant: docking failure 5 April						26.47 Sep	295-356
		5.83 Apr	350-366	Progress 32: redocked 10 November			
5.83 Apr	350-366	6.15 Apr	344-364			10.85 Nov	288-344
Kvant: docked 9 April				15.83 Nov	287-342	16.21 Nov	326-344
		10.22 Apr	344-364	Progress 32: undocked 17 November			
Kvant Service Module: undocked 12 April						17.92 Nov	327-343
		13.33 Apr	344-364	Progress 33: docked 22 November			
Progress 29: docked 23 April						22.87 Nov	326-343
		23.76 Apr	343-363	12.54 Dec	323-340	12.79 Dec	334-360
27.83 Apr	343-364	28.85 Apr	344-365	Progress 33: undocked 19 December			
Progress 29: undocked 11 May						19.90 Dec	334-359
		11.31 May	343-365	Soyuz-TM 4: docked 23 December			
Progress 30: docked 21 May						23.78 Dec	334-359
		21.36 May	343-366	Soyuz-TM 3: undocked 29 December			
12.86 Jun	342-366	13.31 Jun	342-367	Soyuz-TM 4: redocked 30 December			
8.80 Jul	341-365	9.56 Jul	312-361			30.88 Dec	335-358

Notes: All the major orbital manoeuvres completed by the Mir complex are noted above. However, there seem to have been additional manoeuvres every few days to adjust the orbit; these generally involved shifting the orbit by only 1-2km and they are on the borderline of what can be detected using the Two-Line Orbital Elements.

expendables, with a view to placing orders for the next Progress freighter mission. They serviced the Elektron life support system, checked the computer units, continued research with the Mariya equipment and prepared the Korund 136kg materials processing unit. On 9 January the cosmonauts prepared for an emergency evacuation drill; Moscow Radio noted that such drills were monthly events.

The new Progress 34 freighter was launched on 20 January (Moscow Time, 21 January) and this docked at the rear of the Kvant module on 23 January at 00.39. The new spacecraft was carrying propellant for the Mir complex, food, water, "equipment and apparatus" (to quote Moscow Radio) and mail for the cosmonauts. In total, more than two tonnes of freight was delivered. The two cosmonauts spent much of the next few days both unloading the Progress, and working with the Kvant astrophysical observatory.

In mid-February propellant was transferred from Progress to Mir. During this period, the cosmonauts prepared for their first spacewalk. On 22 February they checked their spacesuits, and during the next day transferred the equipment for the spacewalk to the airlock which they would use. On 24 February the suits were further tested and, after a day of rest, on 26 February at 09.30, a hatch opened and the cosmonauts floated into space to begin what was scheduled to be a spacewalk lasting for 4h 22m. They moved their equipment outside and went over to the third solar array which had been assembled by Romanenko and Leveykin in 1987. They folded the lower part of the panel assembly, removed one of the two sections and installed a replacement; not only was this intended to replace faulty equipment, but the new section had an improved design. This work done, the two men removed some samples from outside Mir for later return to Earth. The spacewalk lasted for 4h 25m.

With this work complete, most of the next day was spent storing the suits and other equipment which had been used for the spacewalk. Progress 34 was gradually being filled with rubbish and on 4 March at 03.40 the freighter undocked from Mir and was de-orbited later the same day. The orbital routine continued with the launch of a new freighter on 23 March and the docking of Progress 35 at the rear of Kvant two days later at 22.22. The standard load of supplies was carried, with 400kg of food which included fresh vegetables and fruit. Progress remained attached to Mir until 5 May at 01.36, and it was de-orbited later the same day. The freighter was quickly replaced by Progress 36, launched on 13 May with docking on 15 May at 02.13. In addition to the normal supplies, Progress was carrying experiments to support the first international mission of 1988.

The Bulgarian Mission

In 1979 the Soviet-Bulgarian Soyuz 33 spacecraft had failed to dock with Salyut 6, making Bulgaria the only Intercosmos nation not to boast a man with experience on an orbital station. Because of this, the Soviets offered Bulgaria the opportunity of

09.29 the cosmonauts redocked it at the front of Mir. Their long stay in space was just beginning.

Titov and Manarov quickly settled into their planned long stay aboard the Mir complex, with their routine of medical checks, photography, materials processing and experimental work being undertaken in much the same way that the long Soyuz-TM 2 residency had operated. At the beginning of 1988, the cosmonauts drew up an inventory of the station's

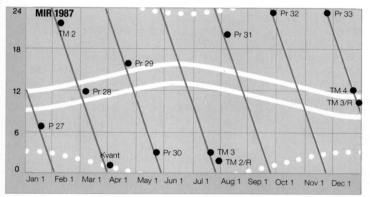

Launch and Landing Windows
The launch of Soyuz-TM 2 towards Mir was probably delayed for nearly a month because of the bad weather at Tyuratam. The routine of regular Progress launches was quickly established, with Kvant being docked at the back of Mir before the Progress 29 launch. The Soyuz-TM 3 visit – with a Syrian cosmonaut – showed that the Soviets are no longer restricted to the previously normal landing constraints. Soyuz-TM 4 placed a new resident crew aboard Mir in December 1987.

LAUNCHES IN SUPPORT OF THE SOYUZ-TM 2, SOYUZ-TM 3 AND SOYUZ-TM 4 MISSIONS

LAUNCH DATE AND TIME (G.M.T.)	RE-ENTRY DATE AND TIME (G.M.T.)	SPACECRAFT	MASS (kg)	CREW
1987 16 Jan 06.06	25 Feb 15.17*	Progress 27	7,150 ?	–
5 Feb 21.38	30 Jul 01.04	Soyuz-TM 2	7,100 ?	Y. V. Romanenko, A. I. Leveykin
3 Mar 11.14	28 Mar 02.59*	Progress 28	7,150 ?	–
31 Mar 00.06	In orbit	Kvant	20,600	–
21 Apr 15.14	11 May	Progress 29	7,150 ?	–
19 May 04.02	19 Jul	Progress 30	7,150 ?	–
22 Jul 01.59	29 Dec 09.16	Soyuz-TM 3	7,100 ?	A. S. Viktorenko, A. P. Alexandrov, M. A. Faris
3 Aug 20.44	23 Sep 00.22*	Progress 31	7,150 ?	–
23 Sep 23.44	18 Nov 00.10*	Progress 32	7,150 ?	–
20 Nov 23.47	19 Dec	Progress 33	7,150 ?	–
21 Dec 10.18	17 Jun 10.13	Soyuz-TM 4	7,100 ?	V. G. Titov, M. K. Manarov, A. S. Levchenko

Notes: The comment "In orbit" shows the spacecraft which were still in orbit on 1 July 1988. There were many crew switches between spacecraft. Leveykin returned in Soyuz-TM 2 after 174d 15h 48m along with Viktorenko and Faris (launched in Soyuz-TM 3) who spent 7d 23h 5m in orbit. Levchenko returned in Soyuz-TM 3 after 7d 22h 58m, with Alexandrov (launched in Soyuz-TM 3) after 160d 7h 17m and Romanenko (launched in Soyuz-TM 2) after 326d 11h 38m – the latter being a new manned duration record. Soyuz-TM 3 was the first spacecraft to return with three cosmonauts, all of whom had been launched on different missions. Titov and Manarov were still in orbit on 1 July 1988. Soyuz-TM 4 was recovered in 1988.

a second manned mission. The protocol was signed with the Soviet Glavcosmos organisation on 28 August 1986 and a ten day mission was scheduled. At the beginning of 1987 two cosmonaut-candidates arrived in the Soviet Union to begin training for the flight. The prime mission candidate was A. P. Alexandrov (not to be confused with the *Soviet* cosmonaut who flew Soyuz-T 9 and Soyuz-TM 3, also called Alexander Alexandrov), who had been the back-up for the Soyuz 33 mission (he had been one of the most highly-rated trainees of the Intercosmos candidates). His back-up was K. Stoyanov.

The initial mission details were announced in December 1987. The prime crew would be A. Y. Solovyov (commander), V.P. Savinykh (flight engineer) and Alexandrov, with the back-ups being V. A. Lyakhov, A. Zaitsev and Stoyanov. Launch was scheduled for 21 June 1988. A few months later it was announced that the launch had been brought forward to 7 June because the later launch would have allowed moonlight to interfere with the planned astronomical observations. Additionally, it was noted that A. A. Serebrov was now the back-up flight engineer, with no explanation as to why he replaced Zaitsev.

In preparation for the launch, the Mir complex was manoeuvred (by Progress 36?) to a higher orbit – the highest altitude which it attained in the first half of 1988 – on 1-2 June. Progress was separated on 5 June at 11.12 and was de-orbited at 20.28.

In the full blaze of publicity, with live television coverage available to any country interested in taking it, Soyuz-TM 5 was launched on 7 June at 14.03 with Solovyov, Savinykh and Alexandrov on board. A successful docking came two days later at 15.57 and soon afterwards a Bulgarian finally entered a Soviet orbital station. The eight days spent on Mir were devoted to

Above and below: *The return of the second Bulgarian space mission. Alexandrov (left) was the second Bulgarian in orbit, and he was joined by Solovyov (centre) and Savinykh. Soyuz-TM 4 heads for a landing (below).*

Left: *The second Franco-Soviet manned mission scheduled for the end of 1988 or perhaps early 1989 will be the second spaceflight for Jean-Loup Chretien (left). His back-up for the month-long flight is Michel Tognini.*

the completion of 46 experiments. It was planned that a Soyuz switch would take place, and accordingly the international crew undocked from Mir in Soyuz-TM 4 on 17 June at 06.18, returning to Earth at 10.13. The next day at 10.11 Soyuz-TM 5 with Titov and Manarov undocked from the rear of Kvant, re-docking at the front of Mir at 10.27.

Two further international missions were scheduled for 1988. Following the success of the Soyuz-T 6 mission in 1982, a second flight was offered to the French. Originally the French asked for a flight lasting for two months, rather than another week-long visit, but they eventually accepted a flight lasting for about 38 days. The agreement for the mission was signed on 7 March 1986, and the flight was scheduled for late 1988. In November 1986 the veteran from Soyuz-T 6, J.-L. Chretien, was selected, backed-up by M. Tognini, and soon afterwards they began training. The high point of the flight – probably to be Soyuz-TM 7 – will be the completion of a spacewalk by Chretien, the first non-Soviet, non-American spacewalk.

Because of the EVA work, it is anticipated that the promised Mir service module will be required for the mission because this apparently will include special EVA airlocks. With the delay in the launching of Mir modules, the launch date of the French mission is uncertain: in mid-1988 it was planned for 21 November, but this might slip. As of June 1988 the full Soviet crews have not been announced, but Moscow Radio has indicated that the commander will be A. A. Volkov, backed up by A. S. Viktorenko. Chretien has said that he expects to be launched with a resident crew for Mir. The flight engineers have not been named at the time of writing.

A purely political guest cosmonaut flight was agreed in July 1987, involving a citizen of Afghanistan. Originally the flight was scheduled for mid-1989, but (probably because of the political situation as the Soviet troops prepared to pull out of the country) the announcement in February 1988 that the cosmonaut-candidates had arrived for training said that the flight was scheduled for August *1988*. The two Afghan trainees would only be spending six months at Zvezdny Gorodok before the launch! This is hardly enough time to provide a familiarisation with the basic spacecraft systems, and it has been noted that even the monkeys for biological Cosmos flights are trained for a year.

The two candidates were named as M. D.-G. Masum and A. A.-M. Sarvar. After some months training, the crew assignments were announced: the prime crew would be V. A. Lyakhov (commander), V. Poliakov (physician) and Masum, backed-up by Berezovoi, G. Arzamazov (physician) and Sarvar. The launch date was set for 29 August and a standard eight day visit was scheduled. Interestingly, the

crews seem to have been put together from fragments of other assignments. Lyakhov was still involved with his back-up assignments for Soyuz-TM 5 with the Bulgarian cosmonaut (possibly his assignment to the prime Afghan crew was the reason that Serebrov, an experienced cosmonaut, replaced the inexperienced Zaitsev in the Bulgarian back-up crew?), while both he and Berezovoi – together with Malyshev, Viktorenko and Volkov – had trained for rescue missions to Mir. Additionally, Poliakov – and presumably Arzamazov – was trained for Soyuz-TM 4, but was removed in favour of a shuttle pilot. Therefore, the four Soviets assigned to the mission were already trained for Mir missions at short notice.

Little scientific work is expected from the Afghan visit. It has been stated that Lyakhov and Masum will return to Earth after eight days, leaving Poliakov in orbit with Titov and Manarov. This is to allow a trained physician to be available for medical monitoring as the flight continues beyond twelve months (400 days might be the target). Additionally, he will undertake the more routine duties of Mir management, leaving Titov and Manarov more time to complete the scientific part of the mission – work which had been falling behind because of a lack of time.

If the Afghan Soyuz-TM 6 mission involves a Soyuz switch, then this will allow the French launch to slip by two months to January 1989, should there be problems with the launch of the Mir service module.

The Soviets have said that these three international missions will end the "free rides" of guest cosmonauts; and all future missions will be on a commercial basis. Already, Austria is planning a commercial manned flight in 1992; selection of mission candidates will lead to training for the flight beginning in 1990. An approach has

been made to Indonesia for a similar mission, and other invitations of commercial involvement are anticipated.

In addition to these flights, one can expect all-Soviet visiting crews to the Mir complex, with more than 2-3 men being permanently on the station as modules are attached and the work-load increases. One can also foresee the Soviet equivalent of "payload specialists" being carried for specific experiments; possibly there will be professional geologists and astronomers visiting the complex before it is abandoned.

Essential for the continued operations of Mir are the flights of the dedicated modules. Unfortunately, the dates for these launches are unclear, with virtually every Soviet spokesman giving his own schedule for the launches. Logically, module launches will come immediately after an old Soyuz-TM has been traded for a new one, the new vehicle remaining at the back of Mir. The module would initially dock at the front port and then be transferred to one of the radial ports, at which point the new Soyuz-TM could transfer to the free front port. An essential addition would seem to be the service module, which would include not only extra work and living space, but also a specialist EVA airlock system. An Earth resources module is planned, and this might be based upon the Kvant design, because the Kvant mock-up displayed at the Paris Air Show in 1987 included the instrument (although the 1987-launched Kvant did not carry it).

A leaflet – also available at the 1987 Paris Air Show – discussed the possibility of flying commercial experiments to the Mir complex, and this revealed that such experiments could be mounted or conducted either inside or outside a module based upon the Heavy Cosmos. Such commercial work will result in the Soviets offering to

fly a commercial cosmonaut as a standard option.

Two more dedicated modules are planned for launches towards the Mir complex: a materials processing module and a biomedical module. The Soviets apparently hope for the full complement of modules to be operating in about 1990.

Closing Comments

As this book went to press, the Mir programme was still evolving, and the timetable which is indicated above will be subject to change. However, from Soviet statements, it is clear that the Mir complex is expected to operate into 1992 at least.

Mir will probably be used as the final testing ground for cosmonauts before a final decision is made as to whether to fly men to the Mars system during the 1990s. Following the eleven months flight of Yuri Romanenko to Mir during 1987, the Soviet Union sees no problems in principle which would prevent such a flight. However, the flights to Mir will continue to get longer, with a 360-400 day mission planned for the Soyuz-TM 4 cosmonauts. If no problems arise, the duration will continue to be extended.

One can look upon the Mir complex as the ultimate variant of the late 1960s Salyut orbital station design, and Soyuz-TM may possibly represent the final version of the Soyuz manned spacecraft. Following Mir will come orbital stations based upon modules with masses of 100 tonnes or more – the announced Mir-2 programme which is discussed in the next chapter.

FURTHER READING

It is too soon in early 1988 for detailed books to appear about Mir, but as usual *Spaceflight* magazine has carried regular mission reports written by Neville Kidger.

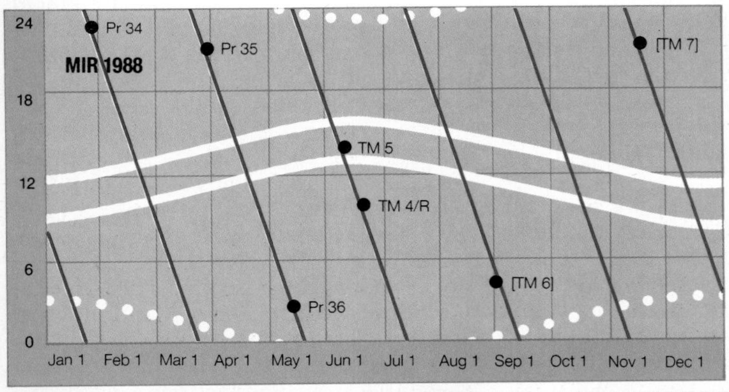

MIR 1988

Launch and Landing Windows
Three manned launches are scheduled for Mir during 1988. The first was Soyuz-TM 5, with a Bulgarian cosmonaut, launched during June with Soyuz-TM 4 being used for the visiting crew's return. A mission with an Afghan cosmonaut (Soyuz-TM 6) is scheduled for August, and this will be followed in November-December by the next 2-man resident crew for Mir accompanied by a French cosmonaut (Soyuz-TM 7). The graph shows the schedule as of June 1988.

LAUNCHES IN SUPPORT OF THE SOYUZ-TM 4 RESIDENCY

LAUNCH DATE AND TIME (G.M.T.)	RE-ENTRY DATE AND TIME (G.M.T.)	SPACECRAFT	MASS (kg)	CREW
1988 21 Jan 22.52	4 Mar	Progress 34	7,150 ?	—
23 Mar 21.05	5 May	Progress 35	7,150 ?	—
13 May 04.30	5 Jun 20.28*	Progress 36	7,150 ?	—
7 Jun 14.03	In orbit	Soyuz-TM 5	7,100 ?	A. Y. Solovyov, V. P. Savinykh, A. P. Alexandrov
29 Aug		Soyuz-TM 6	7,100 ?	V. A. Lyakhov, V. Poliakov, M. D.-G. Masum
21 Nov ?		Soyuz-TM 7	7,100 ?	A. A. Volkov?, Engineer, J.-L. Chretien

Notes: As of mid-June 1988, Soyuz-TM 5 was docked with the Mir/Kvant orbital complex. The launch date and time for Soyuz-TM 6 (with an Afghan cosmonaut) were announced in advance as above. Soyuz-TM 7 is the mission with a French "spationaut", scheduled for November 1988; the full crew has not been announced, as of June 1988, but Moscow Radio announced the names of the prime and back-up commanders. Volkov and his engineer will be the next resident crew to man the Mir complex. The three cosmonauts launched in Soyuz-TM 5 returned to Earth in Soyuz-TM 4 after 9d 20h 10m in orbit.

MIR MANOEUVRES IN SUPPORT OF THE SOYUZ-TM 4 MISSION

PRE MANOEUVRE ORBIT		POST MANOEUVRE ORBIT	
Epoch (1988)	Altitude (km)	Epoch (1988)	Altitude (km)
Progress 34: docked 23 January			
		23.05 Jan	334-356
Progress 34: undocked 4 March			
		4.83 Mar	328-346
Progress 35: docked 25 March			
		26.29 Mar	326-341
22.36 Apr	318-333	22.48 Apr	330-362
Progress 35: undocked 5 May			
		5.24 May	330-360
Progress 36: docked 15 May			
		15.39 May	330-358
1.83 Jun	326-355	2.78 Jun	348-356
Progress 36: undocked 5 June			
		5.71 Jun	349-356
Soyuz-TM 5: docked 9 June			
		9.84 Jun	349-355
Soyuz-TM 4: undocked 17 June			
		17 Jun	347-353
Soyuz-TM 5: redocked 18 June			
		18 Jun	347-353

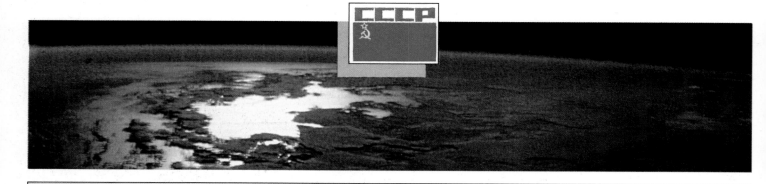

Chapter 14: **The Future of Soviet Manned Spaceflight**

Trying to predict the future of the Soviet space programme is, to say the very least, a difficult task. Any plans are subject to alteration as technology dictates, as priorities change and as budgets are reviewed. During the 1960s the Soviet Union and the United States were in a "race" to land the first man on the Moon. In July 1969 the American Apollo 11 lunar module landed Neil Armstrong and Edwin Aldrin on the Sea of Tranquillity while the Soviet cosmonauts could only watch. Within six years the last Apollo had flown and its accumulated technology thrown away. The American public interest in manned lunar missions had dwindled to virtually zero and the emphasis was to be on the development of a reusable manned Space Shuttle as the spaceflight budget was cut back to the bare minimum required to sustain a space programme.

The Soviet Union's plans (as detailed earlier in this book) for a manned lunar landing were cancelled because of the multiple failures of the giant SL-15 booster. The proposed manned lunar spacecraft (Soyuz) was adapted to serve as a ferry to Earth orbiting space stations in the Salyut and later Mir programmes, with Soyuz-T and Soyuz-TM variants appearing as time passed. The lunar landing and trans-lunar injection stages of the manned lunar complex vehicle have continued flying as the second and third stages of the Proton booster. Therefore, to a great extent the Soviet Union retained the technology of its manned lunar programme.

During the 1970s the Soviet goal seemed to be the development of a permanently manned space laboratory based upon Salyut technology (which itself might originally have been part of a manned lunar programme). However, statements now available which Soviet space planners and politicians made in the mid-1970s show that a major rethink of the space programme took place at that time. Improving the reliability of space systems became a priority and the ultimate goal of the space programme became the economic exploitation of space.

The Salyut programme would continue to expand with improved versions of the Soyuz ferries, but it would push forward the manned duration records in space; the Soviets wanted to find both the optimum duration for space station residencies and how the human body withstood the rigours of really long missions. This work

is continuing today with the Mir space station complex.

Although the Soviet Union has not managed to put a man on the Moon, a programme basically using modifications of existing or soon-to-be-developed standard hardware was instigated with the goal of landing a man on Mars. There was initially no deadline for the mission, although it was perhaps hoped that men could visit the Martian system during the 1990s.

The boldest decision to be made by the Soviet policy-makers during the mid-1970s was to exploit space for the benefit of the Soviet Union's economy. It had long been known that new materials could be processed in space and that space-borne photography was essential for the exploitation and monitoring of the Earth's natural resources, but the Soviet plan went far beyond this. If statements made by Soviet scientists can be believed, the Soviet Union hopes to begin a programme to exploit solar power for home consumption before the end of the twentieth century.

In order even to consider the latter programme, the Soviet Union has had to develop a completely new launch vehicle system which will be fully reusable, economic and capable of lifting heavy payloads into Earth orbit. This is the Soviet space shuttle system which will be flown as either an unmanned heavy lift launch vehicle or with a large shuttle orbiter carrying a crew.

This chapter will detail the various space-shuttle related tests which took place during the 1970s and 1980s (to the end of 1987), first involving sub-scale payloads and then with the new launch vehicle family. The future of the Soviet space station programme will then be discussed, followed by a study of the requirements for the projected manned mission to Mars. It will conclude with the projected space power programmes.

The Soviet Space Shuttle

Almost as soon as the United States began to develop the Space Shuttle, rumours began that the Soviet Union was planning a manned reusable spacecraft which would rival it. According to the various reports, this could be a small spaceplane carried atop a throwaway Proton booster, or a configuration like the American Shuttle, or a fully reusable manned first and second stage system (like

the original American Space Shuttle designs). One can assume that the Soviet designers looked at the American plans with great interest, but the Soviet effort lagged some years behind the Americans.

The first flights which are generally credited as being connected with the Soviet shuttle programme are four launches, each involving two Cosmos satellites (see table). The first flight was the pair Cosmos 881-882 and it is reported by western sources that both payloads re-entered within one orbit. They could have been de-orbited towards destruction, but it is generally assumed that they were actually recovered inside the Soviet Union. The launch was from Tyuratam and used the Proton booster, possibly in the SL-13 three-stage variant. When the Soviet Union announced the launch, they gave the orbital parameters as 51.6°, 202-248km; significantly, they announced no orbital period, an omission which paralleled the launch announcements of the FOBS (Fractional Orbit Bombardment System) test flights in the late 1960s and early 1970s where the payloads had re-entered after less than one orbit. Additionally, the launch announcements stated that the satellites had completed their missions — a statement which had previously been applied regularly to the active weapons in the anti-satellite programme.

Two similar flights followed: Cosmos 997-998 in 1978 (51.6°, 200-230km announced) and Cosmos 1100-1101 (51.6°, 199-230km announced). In addition, there have been western rumours of a launch failure in August 1977, although this has not been confirmed by the Soviet Union.

At the time of these missions, western analysts were carefully watching the Cosmos programme for possible tests of a shuttle vehicle, and these pairs of Cosmos flights seemed to fit such a test programme. They were launched by the Proton — the largest booster then operating in the Soviet programme — and apparently they were being recovered. Almost by default (no-one could think of any other mission which they could plausibly be completing), these flights were identified as the long sought-after Soviet space shuttle flights, the Soviet shuttle being akin to the American Dyna-Soar research vehicle which was abandoned in the early 1960s.

Perhaps the ultimate in such speculation was published in an article by Anthony Lawton, who called the config-

Above: *The spectacular night launch of the first Energiya shuttle booster on 15 May 1987. The large towers on either side* *of the vehicle are about 200 metres tall and might carry cables between them to prevent electrical discharges at the* *launch complex. This launch pad is some 20km away from the two pads which will be used for operational Energiya flights.*

The Cosmos Spaceplane
This is a drawing of the small Cosmos spaceplane model which made four flights during 1982-1984. It was launched from the Kapustin Yar site using the smallest booster currently in use (left) – the SL-8. It is possible that if shuttle hardware was being tested, then the same spaceplane was launched at least twice in the series.

LAUNCHES IN THE DUAL COSMOS PROGRAMME					
LAUNCH DATE AND TIME (G.M.T.)	**COSMOS SATELLITES**	**OBJECT**	**INCL** (deg)	**PERIOD** (min)	**ALTITUDE** (km)
1976 15 Dec 01.29	881	A	51.6	88.91	200-242
	882	B	51.6	88.53	190-214
	Rocket	C	51.6	88.83	197-237
	Fragments	D-F			
1977 4 Aug			Rumoured launch failure		
1978 30 Mar 00.00	997	A	No orbit available		
	998	B	No orbit available		
	Rocket	C	51.6	88.51	189-213
	Fragment	D			
1979 22 May 23.00	1100	A	51.6	88.66	193-223
	1101	B	No orbit available		
	Rocket	C	51.6	88.62	182-231
	Fragments	D-E			

Notes: The launch times quoted above are calculated from the Two-Line Orbital Elements, and should be correct to ±1 minute. The orbits are also derived from the Two-Lines, although those for objects A and B (the payloads) from each launch are normally for epochs after the presumed re-entry of the payloads, and therefore they may represent some fragment objects which are mis-identified. The presence of fragments is noted, but orbital data are not given for the fragments. Apart from the rumoured date of launch, there is no further information available for the launch failure in 1977.

uration "Kosmolyot 1" ("Kosmolyot" is supposedly the generic name that the Soviets use for their space shuttle). This had a manned spaceplane with a throwaway orbital tank launched atop a four-stage Proton which Lawton calls "SL-13/Mod-X"; the six external tanks forming the Proton first stage each had a fin attached to aid stability.

The evidence for this configuration being the rationale for the dual Cosmos missions is zero, and such analysis shows how easy it is to build an argument or launch vehicle programme on the flimsiest of evidence. It is now believed that the Proton booster was probably impossible to man-rate because of vibrations induced by the first stage (the second and third stages were man-rated since they formed part of the manned lunar programme), and this was known by the Soviets in the late 1960s. Therefore, if the dual Cosmos missions were shuttle-related then they cannot have formed part of a manned configuration; they could only have been research vehicles.

One can hypothesise over the number of flights in the programme, assuming that there was a launch failure in 1977. If shuttle-related hardware was being tested, then a prime requirement would be to fly the hardware more than once. Therefore, the launch failure could have been a second launch attempt of Cosmos 881-882, and as the payloads were lost on the failed mission, another pair of payloads had to be flown twice (Cosmos 997-998 then Cosmos 1100-1101) to prove the technology. It is interesting to note that the intervals between the Cosmos 881/882 launch and the presumed launch failure (232 days), and the presumed launch failure and Cosmos 997/998 (238 days) are almost identical. However, the interval between Cosmos 997/998 and Cosmos 1100/1101 was almost twice this value (418 days). The reason for the apparently connected intervals is unknown – especially since the payloads might well have been lost in the launch failure.

Following the dual Cosmos programme, there was no activity in the supposed Soviet shuttle programme, as far as the flight record was concerned, until 1982. On 3 June Cosmos 1374 was launched from the Kapustin Yar launch site and it was reportedly recovered in the Indian Ocean, about 560km south of the Cocos Islands, some 109 minutes after its launch. The whole recovery was said to have been monitored by the Royal Australian Air Force. The launch announcement on 4 June stated that:

"Another artificial Earth satellite, Cosmos 1374, was launched in the Soviet Union today. The satellite carried scientific apparatus to continue space research. The satellite was placed in a circular orbit with the following parameters: distance from the Earth's surface – 225km; orbital inclination – 50.7°. The satellite carries a radio system for precise measurement of orbital elements and a radio telemetry system for transmitting to Earth data on the operation of instruments and scientific apparatus.

The scientific research envisaged by the programme has been carried out. The information received is being processed at the Co-ordinating and Computing Centre."

This was a standard Cosmos launch announcement, but two interesting features are apparent: no orbital period was given and it was noted that the research programme had been completed. Exactly the same comments had been made about the dual Cosmos flights. However, it was thought in some quarters that Cosmos 1374 might be weapons-related; the lack of orbital period suggested a connection with the FOBS programme, while the early completion of the programme implied a connection with the ASAT (antisatellite) weapon flights. However, the consensus opinion was that Cosmos 1374 was a test of shuttle-related hardware.

Kapustin Yar, in the 1980s, was running at a very low level in the satellite programme with perhaps one or two launches each year; since it was the most "minor" of the three cosmodromes, (Tyuratam/Baykonur and Plesetsk are the others), it had only fielded the two smallest satellite-carriers. The small SL-7 vehicle had been retired, while the SL-8 was still in use, but

Above: *The recovery of Cosmos 1445 in the Indian Ocean was photographed by the Royal Australian Air Force. The released photographs confirmed speculation that the mission (and earlier Cosmos 1374) had been shuttle-related.*

it only had a payload capacity of 1-1.25 tonnes to low Earth orbit; hardly enough for an actual shuttle spacecraft!

Most of the questions raised by Cosmos 1374 were answered nine months later when Cosmos 1445 was flown as an almost carbon-copy of the Cosmos 1374 mission. However, there was a major difference in the Cosmos 1445 launch announcement; for the first time in the Cosmos programme, no orbit was announced. Once more, the Royal Australian Air Force was monitoring the recovery, but this time, after some delay, the photographs of the recovery were released to the world by the Australian authorities. Cosmos 1445 was revealed to be a small winged vehicle, 3.4m long, 1.4m in diameter and with a wingspan of 2.6m. There was an upright fin in the back and each wing was curved upwards. A small mock-up crew cabin could be seen.

The size of the payload confirmed that

LAUNCHES IN THE COSMOS SPACEPLANE PROGRAMME					
LAUNCH DATE AND TIME (G.M.T.)	**COSMOS SATELLITES**	**OBJECT**	**INCL** (deg)	**PERIOD** (min	**ALTITUDE** (km)
1982 3 Jun 21.29	1374	A	49.6	88.69	190-232
	Rocket	B	50.7	88.39	167-222
	Fragment	C	–	–	–
1983 15 Mar 22.30	1445	A	50.7	88.14	158.208
	Rocket	B	50.7	88.48	178-221
	Fragment	C	–	–	–
27 Dec 10.20	1517	A	50.7	88.51	180-221
	Rocket	B	50.7	88.53	181-222
1984 19 Dec 03.53	1614	A	No orbit available		
	Rocket	B	50.7	88.5	176-223
	Fragments	C-D	–	–	–

Notes: These four launches were all from the small Kapustin Yar launch site, using the SL-8 booster. The orbits of object A on each mission are open to doubt (especially for Cosmos 1374) since the objects may not have been tracked properly. The first two flights were recovered in the Indian Ocean, while the second two flights were recovered in the Black Sea.

the launch vehicle was the SL-8 booster, since it could just fit underneath the standard payload shroud. Once more, if reusable spacecraft systems were being tested, then Cosmos 1445 could have been the second flight of Cosmos 1374.

Presumably the Soviet authorities were none too pleased that the photographs of the recovery reached the public domain. It is one thing to have intelligence sensors and personnel watching you, but another when their data are released for public consumption. When the next launch in the series took place, Cosmos 1517, the spacecraft returned to Earth in the Black Sea where the Soviets could recover the spacecraft knowing that no unauthorised photographs would be published. Once more, no orbit was announced. Interestingly, the launches of Cosmos 1374 and Cosmos 1445 had been 285 days apart, and those of Cosmos 1445 and Cosmos 1517 were 286 days apart; was this simply the turnaround time of the model shuttle?

The last flight in the series came nearly a year after Cosmos 1517. Cosmos 1614 again returned to Earth in the Black Sea and no orbit was announced. Apparently the technology had been proven as best it could be with small spaceplanes, and it would be applied to larger vehicles, capable of carrying men.

Lawton in his 1987 paper has called these small Cosmos satellites "Mischka" (or "fly") and has speculated that they are intended to be one-manned shuttles, to be used for military observation work and perhaps even space defence work. The present author does not agree with these suggestions.

The New Family of Launch Vehicles

In 1981 the US Department of Defense published the first edition of what would later become an annual report: "*Soviet Military Power*". In the second edition (March 1983) some details were given of three new space vehicle systems which the Soviet Union was thought to be developing:

Medium Lift Launch Vehicle:
  launch mass 500 tonnes
  launch thrust 1,300 tonnes
  payload mass 13 tonnes

Heavy Lift Launch Vehicle:
  launch mass not stated
  launch thrust 8-9,000 tonnes
  payload mass 130-150 tonnes

HLLV/Shuttle:
  launch mass 1,500 tonnes
  launch thrust 4-6,000 tonnes
  payload mass 60 tonnes

The payload masses were "normalised" to a 180km orbit, but the assumed inclination was not stated.

The original reconstructions showed the Medium Lift Launch Vehicle (or MLLV, designated SL-16 when it was introduced) as a conventional two-stage tandem vehicle. The Heavy Lift Launch Vehicle (or HLLV, initially designated SL-W but later SL-17) had a first stage which could be augmented by 2-3 strap-on boosters; it appeared that the strap-ons might have the same diameter as the first stage of the MLLV, suggesting a commonality of systems. The HLLV/Shuttle seemed to use the first stage of the HLLV with two strap-on boosters and the shuttle orbiter mounted in parallel with the first stage core.

Over the years, the DoD estimates of the launch vehicle reconstructions and performances varied — presumably as more intelligence data became available. The March 1987 edition included these data:

MLLV (designated SL-16):
  launch mass 400 tonnes
  launch thrust 600 tonnes
  payload mass 15 tonnes +

HLLV (designated SL-W):
  launch mass 2,000 tonnes
  launch thrust 3,000 tonnes
  payload mass 100 tonnes +

HLLV/Shuttle:
  launch mass 2,000 tonnes
  launch thrust 3,000 tonnes
  payload mass 30 tonnes

The SL-16 vehicle was now clearly shown as a two-stage booster, but it was noted that a manned spaceplane payload was thought to be in development for the vehicle. The SL-W appeared to be outwardly similar to the American Space Shuttle system, having a large central tank second stage with strap-ons forming the first stage; the payload was side-mounted, possibly with its own propulsion system. Unlike the American Shuttle, the Soviet vehicle's central tank carried its own main rocket motors. Finally, the SL-W/Shuttle was shown to be the standard SL-W with the shuttle orbiter replacing the side-mounted payload. With all of the payload masses which the DoD quotes for the SL-W/Shuttle combination, it would seem that the shuttle payload is being quoted.

Below: *A US Department of Defense concept of the Soviet shuttle orbiter strapped to the side of the Energiya launch vehicle. Prepared before the first Energiya launch, the vehicle itself is quite accurately portrayed.*

LAUNCHES OF THE MEDIUM LIFT LAUNCH VEHICLE				
LAUNCH DATE AND TIME (G.M.T.)	**SATELLITE**	**INCL** (deg)	**PERIOD** (min	**ALTITUDE** (km)
1985 13 Apr		Sub-orbital test		
21 Jun 08.30	1985-053A-C	64.4	89.88	197-340
22 Oct 07.08	Cosmos 1697	71.0	101.97	850-854
28 Dec 09.19	Cosmos 1714	71.0	94.78	163-853
1986 10 Jul 08.30	Cosmos 1767	64.9	88.52	196-207
22 Oct 08.14	Cosmos 1786	64.9	113.29	190-2,564
1987 14 Feb 08.30	Cosmos 1820	64.8	88.82	180-252
18 Mar 08.37	Cosmos 1833	71.0	101.94	849-852
13 May 05.47	Cosmos 1844	71.0	101.95	849-853
1 Aug 03.59	Cosmos 1871	97.0	88.27	179-199
28 Aug 08.20	Cosmos 1873	64.8	88.82	177-255

Notes: This table provides a list of the launches by the SL-16 Medium Lift Launch Vehicle to the end of 1987. All of the orbital data are derived from the Two-Line Orbital Elements and relate to the main object tracked in orbit from each mission. The launch times are calculated from the initial Two-Line data, and should be correct to perhaps ±1 minute.

To the end of 1987 both the SL-16 and the SL-W (redesignated SL-17 following its first flight) had been launched, although the large booster's maiden flight had been less than successful.

Introduction of the SL-16

Of the new space vehicles reported to be under development, the Medium Lift Launch Vehicle was the first to reach flight status, although its launches were not immediately recognised as being those involving the new vehicle. When discussing the development of the RD-170 engine which is used for the strap-ons on Energiya (the Soviet name for the SL-17 booster), Glushko has noted that the first use of the new engine came when it was used in the first stage of a carrier rocket, launched on 13 April 1985. This represents the first flight of the SL-16 booster, and also confirms western speculation that the SL-16 first stage and the Energiya strap-ons share the same engine technology.

The first signs of a new vehicle surfaced on the morning of 21 June 1985 when three small pieces of debris were tracked in orbit and designated in the West as 1985-053A, 1985-053B and 1985-053C (objects A, B and C from the 53rd launch in 1985). From the orbital data, it was possible to backtrack and discover that the debris related to a launch out of the Tyuratam launch site; however, the orbital inclination of 64.4° was not in use by any launch vehicle at that time.

The first orbiting of a payload followed in October 1985 when Cosmos 1697 was launched. The orbit used was similar to the final ones attained on two earlier Cosmos flights – Cosmos 1603 and Cosmos 1656 – which had been launched by the four-stage Proton SL-12 and which had to manoeuvre extensively to reach the final orbits. An attempt to repeat the Cosmos 1697 mission failed with Cosmos 1714 when the final stage of the launch vehicle failed to re-ignite and circularise the orbit. The payload and rocket stage remained attached and were destroyed upon re-entry into the atmosphere.

There followed a pause in the flights of the SL-16 vehicle and when they resumed the missions in 1986 and 1987 fell into four groups:

Group 1: low orbit with the payload decaying with no apparent activity (Cosmos 1767, Cosmos 1820, Cosmos 1873).

Group 2: very eccentric orbit with no apparent activity after orbital injection (Cosmos 1786).

Group 3: high circular orbits (Cosmos 1833, Cosmos 1844).

Group 4: retrograde orbit (Cosmos 1871).

The flight of June 1985 could possibly fall within the Group 1 missions, while the Cosmos 1697 and Cosmos 1714 flights should be classified as Group 3.

The series of flights at 71° is believed to consist of ELINT (electronic intelligence) satellites in the unmanned military programme, and therefore is not considered further here.

The flights of Cosmos 1767, Cosmos 1820 and Cosmos 1873 seem to be test flights of some kind, although what is being tested remains unclear at the time of writing. According to Soviet admissions, Cosmos 1871 should have manoeuvred to a higher altitude, but a propulsion system (whether the last stage of the rocket or that of the satellite was not stated) failed to ignite. The orbital inclination of 97° suggests that the orbit should have been of a special type called Sun-synchronous, and if this were the case then Cosmos 1871 should have manoeuvred to an altitude of about 391km. Most interesting was a comment in the launch announcement for Cosmos 1873: "The Cosmos 1873 satellite's flight programme is analogous to the flight programme of Cosmos 1871". When commenting on the re-entry of Cosmos 1871, the Soviets indicated that its mass was about 10 tonnes – a figure which would square only with the SL-16.

The purpose of the test flights of the Cosmos 1767 series and Cosmos 1871 also remain obscure since there were no in-orbit manoeuvres. Unconfirmed western reports connect these flights to possible spaceplane development.

The Introduction of Energiya

During the mid-1980s rumours abounded that the new SL-W Heavy Lift Launch Vehicle was on the verge of its maiden flight. The winter of 1986-1987 saw on-the-pad testing of the vehicle, with the central core stage engines being tested twice in a launch configuration. There were western reports that the first orbital flight was scheduled for April 1987.

On 13 May 1987 it was announced by Moscow Radio that Mikhail Gorbechev – the General Secretary of the Communist Party of the Soviet Union (CPSU) whose policies advocating greater "openness" resulted in the immediate and even premature release of spaceflight plans – had paid a visit to the "Baikonur Cosmodrome" and Leninsk during 11-13 May. During this time he may have witnessed the launches of Gorizont 14 (a communications satellite launched by a Proton SL-12 booster) and Cosmos 1844 (using the SL-16 booster), but it was also stated that he saw

"work . . . in progress at the cosmodrome in preparation for the launch of a new general-purpose carrier rocket capable of putting both reusable orbital craft and large spacecraft for scientific and economic purposes into low orbit, including modules for a long-term space station."

This was the first Soviet admission that they had a large booster under development and, more importantly, it confirmed that a space shuttle was in development.

On 14 May a further announcement was made, indicating that the new launch vehicle was on the pad and that the countdown ready for its maiden flight was underway. The launch of the new booster, called Energiya, came on 15 May at 17.30 and full television coverage (albeit retrospective rather than "live") was given of the event by Moscow Television. The ignition of the vehicle happened in a number of distinct phases: possibly the core ignited first and after some 12 seconds the four strap-on boosters ignited resulting in an immediate lift-off. The acceleration of the vehicle indicated that the sea level thrust:mass ratio of the vehicle was 1.7:1. The first and second stages of the vehicle seem to have performed almost perfectly, but a "kick" stage which should have put the (unspecified) payload into orbit failed to operate and the payload re-entered to destruction over the Pacific Ocean.

The Soviet Union has described Energiya in some detail – something which is surprising considering that the first flight was less than successful. The basic vehicle

Below: *The launch of the first Energiya came on the 30th anniversary of the first R-7 flight, the missile which was adapted as the Vostok first stage. Energiya is equally important to the Soviets.*

has four strap-on boosters which use liquid oxygen and kerosene (a propellant combination which was used on the Sputnik-Vostok-Soyuz family of launch vehicles, the most widely used vehicles in the world) and each strap-on is powered by an RD-170 engine with a thrust of 806 tonnes, making these the most powerful engines ever flown. Each engine has four main thrust chambers. The second stage central core uses liquid oxygen and liquid hydrogen and has four engines (each with a single thrust chamber) of 200 tonnes thrust each.

On the first flight, the payload and "kick" stage were mounted on the side of the vehicle, in much the same way that a shuttle orbiter will be. The placement of the four strap-ons was asymmetric, and this allows for the protrusion of the orbiter wings and the separation of the strap-ons without endangering the orbiter. It was later stated that the payload on the first flight had included a test of the shuttle orbiter's in-orbit manoeuvring system, although the test could not be completed because of the failure of the payload to reach orbit.

On the first flight it was said that the first stage strap-ons landed in the western Soviet Union, but they were not recovered for re-use. It is believed that the strap-on boosters separate individually. The core of the booster splashed down in the Pacific Ocean and was designed to split into three sections for recovery (it is unclear what the three sections are and whether the separation did take place on the first flight). It is believed that Energiya's first and second stage assembly will eventually represent a fully-reusable launch vehicle, possibly with the core initially going into orbit on later missions and then being de-orbited some 3-4 orbits later for a recovery within the Soviet Union.

After the first flight of Energiya, it was reported that the launch pad had been

Above: *The first Energiya sits on the pad. The giant central core second stage marked the first use of liquid oxygen and liquid hydrogen by a launch vehicle in the Soviet space programme.*

damaged somewhat more than expected, but this seems actually to have involved the launch simply dislodging loose structures at the pad rather than any more serious structural damage. In autumn 1987 an Energiya test vehicle was rolled out to the launch pad and erected, while further tests of the facilities were undertaken. However, no flight was planned for that time because even six months after the May launch Soviet engineers were still poring over data returned from the first flight which had been carrying a full load of test equipment to take measurements of the vehicle in flight.

The Soviets announced that the launch mass of Energiya was about 2,000 tonnes

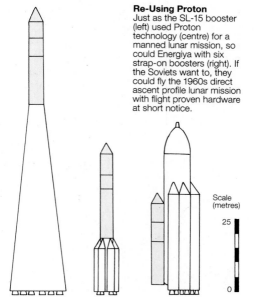

Re-Using Proton
Just as the SL-15 booster (left) used Proton technology (centre) for a manned lunar mission, so could Energiya with six strap-on boosters (right). If the Soviets want to, they could fly the 1960s direct ascent profile lunar mission with flight proven hardware at short notice.

Scale (metres)

25

0

and that the payload had been about 100 tonnes. From the thrust levels of the vehicle at launch and the observed acceleration, the actual mass seems to have been close to 2,150 tonnes. However, scaling the launch vehicle suggests that its fully-fuelled mass would be closer to 2,700 tonnes and therefore the first flight must have represented a test mission using a partially-fuelled booster.

If the core of Energiya can be placed into orbit on later flights, a fully-fuelled booster is capable of orbiting 140-150 tonnes, while it is probable that Energiya will also be launched using six strap-on boosters and this could orbit 220-230 tonnes.

The Space Shuttle Programme

For many years Soviet statements about a possible space shuttle programme in the planning stage were contradictory; sometimes it was said that there were no plans to develop a shuttle system, while at other times the Soviets said that they were undertaking studies to ascertain whether a shuttle would be an economic method of flying men and payloads into space. The final official Soviet confirmation of the programme came with the launch of the first Energiya when it was stated that the booster would be used for reusable space systems.

During the early 1980s the first Soviet shuttle orbiter was rolled out and after some initial testing it began flight trials. It has been reported that in the mid-1980s a Bison aircraft carrying the orbiter ran off the runway while landing at the Ramenskoye research centre.

It would appear that in 1978 a group of trainee cosmonauts was selected which included Volk, Levchenko (who flew on Soyuz-TM 4) and Shchukin (Levchenko's back-up for Soyuz-TM 4), among others, specifically for the testing of the space shuttle system. Certainly, Levchenko had not been based at Zvezdny Gorodok with the other cosmonauts (Romanenko on Mir did not meet him until after the Soyuz-TM 4 docking!), and therefore one may speculate that the shuttle cosmonauts are based elsewhere, only joining the Zvezdny cosmonauts as they are assigned to specific missions.

The shuttle orbiter is believed to be the same general size and shape as the American Space Shuttles, although there are major design changes. The main propulsion system is carried on the core of Energiya rather than on the shuttle orbiter (as is the case with the American design) and therefore the Soviet orbiter only carries an in-orbit manoeuvring system, as well as attitude control thrusters. The most important difference is possibly the inclusion on the Soviet orbiter of four jet engines (derived from the AL-21 engine, with a thrust of 9 tonnes?) which allow the craft to be controlled as it comes down for a runway landing. If necessary, the orbiter can fly around and make second or third landing attempts; by contrast the American orbiter is described as a "flying brick" during the landing and there is no opportunity for the crew to make a second landing attempt.

Two orbiters have been used for the landing tests: one has been carried to altitude and dropped for glide landing tests in much the same way that the American *Enterprise* orbiter conducted drop tests from a converted Boeing 747 in 1977, while the second has been conducting jet powered landing tests. During 1987 five landing tests took place which involved the orbiter taking off under the power of its jet engines and coming down to land, again using the jets. In charge of these tests was Georgi Shonin, the Soyuz 6 cosmonaut, while the pilot who conducted at least some of the flights was Igor Volk, who had flown on Soyuz-T 12 and who was already rumoured to have been a shuttle crew candidate prior to this announcement.

The Soviet shuttle will be launched by Energiya with four strap-ons, allowing for the orbiter having a mass of about 70 tonnes and a payload capacity of 60 tonnes (twice the capacity of the American orbiter). More realistically, a maximum payload mass would be about 40 tonnes, since anything higher than this would imply a very dense payload. Assuming that the core of Energiya is recovered after a suborbital trajectory, the shuttle orbiter manoeuvring system will have to perform the final orbital injection manoeuvre.

However, unlike the American Shuttle system, the Soviet space shuttle programme does not call for the replacement of expendable launch vehicles with reusable ones. The Soviet shuttle will be used for special missions which require the launching of heavy modules and spacecraft or the return to Earth of spacecraft for repair and later reflight. One possible mission will be a visit to the Salyut 7/Cosmos 1686 complex, the orbit of which will have reduced to about 300km in the 1990s. The station could even be returned to Earth by the shuttle so that ground inspection of the vehicle may be completed to gain a further insight into the wear-and-tear experienced by a spacecraft which has been in space for a decade or more.

In addition to the large shuttle programme, there is rumoured to be a second manned reusable spacecraft under development which would be launched using the SL-16 booster. Such a programme has not been acknowledged by the Soviet Union and there are no rumours of the "spaceplane" orbiter being seen during roll-out or atmospheric tests. Therefore, at the time of writing the reality of this programme is open to doubt. Such a spaceplane might be used for quick-reaction orbital reconnaissance or even as a Soyuz-TM replacement as a space station crew and supplies ferry.

The Mir-2 Complex

During 1987 the Soviet Union began to talk about the post-Mir space station programme. Flights to the Mir complex, which began assembly in 1987, are expected to continue until 1992-1993, when the next generation space station complex should be ready for launch. Known as either Mir-2 or Cosmograd ("Spacetown"), in overall philosophy the new station will adhere to the general Mir-1 practice in that it will be assembled from modules.

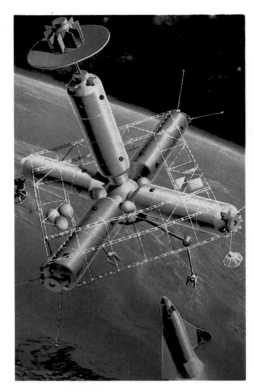

Above: *The next generation Mir-2 space station? This Soviet concept seems to be a reasonable portrayal of the modular space station to be launched by Energiya during the 1990s and serviced by shuttles.*

Mir-2 will be constructed around a large core station to be orbited by an Energiya booster. Depending upon the number of strap-ons used for the Energiya launch, the mass of the core would be in the range 140-230 tonnes. The core would act as the main station, on to which would be docked other specialist modules (either launched by Energiya, the SL-16 or the Proton SL-13 vehicles) and which would have large radiating booms carrying solar panels, antennas and experimental equipment. Mir-2 would be the home for about twenty cosmonauts; not quite a "Space Town", more a "Space Village". This would be the base for a major Soviet expansion in space activities during the 1990s.

A Manned Lunar Mission

The introduction of Energiya will allow the Soviet Union to reconstruct its manned lunar programme, should it so wish. The six strap-on variant of Energiya will match the payload capability of the failed SL-15 giant booster which should have carried men to the Moon in the late 1960s and early 1970s. Although the SL-15 vehicle was abandoned, the Proton rocket stages have been routinely flying unmanned missions since 1965 (second stage) and 1967 (third stage), while the Soyuz spacecraft has been conducting manned missions for more than two decades. Of course, the Soyuz-TM propulsion system does not have the capability to fly a lunar mission, but the 1960s technology could easily be reproduced, updated and re-introduced to the manned programme.

Therefore, for more than twenty years the Soviet Union has been using the essential elements of the manned lunar pro-

gramme individually, with only the heavy lift launch vehicle still being required. Once Energiya has been man-rated for shuttle missions, the "building blocks" of the manned lunar programme could again be brought together and with a minimum of unmanned tests around the Moon a manned lunar expedition could be mounted.

Throughout the Soviet space programme the "production line" philosophy has been applied, since only this can support a programme which sees nearly 100 launches every year to Earth orbit or beyond (the Soviet programme generally accounts for 80 per cent of all space missions each year). Launch vehicles are produced on a production line, with the original Sputnik vehicle still being used for the manned Soyuz-TM missions (plus many more routine unmanned missions) with an added orbital stage more than 30 years after Sputnik 1. Additionally, Soviet spacecraft have used a limited number of satellite "buses" which are produced to a standard design; experiments and equipment for specific missions can be "plugged" on to the standard vehicles as required.

In this way, a programme which would probably drain American space budgets can be mounted by the Soviet Union using modifications of existing spacecraft. The manned lunar programme using Energiya, outlined above, could be flown with existing space equipment, although modifications would have to be made to the standard equipment (eg, the Proton third stage modifications as a lunar landing stage and reconfiguring Soyuz for a lunar mission). This would mean that such a project might be considered as a feasible sideline which could be undertaken without diverting resources from the main thrust of the Soviet space programme – space stations.

A Manned Flight to Mars?

The same philosophy can also be applied to a far more difficult and far more spectacular manned mission: to the planet Mars.

The later long duration missions in Earth orbit, particularly those of Soyuz-T 10 to Salyut 7 and Soyuz-TM 2 (Romanenko) to the Mir complex, have been vital in the obtaining of biomedical data relating to a possible manned flight to Mars. Such a mission would be more than an order of magnitude beyond a manned flight to the Moon, but the evidence suggests that the Soviet Union is thinking of just such a fight. In fact, it is possible that some very preliminary cosmonaut training has begun since Viktor Savinykh has indicated that he is preparing for such a mission. Additionally, a comment has been made that Mir can be used for a manned Mars mission – a statement which should not be taken too literally, although a Mir-derived module could certainly form the basis of cosmonaut quarters for such a flight.

Starting with the Phobos missions, launched in July 1988, the Soviet space planners are looking forward to a major assault on Mars and its environment with automatic spacecraft during the 1990s.

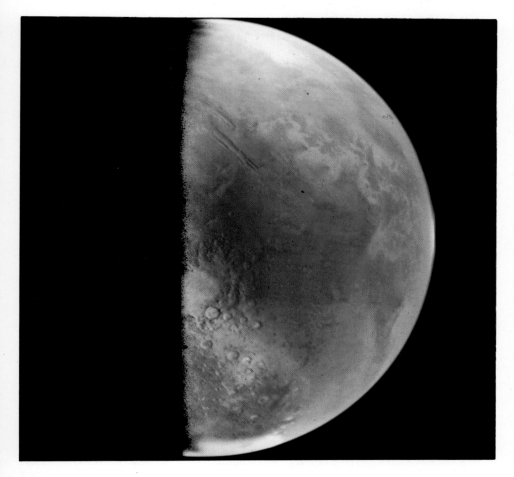

Above: *The Martian system will be the next target for manned exploration, with visits and possibly landing during the late 1990s. How soon before the Red Flag is planted on Mars, the Red Planet?*

Below: *It is too early for actual crews to be selected for Mars missions, but Viktor Savinykh (left) has indicated that he is ready for such a flight. Whether he is a serious candidate is less certain.*

Plans call for Mars rovers and a Mars sample-return mission to be flown, possibly before 1998 (the Soviets have openly discussed the unmanned missions which they would like to fly, while admitting that they have not the hard funding in 1987 for all of them). Such missions could be seen as a reconnaissance of the planet prior to a manned flight.

A possible manned flight to Mars would have to rely on spacecraft production lines which are operating to service the main thrust of the manned space programme. Two complexes would be launched from Earth orbit: one would house the landing spacecraft and the other would carry the manned Mir-derived module in which the cosmonauts would live during the coast to and from Mars and during their period in Mars orbit.

In terms of mission duration, a Mars mission would be the most demanding manned flight to be planned in the whole space programme so far. A flight to the lunar surface and back can be accomplished in a week, while a manned Mars mission would last for more than two-and-a-half years. Launch opportunities to Mars vary greatly because the orbit of Mars around the Sun is far less circular than that of the Earth; a Mars arrival when the planet is close to the Sun requires a smaller launch velocity from Earth orbit than would arrival when Mars is at its furthest from the Sun. Therefore, the figures presented here will represent an average for the spectrum of mission requirements.

The trip to Mars can be split into three sections: the trans-Mars coast, the Mars orbit activities (including the landing), and the trans-Earth coast. Approximate times

for these phases are 290 days, 450 days and 250 days respectively. The long period spent in Mars orbit is required because the spacecraft will have to wait for Earth and Mars to be in the correct relative orbital positions for the minimum velocity required to achieve the manoeuvre out of Mars orbit. Faster trajectories to Earth can be calculated, but these would increase the velocity requirements and thus the propellant load.

The cosmonauts on the mission would have to take all their food, air and water supplies with them; conceivably, these could be launched separately (for example, those for the return trans-Earth coast could be launched on the Mars lander complex, but what would happen if the lander complex failed to get into Mars orbit or if the rendezvous in Mars orbit failed?) However, here it will be assumed that they will be stored on the manned Mir-derived complex. The consumables are estimated to be 1.875kg of oxygen per man-day, 1.8kg of food per man-day and 3.0kg of water per man-day. The water supply can be reduced, since a 50 per cent recycling of water seems to be a reasonable assumption. These consumables will therefore amount to 5.175kg per man-day or about 15.5 tonnes for the complete mission with three cosmonauts.

As much as possible, the manned Mars mission will be assumed to use hardware derived from existing spacecraft. The Proton second stage could be used for part of the trans-Mars manoeuvre, the Mars orbital injection and the trans-Earth injection. As previously noted, a Mir-derived module could be used for transporting the cosmonauts to Mars orbit and returning them to near-Earth space. An up-rated Soyuz descent module could return the cosmonauts and their supplies to the Soviet Union using a direct re-entry trajectory (that is, without first going into Earth orbit, something which would require a great deal of propellant to be carried all the way to Mars and back). The Mars lander would be based upon a modified Proton third stage with an integral manned module. A specially developed heat shield would be required for the Mars landing.

The major piece of new equipment which the mission would require is the development of a large rocket stage which would be launched as a payload of an Energiya with six strap-on boosters. This would have a fully-fuelled mass of 220 tonnes and a propellant mass of 200 tonnes. This would be used to perform the first part of the trans-Mars injection manoeuvres for both the manned complex and the lander complex. This giant rocket stage would probably use liquid oxygen and liquid hydrogen, like the core of Energiya, and the mission requirements suggest that it could be powered by two engines derived from those which are used on Energiya's core stage.

The Mars Flight Plan

The Mars flight plan which is described here does not purport to be the method which the Soviet Union *will* use to put men on Mars, but it illustrates one option which could be used using existing technology in the main. The tables give the numerical details for the mission; the accuracy which is shown in the velocity requirement tables should not be taken to be actually 0.1 tonne – this is simply the mathematical result of the calculation. However, the figures are presented to illustrate a reasonable method by which the mission could be accomplished. Additionally, it represents the *minimum* requirements for a manned Mars mission. More realistically, a crew of 10-12 men will be required, rather than three.

The Mars flight is built around a minimum of five launches to Earth orbit; additional flights of specialist cosmonauts to assist in the docking and pre-departure checking and preparations are possible. Similarly, the mission could be expanded to allow for Heavy Cosmos derivatives to be placed in Mars orbit as additional supply and experiment bases or unmanned landers to act as cargo carriers and surface shelters. Additionally, it would be advisable to fly two of each spacecraft complex, so that if there was a major failure on one, the cosmonauts could switch to the second. However, here the basic "bare bones" mission only will be described.

The five launches would consist of four using Energiya with six strap-ons and one using a standard Soyuz booster carrying the Mars crew into Earth orbit. (Alternatively, a fifth Energiya launch carrying a shuttle payload could take the crew into orbit.) Two complexes would be assem-

bled in Earth orbit, assumed to be 225km high; one to carry the men to and from Mars orbit and one to carry the Mars landing craft.

The first assembly would probably be the Mars lander. The first Energiya launch would orbit the landing craft (built around a modified Proton third stage) and a Proton second stage, the total payload being 233.2 tonnes. This figure is higher than the assumed Earth orbit capability of the six strap-on Energiya, but bearing in mind that there are many performance uncertainties with the vehicle and that the lander complex is only a concept, this is not a real problem. The lander complex could be scaled down slightly to fall within the 220-230 tonne payload mass for this Energiya variant. Once this complex is in orbit and has been checked out, a further Energiya launch would orbit the 220 tonne rocket stage, fully fuelled. This would dock with the lander complex and once the checks were complete the full complex would begin the journey to Mars.

The injection into a trans-Mars trajectory would be completed in two stages. The 220 tonne stage burns to depletion, leaving the lander complex in a 225-64,000km orbit; the empty stage will be abandoned. After the 220 tonne stage manoeuvre the P-2 stage would ignite for the first time to perform the final part of the trans-Mars injection.

The assembly of the manned Mir-module complex would be completed in a

similar manner. The main complex with a mass of 209 tonnes would be launched first, followed by the 220 tonne stage. Following the docking and check-out, a Soyuz would be launched with the three man crew for the Mars mission (alternatively, the crew could be carried as passengers aboard a shuttle mission, using a fifth Energiya booster but with only four strap-on boosters). The ferry would dock at the front of the Mir-module and the crew would transfer to the module. The Soyuz would then be abandoned in orbit or be used to return instruments and tools to Earth.

Once more, the injection into trans-Mars orbit would take place in two stages. Because the Mir-module complex is lighter than the lander, the 220 tonne stage could place the complex into a higher initial orbit: 225-124,100km. Then the Proton second stage would ignite to place the complex onto its trans-Mars trajectory.

Both complexes are assumed to be capable of course corrections of up to 100m/s during the trans-Mars coast. A major danger for the crew during the mission – other than the general problems related to weightlessness – will be the effects of cosmic radiation. Should a major solar storm begin or a major deep space explosion take place, then the cosmonauts should receive a warning from on-board automatic detectors, giving themselves sufficient time to manoeuvre the Mir-module complex to place the large water supply between themselves and the radiation source, since

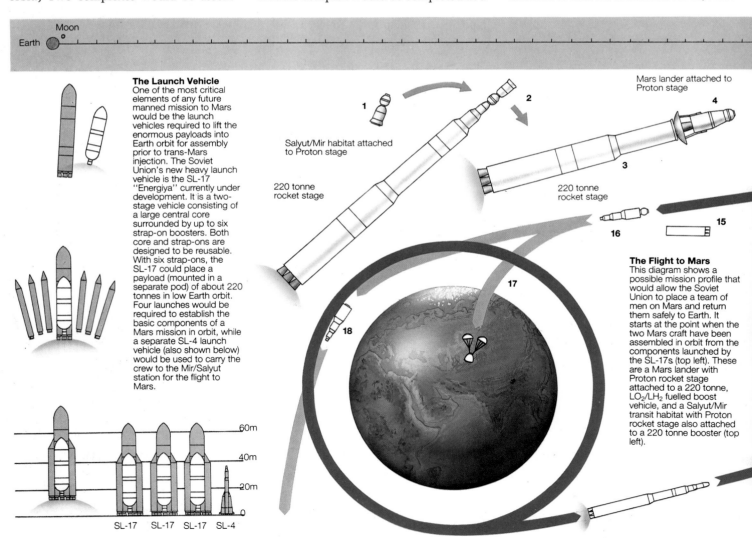

The Launch Vehicle
One of the most critical elements of any future manned mission to Mars would be the launch vehicles required to lift the enormous payloads into Earth orbit for assembly prior to trans-Mars injection. The Soviet Union's new heavy launch vehicle is the SL-17 "Energiya" currently under development. It is a two-stage vehicle consisting of a large central core surrounded by up to six strap-on boosters. Both core and strap-ons are designed to be reusable. With six strap-ons, the SL-17 could place a payload (mounted in a separate pod) of about 220 tonnes in low Earth orbit. Four launches would be required to establish the basic components of a Mars mission in orbit, while a separate SL-4 launch vehicle (also shown below) would be used to carry the crew to the Mir/Salyut station for the flight to Mars.

Salyut/Mir habitat attached to Proton stage

220 tonne rocket stage

Mars lander attached to Proton stage

220 tonne rocket stage

The Flight to Mars
This diagram shows a possible mission profile that would allow the Soviet Union to place a team of men on Mars and return them safely to Earth. It starts at the point when the two Mars craft have been assembled in orbit from the components launched by the SL-17s (top left). These are a Mars lander with Proton rocket stage attached to a 220 tonne, LO_2/LH_2 fuelled boost vehicle, and a Salyut/Mir transit habitat with Proton rocket stage also attached to a 220 tonne booster (top left).

Moon

Earth

60m

40m

20m

0

SL-17   SL-17   SL-17   SL-4

water can act as a radiation shield.

The lander complex would be first to reach Mars orbit, followed by the Mir-module complex. A number of Mars orbits could be chosen for the mission, but here a 2,000-32,052km orbit with a period of 1,477 minutes is chosen. The orbital period is the same as the rotation of Mars on its axis (a Martian "day"), so that if orbital precession is ignored the complexes will pass over almost the same areas of Mars each day, while the orbital altitude, – a minimum of 2,000km, – is high enough to prevent premature orbital decay. Once both complexes were in Mars orbit, they would manoeuvre and dock. When the right time comes, two cosmonauts would transfer to the lander complex to begin their descent to the Martian surface.

After separation, the Proton second stage on the lander would perform a manoeuvre to ensure that the lander enters the Martian atmosphere, and almost immediately the lander separates; this will allow the Proton stage to perform a final manoeuvre which will raise the orbit, preventing the rocket stage from entering the Martian atmosphere and contaminating the Martian surface when it crashes.

The actual landing might be accomplished in three stages: a large aeroshield would use air resistance to reduce the velocity of the lander, at the same time acting as a heat shield; a parachute system would then deploy to slow down the lander further; finally, as the landing site is

A Mars Lander
In the "minimum-mission" described here, the Mars landing module is a derivative of the Kvant module for living quarters, mounted atop a modified Proton third stage.

Experimental equipment could be carried in pods strapped to the exterior of the rocket stage. A single burn of the rocket engine would allow launch from Mars, back into orbit around the planet.

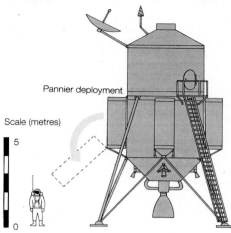

Pannier deployment

Scale (metres)

approached, the lander's rockets could be used to perform the final site selection.

Here, the stay on the Martian surface has been estimated to be about 100 days, but it could range from a few weeks to longer than 100 days. An experimental load of five tonnes has been allowed for, and this would include a manned roving vehicle to allow an extended exploration of the landing site. The lander would presumably require an airlock system, rather than lose the whole of the cabin's atmosphere to the Martian environment each time surface activities were undertaken.

Once the surface exploration has been completed, the complete landing vehicle would be launched from the Martian surface and placed back into Mars orbit; the lander is assumed to be a single stage ascent vehicle (based upon the Proton third stage and a Kvant-type cabin), although credible designs for two stage vehicles could be produced. At least a single stage vehicle would not require a

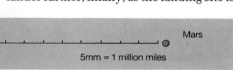

Mars

5mm = 1 million miles

A Long Way Home
Mars is some 48 million miles (78 million km) from the Earth. This diagram

shows this distance in relation to the mere 0.25 million miles (0.4 million km) that Apollo travelled.

1 A Soyuz-type craft containing the crew for the Mars flight docks with Salyut/Mir, the three crew members transfer.
2 After transfer, Soyuz module is discarded.
3 and 4 The 220 tonne transfer stages are fired to put both the Salyut/Mir habitat and the Mars lander craft into a Mars injection trajectory. When their propellants are exhausted, the 220 tonne stages are discarded.
5 The lander arrives at Mars first, and its Proton stage fires to brake it, and put it into Mars orbit.

6 Salyut/Mir assembly enters Mars orbit.
7 Transit habitat docks with the lander, and two crew members transfer to the lander.
8 Lander separates from Salyut/Mir and uses Proton engines to de-orbit. Proton stage is then discarded, and lander drops through Martian atmosphere, its descent being successively braked by an aeroshield, drogue parachutes, and finally its own retro-rocket.

11 Lander docks with Salyut/Mir, crew transfer with soil samples etc.
12 Lander discarded.
13 Salyut/Mir uses remaining Proton fuel to escape from Mars orbit.
14 Salyut/Mir in trans-Earth trajectory.
15 Proton stage discarded.
16 Crew transfer to Soyuz descent module on rear of Salyut/Mir.

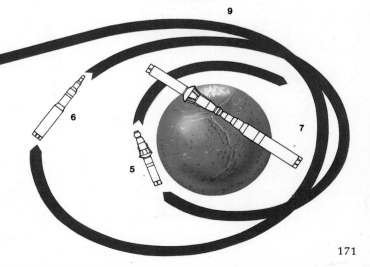

9 During the period of surface exploration, the remaining crew member orbits Mars in the Salyut/Mir habitat.
10 At the conclusion of the surface experiments, lander fires its motor to lift off from Mars.

17 Soyuz re-enters Earth atmosphere and lands in Soviet Union in the normal manner. Crew are recovered and taken for exhaustive medical checks and debriefing.
18 Empty Salyut/Mir habitat continues into solar orbit.

stage to be discarded during the ascent to orbit which might crash onto and contaminate the Martian surface.

The lander would be the active spacecraft in the docking exercise, although the Mir-module complex would have propellant to perform a rendezvous if the lander managed to reach a reasonable orbit. The cosmonauts would transfer the results of the surface work to the Mir module and then the lander would be abandoned in an orbit high enough to prevent an early entry into the Martian atmosphere.

The cosmonauts would have to spend about fifteen months in orbit around Mars (including the period spent on Mars) before the Earth would reach the right orbital position for a launch out of Mars orbit. The Mir-module's Proton stage would perform its final major manoeuvre, to take the cosmonauts out of Mars orbit into the trans-Earth trajectory. Taking about eight months to reach Earth, this part of the mission might seem the most anti-climactic part of the flight. As it approached the Earth, the Proton stage would separate from the Mir-module, revealing for the first time a Soyuz-derived descent craft docked at a rear port of the module. The cosmonauts would have transferred the results of their work into the capsule during the final days aboard the Mir-module. Finally, they too would enter the capsule and separate from the module which would be allowed either to burn up in the atmosphere or perform a small manoeuvre to ensure that it did not enter Earth's atmosphere. The Soyuz capsule with its precious cargo and crew would perform a double-skip re-entry, finally landing within the Soviet Union.

It is not possible to say with certainty in what physical condition the cosmonauts would be in after their 2.7-year mission. The Soviet Union sees no biological problem preventing the cosmonauts from being able to re-adapt to Earth's conditions after being away so long, given that the correct countermeasures were taken. The cosmonauts who reached the Martian surface would require reconditioning to gravity prior to the landing if their flight took place in a weightless or microgravity mode. Since the gravity field of Mars is some 38 per cent that of the Earth, such reconditioning could take some weeks. It may therefore be easier initially to land only on Phobos, one of the two natural Martian

satellites, since this body is so small that it can be considered to be a microgravity environment as far as human biological responses are concerned.

Of course, this minimum mission plan to place men on Mars can be improved upon, especially as new technologies are developed. One innovation which would allow a major saving of propellant would be the use of "aerobraking" to enter orbit around Mars. This requires the addition of a large heat shield to each spacecraft complex. Instead of using rockets to enter orbit around Mars, the complexes direct themselves to the upper atmosphere of Mars and use air resistance to lose sufficient velocity to enter Mars orbit. This done, rockets can be used to trim the orbit to the one actually required. It is known that the Soviet unmanned Mars missions due in 1994 will probably be the first spacecraft to use this technique to gain Martian orbit.

In the same way, when returning to Earth, instead of coming back by a direct entry into the atmosphere, the Mir-derived module could use aerobraking initially to enter orbit around the Earth. The crew could then be transferred to a Soviet orbital station for medical checks and debriefing,

returning to Earth as shuttle passengers.

These are all refinements to the basic mission scenario discussed above and most probably they will be used for a manned Mars mission. The mass of the additional aerobrakes required for the missions will be less than the propellant saved during the manoeuvres, and therefore the total spacecraft mass could be reduced.

Space Industrialisation

When the "Space Age" began in 1957 the primary reasons for developing space pro-

RECONSTRUCTION OF THE MANNED TRANS-MARS COMPLEX	
Earth return capsule (Soyuz derived)	3.5 tonnes
Trans-Mars module (Mir derived)	25.0 tonnes
Consumables	15.5 tonnes
Fuelled Proton second stage	165.0 tonnes
Single Energia payload	209.0 tonnes
Fuelled heavy rocket stage	220.0 tonnes
Total mass of complex	429.0 tonnes

Notes: This assembly is launched into Earth orbit using two Energiya missions, each with the six strap-on variant: one carries the complex with a mass of 209 tonnes, the second the heavy rocket stage. The consumables are calculated for 3 cosmonauts for 990 days at 5.175kg per day per man (as noted in the text).

VELOCITY REQUIREMENTS FOR THE MARS LANDER COMPLEX			
MANOEUVRE	ROCKET STAGE	VELOCITY CHANGE (m/s)	POST-MANOEUVRE MASS (tonnes)
Mass in Earth orbit			453.2
Trans-Mars injection 1	Heavy	2,736	253.2
Trans-Mars injection 2	P-2	964	173.7
Course correction	P-2	100	168.4
Mars orbit injection	P-2	1,900	94.2
Descent orbit burn	P-2	400	83.4
Less Mars lander (mass 68.2 tonnes)			15.2
Raise orbit	P-2	400	14.7
Mars lander			68.2
Less parachute system and heat shield			60.2
Hover over Martian surface	P-3	50	59.3
Less Mars surface experiments (5.0 tonnes)			54.3
Less consumables (1.4 tonnes)			52.9
Ascent and rendezvous	P-3	5,000	12.0

Notes: This table shows how the mass of the spacecraft complex reduces as each manoeuvre takes place. The mid-course corrections are allowed for, should they be required. It is assumed that the P-2 stage will raise its orbit again after the lander itself has separated, thus preventing the uncontrolled crash landing of the stage on the Martian surface. The rocket stages are identified as in the previous table, other than the third Proton stage, the P-3.

RECONSTRUCTION OF THE MANNED MARS LANDER COMPLEX	
Mars lander cabin (P-3 stage + manned capsule)	53.8 tonnes
Mars surface experimental equipment	5.0 tonnes
Consumables	1.4 tonnes
Landing parachute system	1.5 tonnes
Atmospheric entry heat shield	6.5 tonnes
Fuelled Proton second stage	165.0 tonnes
Single Energiya payload	233.2 tonnes
Fuelled heavy rocket stage	220.0 tonnes
Total mass of complex	453.2 tonnes

Notes: This assembly is launched into Earth orbit using two Energiya missions, each with the six strap-on variant: one carries the lander complex with a mass of 233.2 tonnes, the second the heavy rocket stage. The consumables are calculated for 2 cosmonauts for 100 days at 6.675kg per day per man (as noted in the text).

VELOCITY REQUIREMENTS FOR THE MANNED MARS COMPLEX			
MANOEUVRE	ROCKET STAGE	VELOCITY CHANGE (m/s)	POST-MANOEUVRE MASS (tonnes)
Mass in Earth orbit			429.0
Trans-Mars injection 1	Heavy	2,950	229.0
Trans-Mars injection 2	P-2	750	166.2
Course correction	P-2	100	161.2
Less consumables (4.5 tonnes)			156.7
Mars orbit injection	P-2	1,900	87.6
Mars orbit manoeuvres	P-2	250	81.2
Less consumables (7.0 tonnes)			74.2
Trans-Earth injection	P-2	1,400	48.4
Course correction	P-2	100	46.9
Less consumables (3.9 tonnes)			43.0

Notes: This table shows how the mass of the spacecraft complex reduces as each manoeuvre takes place. The mid-course corrections are allowed for, should they be required. The Mars orbit manoeuvres are those required for rendezvous and docking with the Mars lander complex following orbital injection, as well as a contingency for course corrections in Mars orbit. The velocity requirements for the course corrections are on the high side, thus reducing the final mass of 43.0 tonnes to almost the minimum required: the Proton second stage dry (13.5 tonnes), the Mir module (25 tonnes) and the re-entry capsule (3.5 tonnes), a total of 42 tonnes. The rocket stages indicated are those which performed each manoeuvre; "Heavy" is the Energiya 220 tonne stage and P-2 is the Proton second stage.

grammes were, in order of priority: scientific, prestige, military, commercial.

Both the Soviet Union and the United States claimed that their programmes were scientific, although clearly the "prestige" element was something which no politician – who governed the budgets – could ignore. The military reason for going into space quickly became a major application within the first five years of space exploration. Initially, the commercial considerations were small, but during the 1970s the potential market for launching satellites on a commercial basis for countries without their own satellite launching capability was realised. The American Space Shuttle system was geared towards the commercial satellite market (for want of a better mission, since it would have no space station to service for at least a decade-and-a-half after its first flight), while the European Space Agency developed the Ariane launch vehicle which was designed specifically for commercial satellite missions to geostationary transfer orbit.

One important event seems to have struck the Soviet Union with particular force during the 1970s: the energy shortage which affected the western world as oil prices were raised following the Arab defeat in the Yom Kippur War of October 1973. The Soviet Union had its own supplies of energy, but its major oil and gas fields were in Siberia and the harnessing of the resources was pushing the available Soviet technology to its limits.

It has already been mentioned that the mid-1970s saw a major redirecting of the Soviet space programme, and it was during this period that a new consideration was added to the existing list of reasons for going into space. Economic advantage.

Previously, such considerations had been of little consequence, but slowly even politicians became aware of the potential of space to provide new materials to assist technological progress on Earth. The weightless environment of orbital flight allowed the preparation of purely mixed alloys which could not be manufactured on Earth because of differing densities of the component parts. Additionally, space would also allow the preparation of medical serums purer than those which could be made on Earth – again by taking advantage of microgravity.

However, in the Soviet grand scheme of spaceflight, such stratagems were merely trinkets; a far bolder plan seems also to have been hatched: this envisaged being able to build solar power stations and other power-related satellites in Earth orbit which would have a direct economic benefit for the Soviet Union. This is a long-term plan to which the Soviets seem seriously committed.

The idea of developing solar power satellites was originally proposed by the American, Peter Glaser, in 1968, but while the United States undertook feasibility studies and little more, the Soviet Union read his proposals during the 1970s with great interest. Economists and space planners were instructed to look at the ideas to assess whether such a programme could indeed be undertaken. In 1985 S. A. Sarkisyan and others presented a paper to the

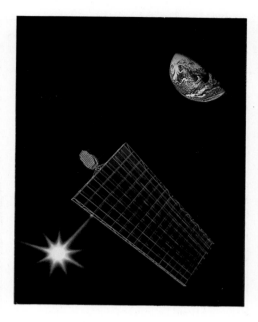

Above: *Looking further into the future, the Soviets are believed to be considering using orbiting power stations to harness the Sun's energy and beam it to Earth for industrial and domestic use.*

IAF meeting which outlined a programme for the period up to 2020 involving the initial testing of and later operational use of space mirrors to illuminate Soviet towns at night, as well as space power stations which would convert solar radiation into microwave energy and beam this down to Earth where it could be converted further into a useful form of energy for distribution throughout the Soviet Union.

Such a plan is bold indeed, and when the paper was presented, Sarkisyan was asked whether it was simply a theoretical piece of work or a funded programme; the reply was "We have been watching your studies in the West and we agree with your main conclusions. We do have a programme, and the first elements are in the current five year plan. In due course, we will be offering collaboration to the West."

The introduction of the Energiya booster in 1987 is probably a major part of the space power programme, because its lifting power is required for the assembly of the power stations and space mirrors. Twelve mirrors with diameters of 1.5km would allow Moscow to be fully illuminated at night, while solar power satellites with sides 4.5km long could use laser guns to beam 500MW of energy back to Earth. Such a solar power station would require the launch of perhaps 40 Energiya boosters to carry the building blocks into orbit. This undertaking would be far from cheap. However, for the Soviet economy the benefits of such an operation cannot be calculated in monetary terms; it would completely restructure the Soviet philosophy of energy conversion.

Although such power bases in space would be unmanned, the human element in their construction is essential. Starting with modest experiments on Salyut 7 during 1986 the Soviet Union is planning the assembly of larger and larger space structures throughout the Mir and Mir-2 programmes. This work is essential for

investigating and proving the viability of assembling orbital power stations for the future.

Conclusions

The Soviet space programme has progressed further in the three decades which separate us from the launch of the first Sputnik in October 1957 than anyone could have realistically imagined. It took less than four years from Sputnik to put a man in orbit and within ten years plans were advanced for a manned lunar landing programme (which was not successful). Compared with the spectacular Apollo lunar programme, the Soviet achievement has seemed comparatively unexciting and has progressed at a rate which has seemed pedestrian by western standards. The American "hare" managed to touch the Moon, but then abandoned the dreams and technology which put the first human beings on another world. The Soviet "tortoise" continued with a slowly evolving space station programme during the 1970s and 1980s to a point where by 1990 it should have developed the Mir complex to be the equal of the American space station programme which will not be complete until the latter half of the 1990s.

When reading spaceflight books from the Soviet Union, it is clear that the Soviet Union's policy makers share the dreams of the space scientists who seem to believe that man (whether Soviet or not) is destined to explore beyond the Earth's cradle, and eventually to colonise and exploit the Solar System. The papers presented at the International Astronautical Federation Congress in 1987 show that Soviet space scientists are not only considering the current programmes and programmes for the near future; one paper presented by L. M. Shkadov was entitled "Possibility Of Controlling Solar System Motion In The Galaxy" (paper IAA-87-613), showing that Soviet scientists are freely considering possibilities which will not be undertaken for millennia.

People have a habit of saying that "the world is your oyster". Soviet space planners are progressing with the belief that "the whole space environment is your oyster". Will there be a serious western response?

FURTHER READING

Anon, "Soviet Space Capability: Big Surprises Coming?" in *Space Markets*, Autumn 1986, pp176-182.

Clark, P. S., "Energiya – Soviet Superbooster" in *Space*, Sep-Oct 1987 (Vol 3, issue 4), pp 36-37, 39.

Clark, P. S., "The Energiya Launch Vehicle" in *TRW Space Log 1987*, TRW Public Relations Dept., 1988.

Lawton, A., "The Soviet Space Shuttle Programme" in *Spaceflight*, vol 29 (Supplement 1, 1987), pp4-7.

Sarkisyan, S. A. et al, *Socio-Economic Benefits Connected with the Use of Space Power and Energy Systems* (paper presented at the International Astronomical Federation Congress, 1985), IAF-85-188, Pergamon Press, 1985.

Appendix 1: **Launch Vehicle Data**

This appendix brings together most of the information available relating to the performances of the launch vehicles which the Soviet Union has flown within its manned space programme. The amount of information available for the vehicles varies: for example, almost all of the numerical data are available for the Vostok and Soyuz class launch vehicles, while at the other end of the spectrum the Soviets have not even acknowledged the existence of the SL-15 or Type-G.

Although details of the launch vehicles used in the manned programme are covered, a note at the end of the appendix lists the other launch vehicle variants which the Soviets have flown to the end of 1987.

In the tables which follow, the figures which have been quoted by Soviet sources are marked with an asterisk. Numbers which have been directly derived from Soviet data are shown with a double asterisk (eg, propellant masses can be calculated if the burn time, thrust and specific impulses are given, or from partial mass breakdowns quoted in the Soviet literature). The rocket engine designations (the Soviets generally use "RD-" numbers, but a single "RO-" designation has been announced by the Soviets) are shown where they have been announced, otherwise the design bureau is shown (Kosberg, usually). The engine thrusts and specific impulses which are shown are the vacuum equivalents: the sea level equivalents of the first stage figures are about 90 per cent of the vacuum values. The abbreviation LOX/LH used in some tables stands for liquid oxygen and liquid hydrogen.

The Soviet Union has not developed a consistent designation system for its launch vehicles, the boosters generally being named after their prime payloads (Vostok, Soyuz, etc). In America the Department of Defense has devised a system of "SL-" designators; this has the advantage of assuming nothing about the actual booster configuration (other than it is distinct from other vehicles), but has the disadvantage of not reflecting the launch vehicle families.

Another alpha-numeric system of launch vehicle designators was devised in 1967 by Charles S. Sheldon II, but this is not generally used in this book. The launch vehicle family was identified by an upper case letter, followed by numbers and lower case letters which indicated the different upper stages. The Vostok and Soyuz family of boosters are the Type-A in Sheldon's system, the Proton the Type-D and the giant lunar booster the Type-G.

The information presented here for the launch vehicles is as accurate as Soviet data allows in mid-1988. In the era of Glasnost and with Soviet drive to sell their launch vehicle capabilities on the world market, new information is regularly being released by Glavcosmos, the Soviet commercial spaceflight organisation.

VOSTOK SL-3

FIRST STAGE DATA, *Blocks B, V, G, D*

Engine	RD-107
Propellants	LOX/Kerosene
Thrust (tonnes)	102*
Burn time (sec)	122
Specific impulse (sec)	314*
Length (metres)	19.8*
Diameter (metres)	2.68*
Dry mass (tonnes)	3.45**
Propellant mass (tonnes)	39.64**

SECOND STAGE DATA, *Block A*

Engine	RD-108
Propellants	LOX/Kerosene
Thrust (tonnes)	96*
Burn time (sec)	304
Specific impulse (sec)	315*
Length (metres)	28.75*
Diameter (metres)	2.95*
Dry mass (tonnes)	6.51**
Propellant mass (tonnes)	92.65**

THIRD STAGE DATA, *Block E*

Engine	RO-7
Propellants	LOX/Kerosene
Thrust (tonnes)	5.6*
Burn time (sec)	430*
Specific impulse (sec)	326*
Length (metres)	3.1*
Diameter (metres)	2.6*
Dry mass (tonnes)	1.44*
Propellant mass (tonnes)	7.39**
PAYLOAD MASS (tonnes)	4.72*
SHROUD MASS (tonnes)	2.5
LAUNCH MASS (tonnes)	287.03
TOTAL LENGTH (metres)	38.36*

Further Details: The original variant of the SL-3 was launched successfully during 1959 with the first three Luna probes. These vehicles used an early version of the Block-E stage with the following details: thrust, 5 tonnes, specific impulse, 316 seconds, dry mass, 1.1 tonnes, estimated propellant mass, 7 tonnes. The total length of the original SL-3 was 33.5m.

There are four strap-on boosters which the Soviets designate Blocks -B, -V, -G and -D (the second to fifth letters of the cyrillic alphabet). The orbital capacity of the vehicle is generally quoted as 5 tonnes to a low Earth orbit, and it has placed payloads of up to 1.3 tonnes into Sun-synchronous orbits. The Soviets generally quote a launch mass of 287 tonnes for the vehicle.

The RD-107 and RD-108 engines were developed by Glushko's Gas Dynamics Laboratory (GDL-OKB) during 1954-1957. The RO-7 engine, along with many upper stage engines, was produced by the Kosberg bureau. The first flight of a Vostok booster to reach space was the first Luna mission (2 January 1959), but the first flight with a Vostok payload was the first Korabl-Sputnik (15 May 1960).

In support of their programme to sell the Vostok booster commercially, the Soviets have given some further burn times for the three stages of the

vehicle: the first stage (strap-on) burn time is about 120 seconds, the second stage core burn time is about 310 seconds and the total burn time about 11 minutes to orbital injection (implying that the third stage burns for 350 seconds). These figures, of course, relate to the current variant, not the one used for the Vostok manned missions.

VOSKHOD SL-4

FIRST STAGE DATA, *Blocks B, V, G, D*

Engine	RD-107
Propellants	LOX/Kerosene
Thrust (tonnes)	102*
Burn time (sec)	122
Specific impulse (sec)	314*
Length (metres)	19.8*
Diameter (metres)	2.68*
Dry mass (tonnes)	3.45**
Propellant mass (tonnes)	39.63**

SECOND STAGE DATA, *Block A*

Engine	RD-108
Propellants	LOX/Kerosene
Thrust (tonnes)	96*
Burn time (sec)	314
Specific impulse (sec)	315**
Length (metres)	28.75*
Diameter (metres)	2.95*
Dry mass (tonnes)	6.51**
Propellant mass (tonnes)	95.70**

THIRD STAGE DATA, *Block I*

Engine	RD-461
Propellants	LOX/Kerosene
Thrust (tonnes)	30*
Burn time (sec)	240
Specific impulse (sec)	330*
Length (metres)	9
Diameter (metres)	2.66*
Dry mass (tonnes)	3.25**
Propellant mass (tonnes)	21.30**
PAYLOAD MASS (tonnes)	5.68*
SHROUD MASS (tonnes)	2.5
LAUNCH MASS (tonnes)	307.26
TOTAL LENGTH (metres)	44.9

Further Details: The SL-4 launch vehicle used the core and strap-ons from the Vostok booster, but with a new third stage added. The basic length of the third stage is 8.1m, but for Voskhod a cylindrical section was added to the top of the stage to cover the conical retro-rocket module on the payload. On the SL-3 vehicle propellant was kept in toroidal tanks in the third stage, and the retro-rocket could "sit" inside the torus, but the tanks on the SL-4 third stage were ellipsoids, leaving no room for the retro-rocket.

In the manned programme, this variant was only used for the two manned Voskhod missions and their Cosmos precursors, as well as the Voskhod-related Cosmos 110 biological satellite in 1966. However, it has been used widely in the Cosmos photo-reconnaissance satellite programme and the series of biological missions which began with Cosmos 605 in 1973.

The RD-461 engine was developed by the

Kosberg bureau. The first use of the SL-4 in this variant was the launch of Cosmos 22 on 23 November 1963, although all three stages were used on the four stage SL-6 Molniya booster which first reached orbit in February 1961.

SOYUZ SL-4

FIRST STAGE DATA, *Blocks B, V, G, D*

Engine	RD-107
Propellants	LOX/Kerosene
Thrust (tonnes)	102*
Burn time (sec)	122
Specific impulse (sec)	314*
Length (metres)	19.8*
Diameter (metres)	2.68*
Dry mass (tonnes)	3.45**
Propellant mass (tonnes)	39.63**

SECOND STAGE DATA, *Block A*

Engine	RD-108
Propellants	LOX/Kerosene
Thrust (tonnes)	96*
Burn time (sec)	314
Specific impulse (sec)	315*
Length (metres)	28.75*
Diameter (metres)	2.95*
Dry mass (tonnes)	6.51**
Propellant mass (tonnes)	95.70**

THIRD STAGE DATA, *Block I*

Engine	RD-461
Propellants	LOX/Kerosene
Thrust (tonnes)	30*
Burn time (sec)	240
Specific impulse (sec)	330*
Length (metres)	8.1*
Diameter (metres)	2.66*
Dry mass (tonnes)	2.4**
Propellant mass (tonnes)	21.3**
PAYLOAD MASS (tonnes)	6.8*
SHROUD MASS (tonnes)	4.5
LAUNCH MASS (tonnes)	309.53
TOTAL LENGTH (metres)	49.3

Further Details: The launch mass of the vehicle is generally announced as 310 tonnes. The lengths of the various Soyuz boosters have varied, because there have been at least six different variants of the payload shroud rocket tower, each of which have been of different lengths (Soyuz 1-9, Soyuz/ASTP, Soyuz 12-24 ferries, Soyuz 25-40 ferries, Soyuz-T and Soyuz-TM).

The masses of the early Soyuz spacecraft were within the range 6.5-6.8 tonnes, the Soyuz-T variant had a mass of 6.85 tonnes and the Soyuz-TM variant was about 7.1 tonnes with its cosmonaut crew (the unmanned first Soyuz-TM was 7,070kg). The launch vehicle, with a modified payload shroud (excluding the four aerodynamic flaps) is used for the Progress unmanned freighter missions; the original Progress missions had masses of about 7.02 tonnes, but the Progress missions to Mir seem to have had masses of about 7.2 tonnes.

The first orbital flight of this SL-4 variant was Cosmos 133, on 28 November 1966. According to Glushko, the RD-108 engine performance during the 1980s was increased to give a specific impulse of 323 seconds, but no other details of the up-rated engine were given.

As with the Vostok booster, the Soyuz is being marketed commercially, and the following burn durations have been announced in support of the marketing venture:

Stage 1 (strap-ons) 118 sec
Stage 2 (core) 285 sec

The final orbital injection comes about 9 minutes after launch, implying a third stage burn time of about 255 seconds.

COSMOS SL-8

FIRST STAGE DATA

Engine	RD-216
Propellants	Nitric acid/UDMH
Thrust (tonnes)	176*
Burn time (sec)	170*
Specific impulse (sec)	291*
Length (metres)	19
Diameter (metres)	2.44
Dry mass (tonnes)	6
Propellant mass (tonnes)	102.8**

SECOND STAGE DATA

Engine: No performance details available	
Length (metres)	7
Diameter (metres)	2.44
Dry mass (tonnes)	1.6*
Propellant mass (tonnes)	12.8
PAYLOAD MASS (tonnes)	1.25
SHROUD MASS (tonnes)	0.5
LAUNCH MASS (tonnes)	124.95
TOTAL LENGTH (metres)	31.2

Further Details: This launch vehicle has only been used in connection with the manned programme for the Cosmos 1374 series of spaceplane model flights. The first use of the SL-8 booster was Cosmos 38-40 on 18 August 1964. The RD-216 engine was developed by the GDL-OKB during 1958-1960. No numerical details are available relating to the second stage of the booster.

Below: *Lift-off of a Soyuz SL-4 launch vehicle from Tyuratam launch centre. The launch mass of this booster variant is estimated at around 310 tonnes.*

PROTON SL-12

FIRST STAGE DATA

Engine	RD-253
Propellants	Nitrogen tetroxide/UDMH
Thrust (tonnes)	167*
Burn time (sec)	130*
Specific impulse (sec)	316*
Length (metres)	20.2
Diameter (metres)	7.4*
Dry mass (tonnes)	43.4
Propellant mass (tonnes)	412.2**

SECOND STAGE DATA

Engine	(Kosberg Bureau)
Propellants	Nitrogen tetroxide/UDMH
Thrust (tonnes)	240*
Burn time (sec)	208
Specific impulse (sec)	333
Length (metres)	13.7
Diameter (metres)	4.15*
Dry mass (tonnes)	13.2**
Propellant mass (tonnes)	152.4

THIRD STAGE DATA

Engine	(Kosberg Bureau)
Propellants	Nitrogen tetroxide/UDMH
Thrust (tonnes)	64*
Burn time (sec)	254
Specific impulse (sec)	344
Length (metres)	6.4
Diameter (metres)	4.15*
Dry mass (tonnes)	5.6**
Propellant mass (tonnes)	47.5

FOURTH STAGE DATA *Block D*

Engine	(Korolyov Bureau)
Propellants	LOX/Kerosene
Thrust (tonnes)	8.7*
Burn time (sec)	600*
Specific impulse (sec)	352*
Length (metres)	5.5*
Diameter (metres)	3.7*
Dry mass (tonnes)	1.8**
Propellant mass (tonnes)	14.8**
PAYLOAD MASS (tonnes)	5.4
SHROUD MASS (tonnes)	3.5
LAUNCH MASS (tonnes)	699.8
TOTAL LENGTH (metres)	58.9

Further Details: The four stage Proton SL-12 is the vehicle which the Soviet Union has been trying to sell as a commercial launch vehicle of geosynchronous satellites. The configuration detailed above is that used to launch the Soyuz-derived Zond missions around the Moon during 1968-1970.

The vehicle was designed by the Chelomei bureau, using – it is believed – two stages from the giant SL-15 vehicle for its second and third stages. The fourth stage was designed by Korolyov's bureau, and might have been transferred from Korolyov's own design for a launch vehicle capable of placing 20 tonnes into low Earth orbit. The second and third stage engines were designed by the Kosberg bureau; the second stage uses four engines of a similar type to the one used on the third stage. The thrust of the third stage includes the four vernier engines which are added to the single-chamber engine.

The first flight of the SL-12 vehicle was Cosmos 146, launched on 10 March 1967. Starting with the launches of the heavy Venus probes in 1975, a modified fourth stage (designated "Block DM") has been used, with the last flight of the original variant being Luna 24 in August 1976.

The first stage of the Proton booster uses six RD-253 engines, these being designed by the GDL-OKB during 1961-1965. The six external tanks on the first stage contain UDMH while the central core tank carries the nitrogen tetroxide.

PROTON SL-13	
FIRST STAGE DATA	
Engine	RD-253
Propellants	Nitrogen tetroxide/UDMH
Thrust (tonnes)	167*
Burn time (sec)	130*
Specific impulse (sec)	316*
Length (metres)	20.2
Diameter (metres)	7.4*
Dry mass (tonnes)	43.4
Propellant mass (tonnes)	412.2**
SECOND STAGE DATA	
Engine	(Kosberg Bureau)
Propellants	Nitrogen tetroxide/UDMH
Thrust (tonnes)	240*
Burn time (sec)	208
Specific impulse (sec)	333
Length (metres)	13.7
Diameter (metres)	4.15*
Dry mass (tonnes)	13.2**
Propellant mass (tonnes)	152.4
THIRD STAGE DATA	
Engine	(Kosberg Bureau)
Propellants	Nitrogen tetroxide/UDMH
Thrust (tonnes)	64*
Burn time (sec)	254
Specific impulse (sec)	344
Length (metres)	6.4
Diameter (metres)	4.15*
Dry mass (tonnes)	5.6**
Propellant mass (tonnes)	47.5
PAYLOAD MASS (tonnes)	19.8*
SHROUD MASS (tonnes)	3.0
LAUNCH MASS (tonnes)	697.1
TOTAL LENGTH (metres)	59.8

Further Details: The Proton SL-13 variant uses the first three stages of the SL-12 vehicle, already described. The first flight of the three stage variant was the launch of Proton 4 on 16 November 1968. The vehicle has been used for the Salyut launches (the figures quoted above relate to Salyut 6), the Mir core, the Heavy Cosmos satellites and Kvant. Additionally, the SL-13 was used for the dual Cosmos launches during 1976-1979.

The heaviest payload announced for the SL-13 vehicle was the Mir core, with a mass of 20.9 tonnes. During the 1980s the first stage engine of the vehicle, the RD-253,was up-rated to give a thrust of 178 tonnes.

(No name) SL-15
NO INFORMATION AVAILABLE

Further Details: This is the abandoned giant booster, also known as the Type-G and the TT-5 in the West. Three vehicles are reported to have

been readied for flight. The first exploded on the launch pad on 3-4 July 1969 (it is claimed that propellant ignited while it was being loaded into an upper stage, causing the massive explosion); the second was launched 23-24 June 1971 but disintegrated at an altitude of about 12km and the third was launched on 24 November 1972 but was destroyed at an altitude of 40km after a multiple engine failure.

From Soviet sources there is nothing known about the SL-15 vehicle, but some rumours of its potential performance have appeared in the western aerospace press. The thrust of the first stage was said to be 10-14 million pounds – equivalent to 4,540-6,350 tonnes – and comprised a cluster of 24-36 engines (the engine thrusts would thus have been up to 265 tonnes). The launch mass seems to have been about 5,300 tonnes, implying the higher thrust limit quoted above was applicable. One can speculate that the vehicle was designed by the Korolyov bureau – since it was intended to support Korolyov's plan to land men on the Moon in the late 1960s – and thus had Glushko's GDL-OKB engines clustered on the first stage and Kosberg's engines on the other stages.

The booster was roughly conical in shape, having a base diameter of about 16m. The upper part of the booster, with the manned lunar complex, was cylindrical, having a diameter of about 4.2m. The overall height of the SL-15 was probably about 110m.

Assuming that the vehicle was optimally designed, it would probably have had three stages to reach Earth orbit, and would have been able to orbit a 220 tonne payload. For the lunar mission, the payload would have been the lunar Soyuz and fully fuelled Proton second and third stages, as described earlier in the lunar programme chapter. Assuming that the propellants used were those for the Vostok-Soyuz family to get into orbit, the first three stages would have used liquid oxygen and kerosene, while the Proton stages for the lunar mission would have carried nitrogen tetroxide and UDMH.

Below: *The four strap-on version of SL-17 Energiya which made its first flight on 15 May 1987. This is the vehicle that will lift the Soviet shuttle to orbit. Parts of it will be reusable.*

ENERGIYA SL-17 *(Four strap-on boosters)*	
FIRST STAGE DATA	
Engine	RD-170
Propellants	LOX/Kerosene
Thrust (tonnes)	806*
Burn time (sec)	156
Specific impulse (sec)	336*
Length (metres)	40
Diameter (metres)	4*
Dry mass (tonnes)	25
Propellant mass (tonnes)	375
SECOND STAGE DATA	
Engine	(Not known)
Propellants	LOX/LH
Thrust (tonnes)	200*
Burn time (sec)	470
Specific impulse (sec)	470
Length (metres)	60*
Diameter (metres)	8*
Dry mass (tonnes)	50
Propellant mass (tonnes)	821
THIRD STAGE DATA	
Engine	(Not known)
Propellants	(Not known)
Thrust (tonnes)	(Not known)
Burn time (sec)	(Not known)
Specific impulse (sec)	350
Length (metres)	(Not known)
Diameter (metres)	4
Dry mass (tonnes)	5
Propellant mass (tonnes)	25
PAYLOAD MASS (tonnes)	153
LAUNCH MASS (tonnes)	2654
TOTAL LENGTH (metres)	60

Further Details: The first flight of the Energiya shuttle booster was completed on 15 May 1987. This used four strap-on first stage boosters, although a later variant of Energiya should be capable of carrying six strap-ons. Some performance data were released for the first stage engine, designed by the GDL-OKB and designated RD-170. The second stage uses four engines, each with a thrust of 200 tonnes, clustered together; no further numerical details are available for these engines.

The third "kick" stage was not described for the vehicle, but it is mounted underneath the payload and strapped to one side of the core booster. It was stated that the "kick" stage used the same orbital manoeuvring system as the shuttle orbiter. The numerical data quoted above for the stage are based on the assumption that storable propellants are used.

The amount of reuse built into the Energiya vehicle is uncertain, and Soviet statements range from there being no reuse intended to the vehicle being fully reusable. It has been stated that the core splits into three sections for re-entry and landing, and another Soviet source has specifically said that the engines are reusable.

On its first flight, Energiya was only partially fuelled and carried a payload of 100 tonnes. The "kick" stage failed to ignite (reportedly, due to human error) and operate properly, and American sensors observed the re-entry of the combined "kick" stage and payload, estimating that the combined mass was 130 tonnes or so.

The Soviets have specifically stated that Energiya (presumably the four strap-on version) can launch 32 tonnes of payload towards the

Moon or 27 tonnes towards Mars or Venus. Of course, heavier payloads can be flown with orbital assembly.

The Soviet space shuttle orbiter is to be operated with the four strap-on variant of the launch vehicle. According to western estimates, the orbiter will be about 33m long and have a wing span of about 23m. The fuselage is believed to have a diameter of about 5.5m. The dry mass of the orbiter should be about 65-70 tonnes: this would theoretically allow for more than 80 tonnes for the payload in the shuttle cargo bay, but this figure would imply a very dense payload; perhaps a more realistic figure will be 40 tonnes on a one-way trip to orbit. The shuttle should be typically used for the launch of space station modules and for their return to Earth for refurbishment.

In the performance calculations presented for Energiya, it has been assumed that the orbit will be 51.6°, 200-250km, and the velocity loss due to air resistance and gravity will be about 1,500m/s.

The core apparently burns on the pad some 12.3 seconds before launch, and this reduces the vehicle's actual lift-off mass by about 21 tonnes from the figure which is quoted above.

ENERGIYA SL-? *(Six strap-on boosters)*	
FIRST STAGE DATA	
Engine	RD-170
Propellants	LOX/Kerosene
Thrust (tonnes)	806*
Burn time (sec)	156
Specific impulse (sec)	336*
Length (metres)	40
Diameter (metres)	4*
Dry mass (tonnes)	25
Propellant mass (tonnes)	375
SECOND STAGE DATA	
Engine	(Not known)
Propellants	LOX/LH
Thrust (tonnes)	200*
Burn time (sec)	470
Specific impulse (sec)	470
Length (metres)	60*
Diameter (metres)	8*
Dry mass (tonnes)	50
Propellant mass (tonnes)	821
THIRD STAGE DATA	
Engine	(Not known)
Propellants	(Not known)
Thrust (tonnes)	(Not known)
Burn time (sec)	(Not known)
Specific impulse (sec)	350
Length (metres)	(Not known)
Diameter (metres)	5.5
Dry mass (tonnes)	10
Propellant mass (tonnes)	65
PAYLOAD MASS (tonnes)	230
LAUNCH MASS (tonnes)	3576
TOTAL LENGTH (metres)	60

Further Details: The six strap-on variant of the Energiya booster is believed to be the optimum one for the amount of payload which can be placed in orbit, and this variant has been assumed for the manned Mars mission reconstruction presented in Chapter 14. This variant would match the low Earth orbit lift capability of the abandoned SL-15 vehicle. The Soviets have confirmed that a six strap-on Energiya is planned, this being the maximum number of strap-ons envisaged.

Remaining launch vehicles

To complete the record, the Soviet launch vehicles which have not seen an application in the manned space programme are listed below.

The **Sputnik Rocket**, also called R-7 by the Soviet Union, was the basis for the Vostok, Soyuz and Molniya launch vehicles. The first stage comprised four strap-on boosters, each powered by a single RD-107 engine: the performance parameters for the early RD-107 engines are thrust, 100 tonnes; specific impulse, 309 seconds. The core used a single RD-108 engine, with the following parameters: thrust, 93 tonnes; specific impulse, 308 seconds. This variant is designated **SL-1** and **SL-2** in the West, and was used for the first three Sputnik launches. For Sputnik 3 (mass 1,327kg) the launch mass was 267 tonnes and the length was 29.17m.

The **SL-5** is an obscure variant of the SL-3 vehicle, believed to have launched Polyot 1 and Polyot 2; its actual configuration is not known.

The **Molniya Booster** is a four stage variant of the SL-4 vehicle, and it is designated **SL-6** in the West. It takes the Soyuz launch vehicle, and replaces the payload with a Vostok-type payload shroud. Inside the shroud is the small Block-L stage, known in the West as an 'escape stage'. The Block-L has a dry mass of about 1.1 tonnes and can carry a load of 5.6 tonnes of liquid oxygen and kerosene; the specific impulse is 340 seconds and the thrust is 6.7 tonnes. It was used to launch the second series of Luna probes (masses up to 1.65 tonnes) and the early Mars and Venus missions (masses up to 1.2 tonnes), It is still in use for launching the Molniya communications satellites (masses 1.6 tonnes) and military early warning satellites.

The smallest booster flown by the Soviets is the **Cosmos Rocket**, or **SL-7**. Introduced in 1962, it has not been used since 1977. The first stage was powered by an RD-214 engine and the second stage used an RD-119 engine. The launch mass of the vehicle was about 53 tonnes and it could orbit a maximum of about 500kg.

The first **Proton Booster** variant to fly was the **SL-9**: this used only the first and second stages of the SL-13 vehicle, and carried three Proton satellites into orbit (plus a launch failure) during 1965-66. The satellites had masses of 12.2 tonnes.

A further variant of the Vostok-Soyuz series of launch vehicles has been designated **SL-10**, but its actual configuration is not known from the public record. It was probably used for the launches of Cosmos 102 and Cosmos 125 in 1965-1966.

There have been two variants of the **Tsyklon** booster, which is derived from the Scarp SS-9 missile. The **SL-11** is a two stage vehicle, developed for an orbital weapon called "FOBS" (Fractional Orbit Bombardment System) during 1967-1971. The same vehicle has been used for launching anti-satellite weapons and ocean reconnaissance satellites. A three stage variant of the Tsyklon, the **SL-14**, was introduced in 1977 and that has been used for meteorological, oceanographic and other applications and military payloads. Few details are available

about the Tsyklon series, but the three stage variant can place about 4 tonnes into a low orbit. It has been privately confirmed by Soviet officials that the Tsyklon booster uses the RD-219 engine in its second stage (thrust, 90 tonnes; specific impulse, 293 seconds; burn time, 125 seconds; implying a propellant load of 38.4 tonnes), burning a nitric acid and UDMH.

The final launch vehicle to be noted here is the **SL-16** medium lift launch vehicle, discussed in Chapter 14. A two-stage vehicle, the first stage uses the RD-170 engine which is also used on Energiya's strap-ons. The orbital capability is at least 15 tonnes and it might possibly rival the Proton booster's orbital capacity to low Earth orbit in the future.

FURTHER READING

As far as possible, the information included has been drawn from Soviet sources. The Keldysh volume has provided most of the numerical data for the Sputnik-Vostok-Soyuz family of launch vehicles, while the Glushko books have been essential in deriving the rocket engine data.

Clark, P. S., "Soviet Rocket Engines", *JBIS*, (submitted for publication June 1988).

Glushko, V. P., *Development of Rocketry and Space Technology in the USSR*, Novosti Press Agency, Moscow, 1973, 1st edition.

Glushko, V. P., *Rocket Engines GDL-OKB*, Novosti Press Agency, Moscow, 1975.

Glushko, V. P., *Razvitye Raketostroiniya I Kosmonautika B CCCP* (in Russian), Mashinastroyeniye Press, Moscow, 1981, 2nd edition, 1987, 3rd edition.

Glushko, V. P., *Kosmonautika Encyclopediya*, Sovetskaya Encyclopediya, Moscow, 1985.

Keldysh, M. V. (editor), *The Creative Legacy of Academician Sergei Pavlovich Korolyov* (in Russian), Nauka Publishers, Moscow, 1980.

Most of the launch vehicle analysis is based upon reports marketed by Commercial Space Technologies Ltd (for example, *The Proton Launch Vehicle*, published in revised editions in 1986 and 1988) and discussions with specialists at CST Ltd. The following article was an initial analysis of the data available for Energiya.

Clark, P. S., "Energiya: Soviet Superbooster" in *Space*, Sep-Oct 1987 (Vol 3, issue 4), pp36-37, 39.

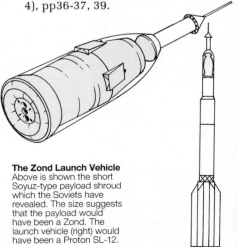

The Zond Launch Vehicle
Above is shown the short Soyuz-type payload shroud which the Soviets have revealed. The size suggests that the payload would have been a Zond. The launch vehicle (right) would have been a Proton SL-12.

Appendix 2: **Soviet Manned Spaceflight Log**

LOG OF ALL ACTUAL AND PLANNED LAUNCHES IN THE SOVIET MANNED SPACE PROGRAMME

Year	MISSION	LAUNCH DATE AND TIME	DOCKING DATE AND TIME	UNDOCKING DATE AND TIME	RECOVERY DATE AND TIME	MASS (kg)	PRIME CREW	BACK-UP CREW
1960	KS 1	15 May			(15 Oct 1965)	4,540	Unmanned Vostok	
	KS 2-1	23 Jul	Failed to reach orbit			4,600 ?	Unmanned Vostok – 2 dogs ?	
	KS 2	19 Aug			20 Aug	4,600	Unmanned Vostok – 2 dogs	
	KS 3	1 Dec			2 Dec	4,563	Unmanned Vostok – 2 dogs	
	KS 4-1	22 Dec ?	Failed to reach orbit			4,575 ?	Unmanned Vostok – 2 dogs?	
1961	KS 4	9 Mar			9 Mar	4,700	Unmanned Vostok – 1 dog	
	KS 5	25 Mar			25 Mar	4,695	Unmanned Vostok – 1 dog	
	Vostok 1	12 Apr 06.07			12 Apr 07.55	4,725	Y. A. Gagarin	G. S. Titov
	Vostok 2	6 Aug 06.00			7 Aug 07.15	4,731	G. S. Titov	A. G. Nikolayev
1962	Vostok 3	11 Aug 08.30			15 Aug 06.52	4,722	A. G. Nikolayev	V. F. Bykovsky
	Vostok 4	12 Aug 08.02			15 Aug 06.59	4,728	P. R. Popovich	V. M. Komarov *then* B. V. Volynov
1963	Vostok 5	14 Jun 12.00			19 Jun 11.00	4,720	V. F. Bykovsky	B. V. Volynov
	Vostok 6	16 Jun 09.30			19 Jun 08.16	4,713	V. V. Tereskhova	I. B. Solovyova
1964	Vostok 7	Summer ?	Planned mission cancelled				B. B. Yegorov	V. G. Lazarev ?
	Cosmos 47	6 Oct 07.15 ?			7 Oct	5,200 ?	Unmanned Voskhod	
	Voskhod 1	12 Oct 07.30			13 Oct 07.47	5,320	V. M. Komarov, K. P. Feoktistov, B. B. Yegorov	B. V. Volynov, G. P. Katys, V. G. Lazarev
1965	Cosmos 57	22 Feb 07.40 ?	Disintegrated in orbit			5,500 ?	Unmanned Voskhod	
	Voskhod 2	18 Mar 07.00			19 Mar 09.02	5,682	P. I. Belyayev, A. A. Leonov	V. V. Gorbatko *then* D. A. Zaikin, Y. V. Khrunov
	Cosmos	2 Jul	Failed to reach orbit			5,700 ?	Unmanned Voskhod – 2 dogs ?	
1966	Voskhod 3	??	Planned mission cancelled			6,000 ?	B. V. Volynov, G. S. Shonin	V. A. Shatalov, G. T. Beregovoi
	Cosmos 110	22 Feb 20.10			16 Mar 14,00 ?	5,700 ?	Unmanned Voskhod – 2 dogs	
	Cosmos 133	28 Nov 11.00			30 Nov 10.21	6,300 ?	Unmanned Soyuz	
1967	Cosmos 140	7 Feb 03.20			9 Feb 02.52	6,300 ?	Unmanned Soyuz	
	Cosmos 146	10 Mar 11.31			18 Mar ?	5,017	Unmanned Zond	
	Cosmos 154	8 Apr 09.00			19 Apr ?	5,020 ?	Unmanned Zond	
	Soyuz 1	23 Apr 00.35			24 Apr 03.23	6,450	V. M. Komarov	Y. A. Gagarin
	Soyuz 2	24 Apr	Planned mission cancelled			6,400 ?	V. F. Bykovsky, A. S. Yeliseyev, Y. V. Khrunov	A. G. Nikolayev, V. N. Kubasov, V. V. Gorbatko
	Cosmos 159	16 May 21.45			(11 Nov 1977)	3,780 ?	Unmanned Soyuz propulsion test	
	Cosmos 186	27 Oct 09.30	30 Oct 09.20	30 Oct 12.50	31 Oct 08.20	6,400 ?	Unmanned Soyuz	
	Cosmos 188	30 Oct 08.12	30 Oct 09.20	30 Oct 12.50	2 Nov 09.10	6,400 ?	Unmanned Soyuz	
	Zond	21 Nov 18.45	Failed to reach orbit			5,140 ?	Unmanned circumlunar mission	
1968	Zond 4	2 Mar 18.30			9 Mar ?	5,140 ?	Unmanned lunar simulation	
	Cosmos 212	14 Apr 10.00	15 Apr 10.21	15 Apr 14.11	19 Apr 08.10	6,400 ?	Unmanned Soyuz	
	Cosmos 213	15 Apr 09.34	15 Apr 10.21	15 Apr 14.11	19 Apr 10.11	6,400 ?	Unmanned Soyuz	
	Zond ?	22 Apr	rumoured launch failure			5,140 ?	Unmanned lunar-related flight	
	Cosmos 238	28 Aug 09.59			1 Sep 09.03	6,400 ?	Unmanned Soyuz	
	Zond 5	14 Sep 21.42			21 Sep 16.08	5,140 ?	Unmanned circumlunar mission	
	Soyuz 2	25 Oct 09.00			28 Oct 07.51	6,450 ?	Unmanned	
	Soyuz 3	26 Oct 08.34			30 Oct 07.25	6,575	G. T. Beregovoi	V. A. Shatalov *and* B. V. Volynov
	Zond 6	10 Nov 19.11			17 Nov 14.10	5,140 ?	Unmanned circumlunar mission	
1969	Soyuz 4	14 Jan 07.29	16 Jan 08.20	16 Jan 12.55	17 Jan 06.53	6,626	V. A. Shatalov	A. V. Filipchenko
	Soyuz 5	15 Jan 07.05	16 Jan 08.20	16 Jan 12.55	17 Jan 08.00	6,585	B. V. Volynov, A. S. Yeliseyev*, Y. V. Khrunov*	G. S. Shonin, V. N. Kubasov, V. V. Gorbatko
	No name	3-4 Jul	Launch pad explosion			200,000 ?	Unmanned giant booster	
	Zond 7	7 Aug 23.55			14 Aug	5,390 ?	Unmanned circumlunar mission	
	Soyuz 6	11 Oct 11.10			16 Oct 09.52	6,577	G. S. Shonin, V. N. Kubasov	A. G. Nikolayev, V. I. Sevastyanov *and* N. N. Rukavishnikov
	Soyuz 7	12 Oct 10.45			17 Oct 09.26	6,570	A. V. Filipchenko, V. N. Volkov, V. V. Gorbatko	A. G. Nikolayev, V. I. Sevastyanov, G. M. Grechko *and* N. N. Rukavishnikov
	Soyuz 8	13 Oct 10.20			18 Oct 09.10	6,646	V. A. Shatalov, A. S. Yeliseyev	A. G. Nikolayev, V. I. Sevastyanov
	Cosmos	16 Nov ?	Rumoured launch failure			18,700 ?	Unmanned Soyuz lunar lander test ?	
1970	Soyuz 9	1 Jun 19.00			19 Jun 11.59	6,590	A. G. Nikolayev, V. I. Sevastyanov	V. G. Lazarev *and* A. V. Filipchenko, V. A. Yazdovski *and* G. M. Grechko
	Zond 8	20 Oct 19.55			27 Oct	5,390 ?	Unmanned circumlunar mission	
	Cosmos 379	24 Nov 05.15			(21 Sep 1983)	5,425 ?	Unmanned Soyuz propulsion test	
	Cosmos 382	2 Dec 16.44			In orbit	18,735 ?	Unmanned Soyuz lunar lander test	
1971	Cosmos 398	26 Feb 05.05			In orbit	5,885 ?	Unmanned Soyuz propulsion test	
	Salyut 1	19 Apr 01.39			11 Oct	18,900	Orbital station	

LOG OF ALL ACTUAL AND PLANNED LAUNCHES IN THE SOVIET MANNED SPACE PROGRAMME – continued

	MISSION	LAUNCH DATE AND TIME	DOCKING DATE AND TIME	UNDOCKING DATE AND TIME	RECOVERY DATE AND TIME	MASS (kg)	PRIME CREW	BACK-UP CREW
1971	Soyuz 10	22 Apr 23.54	24 Apr 01.47 24 Apr 05.47	24 Apr 04.18 24 Apr 07.17	24 Apr 23.40	6,800 ?	V. A. Shatalov, A. S. Yeliseyev, N. N. Rukavishnikov	G. T. Dobrovolsky, V. N. Volkov, V. I. Patsayev
	Soyuz 11	6 Jun 04.55	7 Jun 07.45	29 Jun 18.28	29 Jun 23.17	6,790	A. A. Leonov *then* G. T. Dobrovolsky, V. N. Kubasov *then* V. N. Volkov, P. I. Kolodin *then* V. I. Patsayev	G. T. Dobrovolsky *then* V. A. Shatalov, V. N. Volkov *then* A. S. Yeliseyev, V. I. Patsayev *then* N. N. Rukavishnikov
	No name	23-24 Jun	Failed to reach orbit			200,000 ?	Unmanned giant booster	
	Cosmos 434	12 Aug 05.30	In orbit			5,740 ?	Unmanned Soyuz propulsion test	
1972	Cosmos 496	26 Jun 14.53			2 Jul 13.54	6,675 ?	Unmanned Soyuz	
	Salyut 2-1	29 Jul 03.17	Failed to reach orbit			18,900 ?	Orbital statiion (civilian ?)	
	No name	24 Nov	Failed to reach orbit			200,000 ?	Unmanned giant booster	
1973	Salyut 2	3 Apr 09.00	Failed in orbit		(28 May)	18,900 ?	Military orbital station	
	Cosmos 557	11 May 00.20	Failed in orbit		(22 May)	18,900 ?	Civilian orbital station	
	Cosmos 573	15 Jun 06.00			17 Jun 06.01	6,675 ?	Unmanned Soyuz	
	Soyuz 12	27 Sep 12.18			29 Sep 11.34	6,720	V. G. Lazarev, O. G. Makarov	A. A. Gubarev, G. M. Grechko
	Cosmos 613	30 Nov 05.20			29 Jan 05.29	6,675 ?	Unmanned Soyuz	
	Soyuz 13	18 Dec 11.55			26 Dec 08.50	6,560	P. I. Klimuk, V. V. Lebedev	L. V. Vorobyov, V. A. Yazdovski
1974	Cosmos 638	3 Apr 07.31			13 Apr 05.05	6,575 ?	Unmanned ASTP test	
	Cosmos 656	27 May 07.25			29 May 07.50	6,675 ?	Unmanned Soyuz	
	Salyut 3	22 Jun 22.38			24 Jan 1975	18,900 ?	Military orbital station	
	Soyuz 14	3 Jul 18.51	4 Jul 20.35	19 Jul 09.03	19 Jul 12.21	6,800 ?	P. R. Popovich, Y. P. Artyukhin	B. V. Volynov, V. M. Zholobov
	Cosmos 670	6 Aug 00.01			8 Aug 23.59	6,750 ?	Unmanned Soyuz-T ?	
	Cosmos 672	12 Aug 06.24			18 Aug 05.02	6,575 ?	Unmanned ASTP test	
	Soyuz 15	26 Aug 19.58	Docking failure		28 Aug 20.10	6,760	G. V. Sarafanov, L. S. Demin	B. V. Volynov, V. M. Zholobov
	Soyuz 16	2 Dec 09.40			8 Dec 08.04	6,800	A. V. Filipchenko, N. N. Rukavishnikov	Y. V. Romanenko, A. S. Ivanchenkov
	Salyut 4	26 Dec 04.15			2 Feb 1977	18,900 ?	Civilian orbital station	
1975	Soyuz 17	10 Jan 21.43	12 Jan 01.25	9 Feb 06.08	9 Feb 11.03	6,800 ?	A. A. Gubarev, G. M. Grechko	P. I. Klimuk, V. I. Sevastyanov
	Soyuz 18-1	5 Apr 11.02	Failed to reach orbit			6,830	V. G. Lazarev, O. G. Makarov	P. I. Klimuk, V. I. Sevastyanov
	Soyuz 18	28 May 14.58	29 May 18.44	26 Jul 10.56	26 Jul 14.18	6.825 ?	P. I. Klimuk, V. I. Sevastyanov	V. V. Kovalyonok, Y. Ponomarev
	Soyuz 19	15 Jul 12.20	17 Jul 16.12 19 Jul 12,20	19 Jul 12.02 19 Jul 15.26	21 Jul 10.51	6,790	A. A. Leonov, V. N. Kubasov	V. A. Dzhanibekov, B. D. Andreyev
	Cosmos 772	29 Sep 04.18			2 Oct 04.10	6,750 ?	Unmanned Soyuz-T ?	
	Soyuz 20	17 Nov 14.37	19 Nov 19.20	15 Feb 23.04	16 Feb 02.24	6,700 ?	Unmanned Soyuz	
1976	Salyut 5	22 Jun 18.04			8 Aug 1977	18,900 ?	Military orbital station	
	Soyuz 21	6 Jul 12.09	7 Jul 13,40	24 Aug 15.12	24 Aug 18.33	6,800 ?	B. V. Volynov, V. M. Zholobov	V. D. Zudov, V. I. Rozdestvensky
	Soyuz 22	15 Sep 09.48			23 Sep 07.42	6,510	V. F. Bykovsky V. V. Aksyonov	L. I. Popov *and* Y. V. Malyshev, B. D. Andreyev *and* G. M. Strekalov
	Soyuz 23	14 Oct 17.40	Docking failure		14 Oct 17.46	6,760	V. D. Zudov, V. I. Rozdestvensky	V. V. Gorbatko, Y. N. Glazkov
	Cosmos 869	29 Nov 16.00			17 Dec 10.31	6,750 ?	Unmanned Soyuz-T	
	Cosmos 881 Cosmos 882	15 Dec 01.29 15 Dec 01.29			15 Dec 15 Dec	3,500 ? 3,500 ?	Shuttle model test Shuttle model test	
1977	Soyuz 24	7 Feb 16.12	8 Feb 17.38	25 Feb 06.21	25 Feb 09.38	6,800 ?	V. V. Gorbatko, Y. N. Glazkov	A. N. Berezovoi, M. I. Lisum
	Cosmos 929	17 Jul 08.57			2 Feb 1978	20,000 ?	Heavy Cosmos module	Capsule down 16-17 Aug
	Cosmos Cosmos	4 Aug 4 Aug	Rumoured launch failure Rumoured launch failure			3,500 ? 3.500 ?	Shuttle model test Shuttle model test	
	Salyut 6	29 Sep 06.50			29 Jul 1982	19,825	Civilian orbital station	
	Soyuz 25	9 Oct 02.40	Docking failure		11 Oct 03.25	6,860	V. V. Kovalyonok, V. V. Ryumin	Y. V. Romañenko, A. S. Ivanchenkov
	Soyuz 26	10 Dec 01.19	11 Dec 03.02	16 Jan 08.05	16 Jan 11.25	6,800 ?	Y. V. Romanenko*, G. M. Grechko*	V. V. Kovalyonok, A. S. Ivanchenkov
1978	Soyuz 27	10 Jan 12.26	11 Jan 14.06	16 Mar 08.00	16 Mar 11.19	6,800 ?	V. A. Dzhanibekov*, O. G. Makarov*	V. V. Kovalyonok, A. S. Ivanchenkov
	Progress 1	20 Jan 08.25	22 Jan 10.12	6 Feb 05.53	8 Feb 02.39	7,020	Unmanned freighter	
	Soyuz 28	2 Mar 15.28	3 Mar 17.10	10 Mar 10.23	10 Mar 13.45	6,800 ?	A. N. Gubarev, V. Remek	N. N. Rukavishnikov, O. Pelczak
	Cosmos 997 Cosmos 998	30 Mar 00.00 30 Mar 00.00			30 Mar 30 Mar	3,500 ? 3,500 ?	Shuttle model test Shuttle model test	
	Cosmos 1001	4 Apr 15.07			15 Apr 12.02	6,750 ?	Unmanned Soyuz-T	
	Soyuz 29	15 Jun 20.17	16 Jun 21.58	3 Sep 08.23	3 Sep 11.40	6,800 ?	V. V. Kovalyonok*, A. S. Ivanchenkov*	V. A. Lyakhov, V. V. Ryumin
	Soyuz 30	27 Jun 15.27	28 Jun 17.08	5 Jul 10.15	5 Jul 13.30	6,800 ?	P. I. Klimuk, M. Hermaszewski	N. N. Kubasov, Z. Jankowski
	Progress 2	7 Jul 11.26	9 Jul 12.59	2 Aug 04.57	4 Aug 01.32	7,000 ?	Unmanned freighter	
	Progress 3	7 Aug 22.31	10 Aug 00.00	21 Aug 19.29	23 Aug 17.30	7,000 ?	Unmanned freighter	
	Soyuz 31	26 Aug 14.51	27 Aug 16.38 7 Sep 12.03	7 Sep 10.53 2 Nov 07.46	2 Nov 11.05	6,800 ?	V. F. Bykovsky*, S. Jahn*	V. V. Gorbatko, E. Koellner
	Progress 4	3 Oct 23.09	6 Oct 01.00	24 Oct 13.07	26 Oct 16.28	7,000 ?	Unmanned freighter	
1979	Cosmos 1074	31 Jan 09.00			1 Apr 10.09	6,750 ?	Unmanned Soyuz-T	
	Soyuz 32	25 Feb 11.54	26 Feb 13.30	13 Jun 09.51	13 Jun 16.18	6.800 ?	V. A. Lyakhov*, V. V. Ryumin*	L. I. Popov, V. V. Lebedev
	Progress 5	12 Mar 05.47	14 Mar 07.20	3 Apr 16.10	5 Apr 01.04	7,000 ?	Unmanned freighter	
	Soyuz 33	10 Apr 17.34	Docking failure		12 Apr 16.35	6,860	N. N. Rukavishnikov, G. Ivanov	Y. V. Romanenko, A. Alexandrov
	Progress 6	13 May 04.17	15 May 06.19	8 Jun 08.00	9 Jun 18.51	7,000 ?	Unmanned freighter	

LOG OF ALL ACTUAL AND PLANNED LAUNCHES IN THE SOVIET MANNED SPACE PROGRAMME – continued

	MISSION	LAUNCH DATE AND TIME	DOCKING DATE AND TIME	UNDOCKING DATE AND TIME	RECOVERY DATE AND TIME	MASS (kg)	PRIME CREW	BACK-UP CREW
1979	Cosmos 1100 Cosmos 1101	22 May 23.00 22 May 23.00			23 May 23 May	3,500 ? 3,500 ?	Shuttle model test Shuttle model test	
	Soyuz 34	6 Jun 18.13	8 Jun 20.02 14 Jun 17.48	14 Jun 16.18 19 Aug 09.07	19 Aug 12.30	6,800 ?	Returned with Soyuz 32 crew	
	Progress 7	28 Jun 09.25	30 Jun 11.18	18 Jul 03.50	20 Jul 01.57	7,000 ?	Unmanned freighter	
	Soyuz-T	16 Dec 12.30	19 Dec 14.05	23 Mar 21.04	25 Mar 21.47	6.900 ?	Unmanned Soyuz-T	
1980	Progress 8	27 Mar 18.53	29 Mar 20.01	25 Apr 08.04	26 Apr 06.54	7,000 ?	Unmanned freighter	
	Soyuz 35	9 Apr 13.38	10 Apr 15.16	3 Jun 11.47	3 Jun 15.07	6,800 ?	L. I. Popov*, V. V. Lebedev then V. V. Ryumin*	V. D. Zudov, B. D. Andreyev
	Progress 9	27 Apr 06.24	29 Apr 08.09	20 May 18.51	22 May 00.44	7,000 ?	Unmanned freighter	
	Soyuz 36	26 May 18.21	27 May 19.56 4 Jun 18.09	4 Jun 16.39 31 Jul 11.55	31 Jul 15.15	6,800 ?	V. N. Kubasov*, B. Farkas*	V. A. Dzhanibekov, B. Magyari
	Soyuz-T 2	5 Jun 14.19	6 Jun 15.58	9 Jun 09.20	9 Jun 12.40	6,850 ?	Y. V. Malyshev, V. V. Aksenov	L. D. Kizim, O. G. Makarov
	Progress 10	29 Jun 04.41	1 Jul 05.53	17 Jul 22.21	19 Jul 01.47	7,000 ?	Unmanned freighter	
	Soyuz 37	23 Jul 18.33	24 Jul 20.02 1 Aug 18.20	1 Aug 16.43 11 Oct 06.30	11 Oct 09.50	6,800 ?	V. V. Gorbatko*, Pham Tuan*	V. F. Bykovsky, Bul Thanh Liem
	Soyuz 38	18 Sep 19.11	19 Sep 20.49	26 Sep 12.34 ?	26 Sep 15.54	6,800 ?	Y. V. Romanenko, A. Tamayo-Mendez	Y. V. Khrunov, J. Lopez-Falcon
	Progress 11	28 Sep 15.10	30 Sep 17.03	9 Dec 10.23	11 Dec 14.00	7,000 ?	Unmanned freighter	
	Soyuz-T 3	27 Nov 14.18	28 Nov 15.54	10 Dec 06.10	10 Dec 09.26	6,850 ?	L. D. Kizim, O. G. Makarov, K. P. Feoktistov then G. M. Strekalov	V. G. Lazarev, V. Poliakov, V. P. Savinykh
1981	Progress 12	24 Jan 14.18	26 Jan 15.56	19 Mar 18.14	20 Mar 16.59	7,000 ?	Unmanned freighter	
	Soyuz-T 4	12 Mar 19,00	13 Mar 20.33	26 Mar 09.20 ?	26 May 12.38	6,850 ?	V. V. Kovalyonok, V. P. Savinykh	V. D. Zudov, B. D. Andreyev
	Soyuz 39	22 Mar 14.59	23 Mar 16.28	30 Mar 08.20 ?	30 Mar 11.42	6,800 ?	V. A. Dzhanibekov, J. Gurragcha	V. A. Lyakhov, M. Ganzorig
	Cosmos 1267	25 Apr 02.01	19 Jun 06.52		29 Jul 1982	20,000 ?	Heavy Cosmos module	Capsule down 24 May 13.25
	Soyuz 40	14 May 17.17	15 May 18.50	22 May 10.40 ?	22 May 13.58	6,800 ?	L. I. Popov, D. Prunariu	Y. V. Romanenko, L. Dediu
1982	Salyut 7	19 Apr 19.45	In orbit			19,920	Civilian orbital station	
	Soyuz-T 5	13 May 09.58	14 May 11.36	27 Aug 11.45 ?	27 Aug 15.04	6,850 ?	A. N. Berezovoi*, V. V. Lebedev*	V. G. Titov, G. M. Strekalov
	Progress 13	23 May 05.57	25 May 07.57	4 Jun 06.31	6 Jun 00.05	7,000 ?	Unmanned freighter	
	Cosmos 1374	3 Jun 21.29			3 Jun 23.18	1,250 ?	Shuttle model	
	Soyuz-T 6	24 Jun 16.29	25 Jun 17.46	2 Jul 11.04	2 Jul 14.21	6,850 ?	Y. V. Malyshev then V. A. Dzhanibekov, A. S. Ivanchenkov, J.-L. Chretien	L. D. Kizim V. A. Solovyov, P. Baudry
	Progress 14	10 Jul 19.58	12 Jul 11.41	10 Aug 22.11	13 Aug 01.29	7,000 ?	Unmanned freighter	
	Soyuz-T 7	19 Aug 17.12	20 Aug 18.32 29 Aug	29 Aug 16.47 10 Dec 15.40 ?	10 Dec 19.03	6,850 ?	L. I. Popov*, A. A. Serebrov*, S. Y. Savitskaya*	Y. V. Romanenko then V. V. Vasyutin, V. P. Savinykh, I. Pronina
	Progress 15	18 Sep 04.59	20 Sep 06.12	14 Oct 11.46	16 Oct 17.08	7,000 ?	Unmanned freighter	
	Progress 16	31 Oct 11.20	2 Nov 13.22	13 Dec 15.32	14 Dec 17.17	7,000 ?	Unmanned freighter	
1983	Cosmos 1443	2 Mar 09.37	10 Mar 09.20	14 Aug 14.04	19 Sep 00.34	20,000 ?	Heavy Cosmos module	Capsule down 23 Aug 11.02
	Cosmos 1445	15 Mar 22,30			16 Mar 00.20 ?	1,250 ?	Shuttle model	
	Soyuz-T 8	20 Apr 13.11	Docking failure		22 Apr 13.29	6,850 ?	V. G. Titov, G. M. Strekalov, A. A. Serebrov	V. A. Lyakhov, A. P. Alexandrov, V. P. Savinykh
	Soyuz-T 9	27 Jun 09.12	28 Jun 10.46 16 Aug 14.45 ?	16 Aug 14.25 23 Nov 16.40	23 Nov 19.58	6,850 ?	V. A. Lyakhov, A. P. Alexandrov	L. D. Kizim, V. A. Solovyov
	Progress 17	17 Aug 12.08	19 Aug 13.47	17 Sep 11.44	17 Sep 23.45	7,000 ?	Unmanned freighter	
	Soyuz-T 10-1	26 Sep 19.38	Launch aborted			6,850 ?	V. G. Titov, G. M. Strekalov	L. D. Kizim, V. A. Solovyov
	Progress 18	20 Oct 09.59	22 Oct 11.34	13 Nov 03.08	16 Nov 04.18	7,000 ?	Unmanned freighter	
	Cosmos 1517	27 Dec 10.20			27 Dec 11.50 ?	1,250 ?	Shuttle model	
1984	Soyuz-T 10	8 Feb 12,07	9 Feb 14.43	11 Apr 07.33	11 Apr 10.50	6,850 ?	L. D. Kizim*, V. A. Solovyov*, O. Y. Atkov*	V. V. Vasyutin, V. P. Savinykh, V. Poliakov
	Progress 19	21 Feb 06.46	22 Feb 08.21	31 Mar 08.40	1 Apr 18.18	7,000 ?	Unmanned freighter	
	Soyuz-T 11	3 Apr 13.09	4 Apr 14.31 13 Apr	13 Apr 10.27 2 Oct 08.40	2 Oct 10.57	6,850 ?	Y. V. Malyshev*, N. N. Rukavishnikov then G. M. Strekalov*, R. Sharma*	A. N. Berezovoi, G. M. Grechko, R. Malhotra
	Progress 20	15 Apr 08.13	17 Apr 09.22	6 May 17.46	7 May	7,000 ?	Unmanned freighter	
	Progress 21	7 May 22.47	10 May 00.10	26 May 09.41	26 May	7,000 ?	Unmanned freighter	
	Progress 22	28 May 14.13	30 May 15.47	15 Jul 13.36	15 Jul	7,000 ?	Unmanned freighter	
	Soyuz-T 12	17 Jul 17.41	18 Jul 19.17	29 Jul	29 Jul 12.55	6,850 ?	V. A. Dzhanibekov, S. Y. Savitskaya, I. P. Volk	V. V. Vasyutin, E. A. Ivanova, V. P. Savinykh
	Progress 23	14 Aug 06.28	16 Aug 08.11	26 Aug 16.13	28 Aug 01.28	7,000 ?	Unmanned freighter	
	Cosmos 1614	19 Dec 03.53			19 Dec 05.23 ?	1,250 ?	Shuttle model	
1985	Soyuz-T 13	6 Jun 06.40	8 Jun 08.50	25 Sep 03.58	26 Sep 09.52	6,850 ?	V. A. Dzhanibekov, V. P. Savinykh*	L. I. Popov, A. P. Alexandrov
	Progress 24	21 Jun 00.40	23 Jun 02.34	15 Jul 12,28	15 Jul	7,000 ?	Unmanned Freighter	
	Cosmos 1669	16 Jul 13.05	18 Jul 15.05	28 Aug 21.50	30 Aug	7,000 ?	Unmanned freighter	
	Soyuz-T 14	17 Sep 12.39	18 Sep 14.14	21 Nov	21 Nov 10.31	6,850 ?	V. V. Vasyutin, G. M. Grechko*, A. A. Volkov	A. S. Viktorenko, G. M. Strekalov, Y. Saley
	Cosmos 1686	27 Sep 08.41	2 Oct 10.16		In orbit	20,000 ?	Heavy Cosmos module	
1986	Mir	19 Feb 21.19			In orbit	20,900	Orbital station core	
	Soyuz-T 15 (Docking with Salyut) (Re-docking with Mir)	13 Mar 12.33	15 Mar 13.38 6 May 16.58 26 Jun 19.46	5 May 12.12 25 Jun 14.58 16 Jul 09.07	16 Jul 12.34	6,850 ?	L. D. Kizim, V. A. Solovyov	A. S. Viktorenko, A. P. Alexandrov

LOG OF ALL ACTUAL AND PLANNED LAUNCHES IN THE SOVIET MANNED SPACE PROGRAMME – continued

	MISSION	LAUNCH DATE AND TIME	DOCKING DATE AND TIME	UNDOCKING DATE AND TIME	RECOVERY DATE AND TIME	MASS (kg)	PRIME CREW	BACK-UP CREW
1986	Progress 25	19 Mar 10.08	21 Mar 11.16	20 Apr 19.34	20 Apr	7,150 ?	Unmanned freighter	
	Progress 26	23 Apr 19.40	26 Apr 21.26	22 Jun 18.25	23 Jun 18.40	7,150 ?	Unmanned freighter	
	Soyuz-TM	21 May 08.22	23 May 10.12	29 May 09.23	30 May 06.49	7,070	Unmanned Soyuz-TM	
1987	Progress 27	16 Jan 06.06	18 Jan 07.27	23 Feb 11.29	25 Feb 15.17	7,150 ?	Unmanned freighter	
	Soyuz-TM 2	5 Feb 21.38	7 Feb 22.28	29 Jul 20.34	30 Jul 01.04	7,100 ?	V. G. Titov *then* Y. V. Romanenko*, A. A. Serebrov *then* A. I. Leveykin	Y. V. Romanenko A. I. Leveykin
	Progress 28	3 Mar 11.14	5 Mar 11.43	26 Mar 05.07	28 Mar 02.59	7,150 ?	Unmanned freighter	
	Kvant	31 Mar 00.06	9 Apr 00.36		In orbit	20,600	Mir astrophysics module	
	Progress 29	21 Apr 15.14	23 Apr 17.05	11 May 03.10	11 May	7,150 ?	Unmanned freighter	
	No name	15 May 17.30	Failed to reach orbit			100,000 ?	Energiya shuttle booster	
	Progress 30	19 May 04.02	21 May 05.53	19 Jul 00.20	19 Jul	7,150 ?	Unmanned freighter	
	Soyuz-TM 3	22 Jul 01.59	24 Jul 03.31 30 Jul 23.48	30 Jul 23.28 29 Dec 05.55	29 Dec 09.16	7,100 ?	A. S. Viktorenko*, A. P. Alexandrov, M. A. Faris*	A. Y. Solovyov, V. P. Savinykh, M. Habib
	Progress 31	3 Aug 20.44	5 Aug 22.28	21 Sep 23.58	23 Sep 00.22	7,150 ?	Unmanned freighter	
	Progress 32	23 Sep 23.44	26 Sep 01.08 10 Nov 05.47	10 Nov 04.09 17 Nov 19.25	18 Nov 00.10	7,150?	Unmanned freighter	
	Progress 33	20 Nov 23.47	23 Nov 01.39	19 Dec 08.16	19 Dec	7,150 ?	Unmanned freighter	
	Soyuz-TM 4	21 Dec 11.18	23 Dec 12.51 30 Dec 09.29	30 Dec 09.10 17 Jun 06.18	17 Jun 10.13	7,100 ?	V. G. Titov, M. K. Manarov, A. S. Levchenko*.	A. A. Volkov, A. Y. Kaleri, A. V. Shchukin
1988	Progress 34	20 Jan 22.52	23 Jan 00.39	4 Mar 03.40	4 Mar	7,150 ?	Unmanned freighter	
	Progress 35	23 Mar 21.05	25 Mar 22.22	5 May 01.36	5 May	7,150 ?	Unmanned freighter	
	Progress 36	13 May 00.30	15 May 02.13	5 Jun 11.12	5 Jun 20.28	7,150 ?	Unmanned freighter	
	Soyuz-TM 5	7 Jun 14.03	9 Jun 15.57 18 Jun 10.27	18 Jun 10.11 5 Sept 22.55	7 Sept 00.50	7,100 ?	A. Y. Solovyov*, V. P. Savinykh*, A. P. Alexandrov*	V. A. Lyakhov, A. Zaitsev *then* A. A. Serebrov, K. Stoyanov
	Soyuz-TM 6	29 Aug 04.23	31 Aug 05.41			7,100 ?	V. A. Lyakhov, V. Poliakov, A. A-M. Sarvar	A. N. Berezovoi, G. Arzamazov, M. D-G. Masum

FURTHER MANNED MISSIONS SCHEDULED FOR 1988

	MISSION	LAUNCH DATE AND TIME	DOCKING DATE AND TIME	UNDOCKING DATE AND TIME	RECOVERY DATE AND TIME	MASS (kg)	PRIME CREW	BACK-UP CREW
	Soyuz-TM 7	21 Nov ?				7,100 ?	A. A. Volkov, (Engineer), J-L. Chretien	A. S. Viktorenko, (Engineer), M. Tognini

Notes: This listing includes all of the launches which have taken place – or which are reported to have taken place – in the Soviet manned space programme. All of the times are given in GMT, although a few times are not available from Soviet sources. Where a landing takes place in the year following the launch, this is not normally specifically indicated (by showing the new year) but it can be deduced from the month of the re-entry. For the unmanned Progress missions and Cosmos 1443 the time of retrofire is given, rather than a landing time. Spacecraft which re-entered the atmosphere naturally have the landing date shown in parentheses. ''In orbit'' means that the spacecraft was still in orbit on 1 July 1988.

Different sources give different times for the mission events, generally with errors of a minute or two. This listing has tried to give the most widely quoted announced figures, although total accuracy cannot be guaranteed. Part of the problem is that the Soviets sometimes do not round numbers, they ''chop'' them (eg, 1 minutes 50 seconds may be shown as 1 minute instead of the correctly rounded 2 minutes).

Full names are given for the spacecraft, other than for the Korabl Sputniks (''KS'') which were unmanned Vostok tests. Space does not permit details of all of the crew/spacecraft switches which have taken place, but cosmonauts who were launched in one spacecraft and returned to Earth in another are indicated by an asterisk. Normally, changes in crew assignments are noted, with the original choice shown first followed by ''*then*'' and the replacement crewman; the most complex case was Soyuz 11, where the prime crew were replaced by the back-ups, and a replacement back-up crew was assigned. The back-up crew assignments reflect new information released by the Soviets to Bert Vis as this book went to press; the remainder of the book has not been updated and therefore where differences in assignments occur, this appendix should be taken as correct.

Launch vehicles are not shown here, but their uses in the manned programme are:
SL-3   Korabl Sputnik, Vostok
SL-4   Voskhod, Propulsion Tests, Soyuz, Soyuz-T, Soyuz-TM, Progress
SL-8   Small spaceplane tests (Cosmos 1374, etc)
SL-12   Zond (Soyuz-derived lunar craft)
SL-13   Soyuz lunar lander variant, Salyut, Mir, Cosmos Modules, Kvant, Dual Cosmos (Cosmos 881/882, etc)
SL-15   Abandoned lunar booster
SL-17   Space shuttle booster

Appendix 3: **Cosmonaut Data**

The mission assignment dates relate to the year that the mission was launched, rather than the year that training began (which in many cases is not known). The crew assignments are identified as: Cdr, commander; FE, flight engineer; RE, research engineer. Where a Soyuz mission is associated with a Salyut or Mir station, then the station is identified. The foreign "guest cosmonauts" are not listed here.

The vast majority of mission back-up assignments are known from Soviet statements, but there are a few which have not been confirmed; in such cases, the most probable assignments are listed, but are indicated to be speculative. In general, only prime and back-up crew assignments are listed, but in a few cases the support roles of a cosmonaut for specific missions are noted. This is done particularly where the cosmonaut has not made a subsequent flight.

AKSYONOV, Vladimir Viktorovich

Date of Birth:	1 February 1935		
Selection Group:	1973	Engineer	
Assignments:	1976	FE	Soyuz 22
	1980	FE	Soyuz-T 2/Salyut 6

ALEXANDROV, Alexander Pavlovich

Date of Birth:	20 February 1943		
Selection Group:	1978 December	Engineer	
Assignments:	1983 Back-up	FE	Soyuz-T 8/Salyut 7
	1983	FE	Soyuz-T 9/Salyut 7
	1985 Back-up	FE	Soyuz-T 13/Salyut 7 **(1)**
	1986 Back-up	FE	Soyuz-T 15/Mir/Salyut 7
	1987	FE	Soyuz-TM 3/Mir/Syria **(2)**

(1) The original Soyuz-T 13 long duration crew was Vasyutin, Savinykh and A. A. Volkov, backed up by Viktorenko, Alexandrov and Saley. Following the failure of Salyut 7, Savinykh and Alexandrov were re-assigned to new commanders for a new Soyuz-T 13 mission to rescue Salyut 7: the other cosmonauts were assigned to Soyuz-T 14.
(2) After being launched on the Soyuz-TM 3 Soviet-Syrian mission, Alexandrov remained on Mir complex to replace Leveykin whose health was causing concern.

ANDREYEV, Boris Dmitrievich

Date of Birth:	6 October 1940		
Selection Group:	1970	Engineer	
Assignments:	1975 Back-up	FE	Soyuz 19/ASTP
	1980 Back-up	FE	Soyuz 35/Salyut 6
	1981 Back-up	FE	Soyuz-T 4/Salyut 6

Believed to have been disqualified from the flight team due to physical difficulties.

ANIKEYEV, Ivan Nikolayevich

Date of Birth:	?	
Selection Group:	1960 March	Military Pilot

Left the cosmonaut team following the railway incident in late 1961 involving Filatyev and Nelyubov (qv).

ANOKHIN, Sergei Nikolayevich

Date of Birth:	1910 ?	
Date of Death:	15 April 1986	
Selection Group:	1966	Military Engineer

Left the cosmonaut team soon after joining, due to physical difficulties.

ARTYUKHIN, Yuri Petrovich

Date of Birth:	22 July 1930		
Selection Group:	1963 January	Military Engineer	
Assignments:	1967 Lunar training group		
	1974	FE	Soyuz 14/Salyut 3

ARZAMAZOV, German

Date of Birth:	?		
Selection Group:	1977	Doctor	
Assignments:	1987 Back-up	RE	Soyuz-TM 4/Mir **(1)**
	1988 Back-up	FE	Soyuz-TM 6/Mir/Afghanistan

(1) There are reports that Poliakov might have been an original crew member of Soyuz-TM 4, and Arzamazov would be a logical choice as his back-up since he was Poliakov's back-up for Soyuz-TM 6.

ASANIN

Date of Birth:	?	
Selection Group:	?	Military Pilot

Identified in April 1966 as a trainee cosmonaut by Novosti when reporting a month-long spaceflight simulation during January 1966.

ATKOV, Oleg Yuryevich

Date of Birth:	9 May 1949		
Selection Group:	1977	Doctor	
Assignments:	1984	RE	Soyuz-T 10/Salyut 7

BARSUKOV, S.

Date of Birth:	?	
Selection Group:	?	Military

Pictured in a Soyuz simulator in *Aviation and Cosmonautics* magazine, May 1977; uncertain whether a cosmonaut or a ground trainer.

BELYAYEV, Pavel Ivanovich

Date of Birth:	26 June 1925		
Date of Death:	10 January 1970		
Selection Group:	1960 March	Military Pilot	
Assignments:	1965	Cdr	Voskhod 2
	1967 Lunar training group (manager from late 1967)		

Died following an operation for ulcers.

BEREGOVOI, Georgi Timofeyevich

Date of Birth:	15 April 1921		
Selection Group:	1964 February	Military Pilot	
Assignments:	1966 Back-up	Pilot	Voskhod 3 **(1)**
	1968	Cdr	Soyuz 3

(1) This mission was cancelled in late 1965; it is uncertain whether Beregovoi would have been the back-up pilot or commander for Voskhod 3.

BEREZOVOI, Anatoly Nikolayevich

Date of Birth:	11 April 1942		
Selection Group:	1970 May	Military Pilot	
Assignments:	1977 Back-up	Cdr	Soyuz 24/Salyut 5
	1982	Cdr	Soyuz-T 5/Salyut 7
	1984 Back-up	Cdr	Soyuz-T 11/Salyut 7/India
	1988 Back-up	Cdr	Soyuz-TM 6/Mir/Afghanistan

BOGDASHEVSKY, Rotislav

Date of Birth:	?	
Selection Group:	1965 ?	Military Engineer

Filmed in 1968 by NBC and identified as a trainee cosmonaut.

BONDARENKO, Valentin Vasilyevich

Date of Birth:	1937?	
Date of Death:	23 March 1961	
Selection Group:	1960 March	Military Pilot

Died due to a fire in the pure oxygen atmosphere of a spacecraft simulator.

BONDAREV, K.

Date of Birth:	?	
Selection Group:	?	Military

Pictured in a Soyuz simulator in *Aviation and Cosmonautics* magazine, May 1977; uncertain whether a cosmonaut or a ground trainer.

BYKOVSKY, Valery Fyodorovich

Date of Birth:	2 August 1934		
Selection Group:	1960 March	Military Pilot	
Assignments:	1962 Back-up	Pilot	Vostok 3
	1963	Pilot	Vostok 5
	1967	Cdr	Soyuz 2 **(1)**
	1967 Lunar training group		
	1973 Back-up	Cdr	Soyuz 13?
	1976	Cdr	Soyuz 22
	1978	Cdr	Soyuz 31/Salyut 6 **(2)**
	1980 Back-up	Cdr	Soyuz 37/Salyut 6

(1) The Soyuz 2 mission was cancelled following the in-flight problems with Soyuz 1; the name Soyuz 2 was later given to an unmanned mission.
(2) Replaced Dzhanibekov as Soyuz 31 commander.

DEMIN, Lev Stepanovich

Date of Birth:	11 June 1926		
Selection Group:	1963 January	Military Engineer	
Assignments:	1974	FE	Soyuz 15/Salyut 3

DOBROVOLSKY, Georgi Timofeyevich

Date of Birth:	1 June 1925		
Date of Death:	29 June 1971		
Selection Group:	1963 January	Military Pilot	
Assignments:	1967 Lunar training group		
	1971 Back-up	Cdr	Soyuz 10/Salyut 1
	1971 (Back-up)	Cdr	Soyuz 11/Salyut 1 **(1)**

(1) Originally the back-up commander for Soyuz 11, but the back-up crew replaced the prime crew a few days before the launch. Died when Soyuz 11 lost pressure during its descent (30 June Moscow Time).

DZHANIBEKOV, Vladimir Alexandrovich

Date of Birth:	13 May 1942		
Selection Group:	1970 May	Military Pilot	
Assignments:	1975 Back-up	Cdr	Soyuz 19/ASTP
	1978	Cdr	Soyuz 27/Salyut 6
	1978	Cdr	Soyuz 31/Salyut 6/GDR **(1)**
	1980 Back-up	Cdr	Soyuz 36/Salyut 6/Hungary
	1982	Cdr	Soyuz-T 6/Salyut 7/France **(2)**
	1984	Cdr	Soyuz-T 12/Salyut 7
	1985	Cdr	Soyuz-T 13/Salyut 7 **(3)**

(1) Reportedly, he was originally scheduled as the commander for the Soyuz 3 mission, but re-assigned in late 1977 and replaced by Bykovsky.
(2) Trained in the Salyut 7 resident crew group, and should have flown Soyuz-T 9, but re-assigned to Soyuz-T 6 to replace Malyshev.
(3) Originally, Soyuz-T 13 was to be a long duration mission, but after the failure of Salyut 7 the crews were re-assigned and Dzhanibekov commanded the rescue mission.
His real last name is "Krysn" which means "rat"; he has adopted his wife's last name.

FEOKTISTOV, Konstantin Petrovich

Date of Birth:	7 February 1926		
Selection Group:	1964 February	Voskhod Engineer	
Assignments:	1964	Eng.	Voskhod 1
	1980	RE	Soyuz-T 3/Salyut 6 **(1)**

(1) Trained for Soyuz-T 3 mission, but replaced, following medical problems, by Strekalov.
Senior spacecraft designer, having been involved in Soyuz, Salyut and Mir designs.

FILATYEV, Valentin Ignatyevich

Date of Birth:	1925	
Selection Group:	1960 March	Military Pilot

Left the cosmonaut team, following the late 1961 railway incident involving Anikeyev and Nelyubov (qv).

FILIPCHENKO, Anatoli Vasilyevich

Date of Birth:	26 February 1928		
Selection Group:	1963 January	Military Pilot	
Assignments:	1969 Back-up	Cdr	Soyuz 4
	1969	Cdr	Soyuz 7
	1970 Back-up	Cdr	Soyuz 9 **(1)**
	1974	Cdr	Soyuz 16

(1) Trained with Kubasov; Lazarev and Grechko have also been identified as Soyuz 9 back-ups.

GAGARIN, Yuri Alexeyevich

Date of Birth:	9 March 1934		
Date of Death:	27 March 1968		
Selection Group:	1960 March	Military Pilot	
Assignments:	1961	Pilot	Vostok 1 **(1)**
	1967 Back-up	Cdr	Soyuz 1

(1) First man in space.
Killed in an air crash.

GLAZKOV, Yuri Nikolaievich

Date of Birth:	2 October 1939		
Selection Group:	1965 October	Military Engineer	
Assignments:	1976 Back-up	FE	Soyuz 23/Salyut 5
	1977	FE	Soyuz 24/Salyut 5

GOLOVANOV, Yaroslav

Date of Birth: ?
Selection Group: 1965 July — Journalist

Left cosmonaut training group soon after Korolyov's death, implying that the flight with a journalist was to be a "political first".

GORBATKO, Viktor Vasilyevich

Date of Birth: 3 December 1934
Selection Group: 1960 March — Military Pilot

Assignments:			
1965	Back-up	Cdr	Voskhod 2 (1)
1967	Back-up	RE	Soyuz 2 (2)
1969	Back-up	RE	Soyuz 5
1969		RE	Soyuz 7
1976	Back-up	Cdr	Soyuz 23/Salyut 5
1977		Cdr	Soyuz 24/Salyut 5
1978	Back-up	Cdr	Soyuz 31/Salyut 6/GDR (3)
1980		Cdr	Soyuz 37/Salyut 6/Vietnam

(1) Replaced by Zaikin because of medical reasons.
(2) Mission cancelled due to the in-flight Soyuz 1 difficulties; Soyuz 2 was subsequently flown as an unmanned mission.
(3) Replaced Makarov in late 1977 when Makarov and Dzhanibekov were assigned to Soyuz 27.

GRECHKO, Georgi Mikhailovich

Date of Birth: 25 May 1931
Selection Group: 1966 September — Engineer

Assignments:			
1967	Lunar training group		
1969	Back-up	RE	Soyuz 7 (1)
1970	Back-up	FE	Soyuz 9 (2)
1973	Back-up	FE	Soyuz 12 (3)
1975		FE	Soyuz 17/Salyut 4
1977		FE	Soyuz 26/Salyut 6
1984	Back-up	FE	Soyuz-T 11/Salyut 7/India
1985		FE	Soyuz-T 14/Salyut 7 (4)

(1) Trained in the back-up role with Rukavishnikov.
(2) Trained with Lazarev; Filipchenko and Kubasov also trained as Soyuz 9 back-ups.
(3) Probably also trained for the Salyut/Cosmos 557 mission in 1973.
(4) Following the failure of Salyut 7, Grechko (and Strekalov as his back-up) was added to the original Soyuz-T 13 crew, Vasyutin and A. A. Volkov, to form the Soyuz-T 14 crew.

GUBAREV, Alexei Alexandrovich

Date of Birth: 29 March 1931
Selection Group: 1963 January — Military Pilot

Assignments:			
1973	Back-up	Cdr	Soyuz 12 (1)
1975		Cdr	Soyuz 17/Salyut 4
1978		Cdr	Soyuz 28/Salyut 6/Czech

(1) Probably also trained for Salyut/Cosmos 557 in 1973.

ILLYIN, Yevgeni A.

Date of Birth: 17 August 1937
Selection Group: 1965 — Voskhod Physician

Assignments:		
1966	? Candidate Physician	Bio-Voskhod

Selected as a candidate for a biological Voskhod mission, plans for which were cancelled in 1965.

ILLARIANOV, Valery Vasilyevich

Date of Birth: 2 July 1939
Selection Group: ? — Military Pilot?

Assignments:			
1980	Back-up	Cdr	Soyuz 35/Salyut 6 (1)

(1) Replaced by Zudov due to physical problems.
Acted as a CapCom during the ASTP preparations at Houston, and identified as a trainee cosmonaut.

IVANCHENKOV, Alexander Sergeievich

Date of Birth: 25 September 1940
Selection Group: 1970 — Engineer

Assignments:			
1974	Back-up	FE	Soyuz 16
1977	Back-up	FE	Soyuz 25/Salyut 6 (1)
1977	Back-up	FE	Soyuz 26/Salyut 6
1978	Back-up	FE	Soyuz 27/Salyut 6
1978		FE	Soyuz 29/Salyut 6
1982		FE	Soyuz-T 6/Salyut 7/France

(1) Prior to the Soyuz 25 failure, Ivanchenkov was slated to be the prime flight engineer for Soyuz 26, but it was decided not to fly any further all-rookie crews.

KALERI, Alexander Yurevich

Date of Birth: 13 May 1956
Selection Group: 1987 February — Engineer

Assignments:			
1987	Back-up	FE	Soyuz-TM 4/Mir

KARTASHOV, Anatoli Yakovlevich

Date of Birth: ?
Selection Group: 1960 March — Military Pilot

Retired from the cosmonaut team after being taken ill following centrifuge training.

KATYS, Georgi Petrovich

Date of Birth: 1927
Selection Group: 1964 February — Voskhod Engineer

Assignments:			
1964	Back-up	Eng	Voskhod 1

KHRUNOV, Yevgeny Vasilyevich

Date of Birth: 10 September 1933
Selection Group: 1960 March — Military Pilot

Assignments:			
1965	Back-up	Pilot	Voskhod 2
1967		RE	Soyuz 2 (1)
1969		RE	Soyuz 5
1980	Back-up	RE	Soyuz 38/Salyut 6/Cuba (2)

(1) Following the in-flight problems with Soyuz 1, the planned Soyuz 2 launch was cancelled; Soyuz 2 was later flown unmanned.
(2) After this assignment, Khrunov was offered the command of the Soyuz 40/Salyut 6 Romanian mission, but he turned it down.

KISILEV, Aleksandr S.

Date of Birth: 13 June 1935
Selection Group: 1965 — Voskhod Physician

Assignments:		
1966? Candidate Physician	Bio-Voskhod

Selected as a candidate for a biological Voskhod mission, plans for which were cancelled in 1965.

KIZIM, Leonid Denisovich

Date of Birth: 5 August 1941
Selection Group: 1965 October — Military Pilot

Assignments:			
1980	Back-up	Cdr	Soyuz-T 2/Salyut 6
1980		Cdr	Soyuz-T 3/Salyut 6
1982	Back-up	Cdr	Soyuz-T 6/Salyut 7/France
1983	Back-up	Cdr	Soyuz-T 9/Salyut 7
1983	Back-up	Cdr	Soyuz-T 10-1/Salyut 7 (1)
1984		Cdr	Soyuz-T 10/Salyut 7
1986		Cdr	Soyuz-T 15/Mir/Salyut 7

(1) Launch abort.

KLIMUK, Pyotr Illyich

Date of Birth: 10 July 1942
Selection Group: 1965 October — Military Pilot

Assignments:			
1967	Lunar training group		
1973		Cdr	Soyuz 13
1975	Back-up	Cdr	Soyuz 17/Salyut 4
1975	Back-up	Cdr	Soyuz 18-1/Salyut 4 (1)
1975		Cdr	Soyuz 18/Salyut 4
1978		Cdr	Soyuz 30/Salyut 6

(1) Spacecraft failed to reach orbit.

KOLODIN, Pyotr Ivanovich

Date of Birth: 23 September 1930
Selection Group: 1963 January — Military Engineer

Assignments:			
1971		RE	Soyuz 11/Salyut 1 (1)

(1) The prime crew, including Kolodin, was replaced by the back-up cosmonauts less than a week before the launch after Kubasov fell ill. Kolodin argued with the mission planners for the prime cosmonauts who were not ill to be re-instated and fly with Kubasov's back-up, but to no avail. Presumably Kolodin has not flown since because of the trouble which he caused at this time, although officially he was disqualified from future flights because of medical problems.

KOMAROV, Vladimir Mikhailovich

Date of Birth: 16 March 1927
Date of Death: 24 April 1967
Selection Group: 1960 March — Military Pilot

Assignments:			
1962	Back-up	Pilot	Vostok 4 (1)
1964		Cdr	Voskhod
1967		Cdr	Soyuz 1 (2)

(1) Replaced by Volynov, after the discovery of a heart irregularity.
(2) Died during the re-entry of Soyuz 1, the first person to die during an actual space mission.

KORNIEV, Ivan

Date of Birth: ?
Selection Group: 1965 October — Military Pilot

Filmed in 1968 by an NBC crew; identified as a trainee cosmonaut.

KOVALYONOK, Vladimir Vasilyevich

Date of Birth: 3 March 1942
Selection Group: 1967 May — Military Pilot

Assignments:			
1975	Back-up	Cdr	Soyuz 18/Salyut 4 (1)
1977		Cdr	Soyuz 25/Salyut 6
1977	Back-up	Cdr	Soyuz 26/Salyut 6
1978	Back-up	Cdr	Soyuz 27/Salyut 6
1978		Cdr	Soyuz 29/Salyut 6
1981		Cdr	Soyuz-T 4/Salyut 6

(1) This assignment followed the launch failure of the intended Soyuz 18; the back-ups for that mission flew the replacement mission (Soyuz 18 itself).

KUBASOV, Valery Nikolayevich

Date of Birth: 7 January 1935
Selection Group: 1966 September — Engineer

Assignments:			
1967	Back-up	FE	Soyuz 2 (1)
1969	Back-up	FE	Soyuz 5
1969		FE	Soyuz 6
1971		FE	Soyuz 11/Salyut 1 (2)
1973		FE	Soyuz/Salyut Cosmos 557 (3)
1975		FE	Soyuz 19/ASTP
1978	Back-up	Cdr	Soyuz 30/Salyut 6/Poland
1980		Cdr	Soyuz 36/Salyut 6/Hungary

(1) Launch cancelled following the in-flight Soyuz 1 problems; the name Soyuz 2 was given to a subsequent unmanned spacecraft.
(2) Fell ill within days of the planned Soyuz 11 launch, resulting in the back-up cosmonauts flying the mission.
(3) Probably due to fly a Soyuz mission to Salyut/Cosmos 557 if the station had not failed in orbit.

KUKLIN, A.

Date of Birth: ?
Selection Group: 1963 January — Military Pilot

KUZNETSOVA, Tatiana Dmitryevna

Date of Birth: ?
Selection Group: 1962 March — Vostok 6 Pilot

LAZAREV, Vasily Grigorievich

Date of Birth: 23 February 1928

Selection Group:		
1963?		Vostok 7 Pilot?
1964 February		Voskhod Doctor
1966		Military Pilot

Assignments:			
1964	Back-up	Pilot	Vostok 7 (1)
1964	Back-up	Dr.	Voskhod 1
1970	Back-up	Cdr	Soyuz 9 (2)
1973		Cdr	Soyuz 12 (3)
1975		Cdr	Soyuz 18-1/Salyut 4 (4)
1980	Back-up	Cdr	Soyuz-T 3/Salyut 6

(1) Vostok 7 was cancelled in 1963, although two cosmonauts were selected in 1963 for the mission: one was Yegorov, while the other might have been Lazarev. When Vostok 7 was cancelled, the two cosmonauts were stood down.
(2) Paired with Grechko; Filipchenko and Kubasov were also back-ups.
(3) Probably also trained for Salyut/Cosmos 557 in 1973.
(4) Spacecraft failed to reach orbit.

LEBEDEV, Valentin Vitalyevich

Date of Birth: 14 April 1942
Selection Group: 1972 — Engineer

Assignments:			
1973		FE	Soyuz 13
1979	Back-up	FE	Soyuz 32/Salyut 6
1980		FE	Soyuz 35/Salyut 6 (1)
1982		FE	Soyuz-T 5/Salyut 7

(1) Injured his knee about six weeks before the Soyuz 35 launch, and had to be replaced by Ryumin.

LEONOV, Alexei Arkhipovich

Date of Birth: 30 May 1934
Selection Group: 1960 March — Military Pilot

Assignments:			
1965		Pilot	Voskhod 2 (1)
1967	Lunar training group		
1971		Cdr	Soyuz 11/Salyut 1 (2)
1973		Cdr	Soyuz/Salyut Cosmos 557 (3)
1975		Cdr	Soyuz 19/ASTP

(1) First man to walk in space.
(2) Prime crew for Soyuz 11 were replaced by the back-ups, following Kubasov's illness a few days before launch.
(3) Probably due to command a Soyuz visit to Salyut/Cosmos 557 if the station had not failed in orbit.

LETUNOV, Yuri

Date of Birth: ?
Selection Group: 1965 July — Journalist

Stood down from cosmonaut training soon after Korolyov's death in January 1966.

LEVCHENKO, Anatoli Semenovich

Date of Birth: 21 May 1941
Date of Death: 6 August 1988
Selection Group: 1978 — Shuttle Pilot

Assignments:			
1987		RE	Soyuz-TM 4/Mir

When announced for the Soyuz-TM 4 crew, it was stated that Levchenko was training for the space shuttle programme.

LEVEYKIN, Alexander Ivanovich

Date of Birth: ?
Selection Group: 1978 December — Engineer

Assignments:			
1987		FE	Soyuz-TM 2/Mir (1)

(1) Romanenko and Leveykin were the back-up cosmonauts for Soyuz-TM 2, but the prime and back-ups were switched in the weeks leading up to the launch. Leveykin had trouble adapting to weightlessness, and was returned to Earth with two of the three cosmonauts launched on the Soyuz-TM 3 Syrian mission; his place on Mir was taken by A. P. Alexandrov.

LYAKHOV, Vladimir Afanasyevich

Date of Birth: 20 July 1941
Selection Group: 1967 May — Military Pilot

Assignments:			
1978	Back-up	Cdr	Soyuz 29/Salyut 6
1979		Cdr	Soyuz 32/Salyut 6
1981	Back-up	Cdr	Soyuz 39/Salyut 6/Mongolia
1983	Back-up	Cdr	Soyuz-T 8/Salyut 7
1983		Cdr	Soyuz-T 9/Salyut 7
1988	Back-up	Cdr	Soyuz-TM 5/Mir/Bulgaria
1988		Cdr	Soyuz-TM 6/Mir/Afghanistan

MAKAROV, Oleg Grigoryevich

Date of Birth: 6 January 1933
Selection Group: 1966 November Engineer
Assignments:
1967 Lunar training group
1973 FE Soyuz 12 **(1)**
1975 FE Soyuz 18-1/Salyut 4 **(2)**
1978 FE Soyuz 27/Salyut 6
1978 Back-up Cdr Soyuz 31/Salyut 6/GDR **(3)**
1980 Back-up FE Soyuz-T 2/Salyut 6
1980 FE Soyuz-T 3/Salyut 6

(1) Probably also trained for Salyut/Cosmos 557 in 1973.
(2) Failed to reach orbit.
(3) Replaced by Gorbatko, when he was paired in late 1977 with Dzhanibekov for Soyuz 27/Salyut 6.

MALYSHEV, Yuri Vasilyevich

Date of Birth: 27 August 1941
Selection Group: 1967 May Military Pilot
Assignments:
1976 Back-up Cdr Soyuz 22 **(1)**
1980 Cdr Soyuz-T 2/Salyut 6
1982 Cdr Soyuz-T 6/Salyut 7/France **(2)**
1984 Cdr Soyuz-T 11/Salyut 7/India

(1) Popov was also trained as a back-up commander.
(2) Removed from mission following a personality clash with the French cosmonaut J.-L. Chretien; he was replaced by Dzhanibekov.

MANAROV, Musa Khiramanovich

Date of Birth: 22 March 1951
Selection Group: 1978 December Engineer
Assignments:
1987 FE Soyuz-TM 4/Mir

NELYUBOV, Grigori Grigoryevich

Date of Birth: ?
Date of Death: 18 February 1966
Selection Group: 1960 March Military Pilot
Assignments:
1961 Support Pilot Vostok 1
1962 Support Pilot Vostok 3 **(1)**

(1) Involved in an incident at a railway station. Nelyubov, Anikeyev and Filatryev were drunk and had an argument with a military patrol; Nelyubov tried to "pull rank" as a cosmonaut to escape any disciplinary action, but the cosmonauts were reported to Korolyov. They were given the option of apologising to prevent further action, but Nelyubov refused to apologise. As a result he and the two other cosmonauts were thrown out of the cosmonaut team. Anikeyev and Filatyev had much sympathy from the other cosmonauts, because the trouble had been caused by Nelyubov. At the time, Nelyubov was training as the support cosmonaut for Vostok 3.
Nelyubov was posted to the Far East where he boasted that he had been Gagarin's back-up (actually he had been the support – next in line after Titov), but he was not believed. This led to an emotional problem and Nelyubov was killed by a passing train on a railway bridge north of Vladivostok while he was drunk.

NIKOLAYEV, Andrian Grigoryevich

Date of Birth: 5 September 1929
Selection Group: 1960 March Military Pilot
Assignments:
1961 Back-up Pilot Vostok 2
1962 Pilot Vostok 3
1967 Back-up Cdr Soyuz 2 **(1)**
1969 Back-up Cdr Soyuz 6
1969 Back-up Cdr Soyuz 7
1969 Back-up Cdr Soyuz 8
1970 Cdr Soyuz 9

(1) Mission cancelled when Soyuz 1 developed problems in orbit; the name was later given to an unmanned mission.

OBRATSOV

Date of Birth: ?
Selection Group: 1963? Military Doctor?

Identified by the Novosti Press Agency as a trainee cosmonaut after taking part in a month-long spaceflight simulation during January 1966.

PATSAYEV, Viktor Ivanovich

Date of Birth: 19 June 1933
Date of Death: 29 June 1971
Selection Group: 1969 Engineer
Assignments:
1971 Back-up RE Soyuz 10/Salyut 1
1971 (Back-up RE) Soyuz 11/Salyut 1 **(1)**

(1) The back-up crew flew the Soyuz 11 mission after the whole prime crew was stood down because of Kubasov falling ill just before the planned launch. Died when pressure was lost during the Soyuz 11 re-entry (30 June Moscow Time).

POLIAKOV, Valeri

Date of Birth: 27 April 1942
Selection Group: 1977 Doctor
Assignments:
1980 Back-up RE Soyuz-T 3/Salyut 6
1984 Back-up RE Soyuz-T 10/Salyut 7
1987 RE Soyuz-TM 4/Mir **(1)**
1988 FE Soyuz-TM 6/Mir/Afghanistan

(1) Reportedly assigned to the prime Soyuz-TM 4 crew in 1987, but was removed from the crew to allow Levchenko to fly the mission.

PONOMAREV, Yuri Anatolyevich

Date of Birth: 1932
Selection Group: 1970 Engineer
Assignments: 1975 Back-up FE Soyuz 18/Salyut 4

Left training group 1980.

PONOMAREVA, Valantina Leonidovna

Date of Birth: ?
Selection Group: 1962 March Vostok 6 Pilot
Assignments: 1963 Support Pilot Vostok 6

POPOV, Leonid Ivanovich

Date of Birth: 31 August 1945
Selection Group: 1970 May Military Pilot
Assignments:
1976 Back-up Cdr Soyuz 22 **(1)**
1979 Back-up Cdr Soyuz 32/Salyut 6
1980 Cdr Soyuz 35/Salyut 6
1981 Cdr Soyuz 40/Salyut 6/Romania
1982 Cdr Soyuz-T 7/Salyut 7
1985 Back-up Cdr Soyuz-T 13/Salyut 7 **(2)**

(1) Malyshev was also announced as back-up commander.
(2) Following the Salyut 7 failure, the original Soyuz-T 13 long duration cosmonauts were re-assigned, and Popov was added to the training group as the back-up commander for the revised Soyuz-T 13 mission to rescue Salyut 7.

POPOVICH, Pavel Romanovich

Date of Birth: 5 October 1930
Selection Group: 1960 March Military Pilot
Assignments:
1962 Pilot Vostok 4
1967 Lunar training group
1974 Cdr Soyuz 14/Salyut 3

PRONINA, Irina Rudolvfovna

Date of Birth: 14 April 1953
Selection Group: 1980 Woman Engineer
Assignments: 1982 Back-up RE Soyuz-T 7/Salyut 7

RAFIKOV, Mars Zakirovich

Date of Birth: ?
Selection Group: 1960 March Military Pilot

Left the cosmonaut team in early 1962, due to training problems.

ROMANENKO, Yuri Viktorovich

Date of Birth: 1 August 1944
Selection Group: 1970 May Military Pilot
Assignments:
1974 Back-up Cdr Soyuz 16
1977 Back-up Cdr Soyuz 25/Salyut 6
1977 Cdr Soyuz 26/Salyut 6
1979 Back-up Cdr Soyuz 33/Salyut 6/Bulgaria
1980 Cdr Soyuz 38/Salyut 6/Cuba
1981 Back-up Cdr Soyuz 40/Salyut 6/Romania
1982 Back-up Cdr Soyuz-T 7/Salyut 7 **(1)**
1987 Cdr Soyuz-TM 2/Mir **(2)**

(1) Replaced by Popov for physical reasons.
(2) Romanenko and Leveykin were the back-up crew for Soyuz-TM 2, but the prime and back-ups were switched in the weeks leading up to the launch.

ROZDESTVENSKY, Valery Illyich

Date of Birth: 13 February 1939
Selection Group: 1965 October Military Engineer
Assignments:
1976 Back-up FE Soyuz 21/Salyut 5
1976 FE Soyuz 23/Salyut 5

RUKAVISHNIKOV, Nikolai Nikolayevich

Date of Birth: 18 September 1932
Selection Group: 1966 November Engineer
Assignments:
1967 Lunar training group
1969 Back-up FE Soyuz 6 **(1)**
1969 Back-up RE Soyuz 7 **(2)**
1971 RE Soyuz 10/Salyut 1
1971 Back-up RE Soyuz 11/Salyut 1 **(3)**
1974 RE Soyuz 16
1978 Back-up Cdr Soyuz 28/Salyut 6/Czech
1979 Cdr Soyuz 33/Salyut 6/Bulgaria **(4)**
1984 FE Soyuz-T 11/Salyut 7/India **(5)**

(1) Trained with Sevastyanov.
(2) Trained with Grechko.
(3) When it was decided that the Soyuz 11 back-up crew would fly the mission, the Soyuz 10 crew was re-cycled as their back-ups.
(4) First non-military Soviet mission commander.
(5) Replaced by Strekalov due to physical problems.

RYUMIN, Valery Viktorovich

Date of Birth: 16 August 1939
Selection Group: 1973 Engineer
Assignments:
1977 FE Soyuz 25/Salyut 6
1978 Back-up FE Soyuz 29/Salyut 6
1979 FE Soyuz 32/Salyut 6
1980 FE Soyuz 35/Salyut 6 **(1)**

(1) Replaced Lebedev at short notice after he suffered a knee injury. On Soyuz 35 Ryumin became the first man to celebrate successive birthdays in orbit.

SALEY, Yevgeni Vladimirovich

Date of Birth: 1 January 1950
Selection Group: 1976 February? Military Pilot?
Assignments:
1985 Back-up RE Soyuz-T 13/Salyut 7 **(1)**
1985 Back-up RE Soyuz-T 14/Salyut 7 **(1)**

(1) Originally, Soyuz-T 13 would have carried Vasyutin, Savinykh and A. A. Volkov, backed-up by Viktorenko, Alexandrov and Saley on a long duration mission, but the in-orbit failure of Salyut 7 resulted in the crew being split. Vasyutin, Volkov, Viktorenko and Saley were re-assigned to the revised Soyuz-T 14 mission.
Saley was named in 1988 by A. A. Volkov as his back-up for Soyuz-T 14. When asked about Saley's selection, Volkov simply said that the men had trained for the same period of time: this might mean that they were from the same cosmonaut selection or simply that they trained for their Salyut 7 mission together. Saley is no longer active, following medical problems.

SARAFANOV, Gennady Vasilyevich

Date of Birth: 1 January 1942
Selection Group: 1965 October Military Pilot
Assignments: 1974 Cdr Soyuz 15/Salyut 3

SAVINYKH, Viktor Petrovich

Date of Birth: 7 March 1940
Selection Group: 1978 December Engineer
Assignments:
1980 Back-up FE Soyuz-T 3/Salyut 6
1981 FE Soyuz-T 4/Salyut 6
1982 Back-up FE Soyuz T 7/Salyut 7
1983 Back-up RE Soyuz T 8/Salyut 7
1984 Back-up FE Soyuz-T 10/Salyut 7
1984 Back-up RE Soyuz-T 12/Salyut 7
1985 FE Soyuz-T 13/Salyut 7 **(1)**
1987 Back-up FE Soyuz-TM 3/Mir/Syria
1988 FE Soyuz-TM 5/Mir/Bulgaria

(1) Originally trained with Vasyutin and A. A. Volkov as prime crew for a long duration Soyuz-T mission to Salyut 7. After the Salyut 7 failure, Savinykh was re-assigned to fly the revised Soyuz-T 13 rescue mission commanded by Dzhanibekov, and after the rescue he was joined by Vasyutin and Volkov in orbit for a long duration mission.

SAVITSKAYA, Svetlana Yevgenyevna

Date of Birth: 8 August 1948
Selection Group: 1980 Woman Engineer
Assignments:
1982 RE Soyuz-T 7/Salyut 7
1984 FE Soyuz-T 12/Salyut 7 **(1)**

(1) First woman to make a second spaceflight: first woman to perform a spacewalk.

SENKEVICH, Yuri A.

Date of Birth: 4 March 1937
Selection Group: 1965 Voskhod Physician
Assignments: 1966? Candidate Physician Bio-Voskhod

Selected as a candidate for a biological Voskhod mission, plans for which were cancelled in 1965.

SEREBROV, Alexander Alexandrovich

Date of Birth: 15 February 1944
Selection Group: 1978 December Engineer
Assignments:
1982 FE Soyuz-T 7/Salyut 7
1983 RE Soyuz-T 8/Salyut 7 **(1)**
1987 (Back-up) FE Soyuz-TM 2/Mir **(2)**

(1) First man to be launched on successive space missions.
(2) Titov and Serebrov trained for the Soyuz-TM 2 mission in 1987, but they were switched with Romanknenko and Leveykin, their back-ups.

SEVASTYANOV, Vitaly Ivanovich

Date of Birth: 8 July 1935
Selection Group: 1967 January Engineer
Assignments:
1967 Lunar training group (Manager until late 1967)
1969 Back-up FE Soyuz 6 **(1)**
1969 Back-up FE Soyuz 7
1969 Back-up FE Soyuz 8
1970 FE Soyuz 9
1973 Back-up FE Soyuz 13?
1975 Back-up FE Soyuz 17/Salyut 4
1975 Back-up FE Soyuz 18-1/Salyut 4 **(2)**
1975 FE Soyuz 18/Salyut 4

(1) Trained with Rukavishnikov.
(2) Failed to reach orbit.

SHATALOV, Vladimir Alexandrovich

Date of Birth: 8 December 1927
Selection Group: 1963 January Military Pilot
Assignments:
1966 Back-up Cdr Voskhod 3 **(1)**
1968 Back-up Cdr Soyuz 3 **(2)**
1969 Cdr Soyuz 4
1969 Cdr Soyuz 8
1971 Cdr Soyuz 10/Salyut 1
1971 Back-up Cdr Soyuz 11/Salyut 1 **(3)**

(1) Beregovoi and Shatalov were the back-ups for the cancelled Voskhod 3 mission, but is it unclear who would have been commander and who the pilot.
(2) Volynov has also been named as back-up commander.
(3) When the Soyuz 11 prime crew stood down and their back-ups assigned to fly the mission, the Soyuz 10 cosmonauts were quickly recycled as the new Soyuz 11 back-ups.

SHCHUKIN, Alexander V.

Date of Birth:	19 January 1946	
Selection Group:	1982 February	Shuttle Pilot
Assignments:	1987 Back-up RE	Soyuz-TM 4/Mir

SHONIN, Georgi Stepanovich

Date of Birth:	3 August 1935	
Selection Group:	1960 March	Military Pilot
Assignments:	1966	Pilot Voskhod 3 (1)
	1969 Back-up	Cdr Soyuz 5
	1969	Cdr Soyuz 6

(1) This mission was cancelled in 1965.

SOLOVYEVA, Irina Bayonovna

Date of Birth:	?	
Selection Group:	1962 March	Vostok 6
Assignments:	1963 Back-up	Pilot Vostok 6

SOLOVYOV, Anatoli Yakovlevich

Date of Birth:	16 January 1948	
Selection Group:	1976 February	Military Pilot
Assignments:	1987 Back-up Cdr	Soyuz-TM 3/Mir/Syria
	1988	Cdr Soyuz-TM 5/Mir/Bulgaria

SOLOVYOV, Vladimir Alexeievich

Date of Birth:	11 November 1946	
Selection Group:	1978 December	Engineer
Assignments:	1982 Back-up FE	Soyuz-T 6/Salyut 7/France
	1983 Back-up FE	Soyuz-T 9/Salyut 7
	1983 Back-up FE	Soyuz-T 10-1/Salyut 7 (1)
	1984 FE	Soyuz-T 10/Salyut 7
	1986 FE	Soyuz-T 15/Mir/Salyut 7

(1) Mission aborted due to launch vehicle explosion.

SOROKIN, Aleksei Vasilyevich

Date of Birth:	30 March 1931	
Date of Death:	23 January 1976	
Selection Group:	1964 February	Voskhod Doctor
Assignments:	1964 Support Doctor Voskhod 1 (1)	

(1) Sorokin was selected and trained for Voskhod 1, although whether he was actually the ''support'' doctor is unclear.

STREKALOV, Gennady Mikhailovich

Date of Birth:	28 October 1940	
Selection Group:	1973	Engineer
Assignments:	1976 Back-up FE	Soyuz 22 (1)
	1980 FE	Soyuz-T 3/Salyut 6 (2)
	1982 Back-up FE	Soyuz-T 5/Salyut 7
	1983 FE	Soyuz-T 8/Salyut 7
	1983 FE	Soyuz-T.10-1/Salyut 7 (3)
	1984 FE	Soyuz-T 11/Salyut 7/India (4)
	1985 Back-up FE	Soyuz-T 14/Salyut 7 (5)

(1) There was another flight engineer training with Strekalov for Soyuz 22, because two back-up commanders have been named. The second flight engineer has not been identified.
(2) Replaced Feoktistov, who had medical problems.
(3) Launch abort.
(4) Replaced Rukavishnikov, who had medical problems.
(5) After the crew re-assignments of the original Soyuz-T 13 cosmonauts (Vasyutin, A. A. Volkov, Viktorenko and Saley), Grechko and Strekalov were added to the four-man training group to provide a three-man crew (and back-ups) for Soyuz-T 14.

TERESHKOVA, Valentina Vladimirovna

Date of Birth:	6 March 1937	
Selection Group:	1962 February	Vostok 6
Assignments:	1963	Pilot Vostok 6 (1)

(1) First woman in space.

TITOV, Gherman Stepanovich

Date of Birth:	11 September 1935	
Selection Group:	1960 March	Military Pilot
Assignments:	1961 Back-up	Pilot Vostok 1
	1961	Pilot Vostok 2

TITOV, Vladimir Georgyevich

Date of Birth:	1 January 1947	
Selection Group:	1976 Februaryr	Militay Pilot
Assignments:	1982 Back-up Cdr	Soyuz-T 5/Salyut 7
	1983 Cdr	Soyuz-T 8/Salyut 7
	1983 Cdr	Soyuz-T 10-1/Salyut 7 (1)
	1987 (Back-up) Cdr	Soyuz-TM 2/Mir (2)
	1987 Cdr	Soyuz-TM 4/Mir

(1) Launch abort: booster exploded on the pad.
(2) Trained with Serebrov as prime crew for Soyuz-TM 2, but they were replaced by Romanenko and Leveykin.

TOKOV

Date of Birth:	?	
Selection Group:	?	Military Pilot

Died in the early 1970s in an air crash, and was disclosed as a cosmonaut trainee to the ASTP crews.

VARLARMOV, Valentin Stepanovich

Date of Birth:	?	
Date of Death:	? October 1980	
Selection Group:	1960 March	Military Pilot

Retired from cosmonaut training in 1960 after fracturing a vertibra. Died of natural causes.

VASYUTIN, Vladimir Vladimirovich

Date of Birth:	8 March 1952	
Selection Group:	1976 February	Military Pilot
Assignments:	1982 Back-up Cdr	Soyuz-T 7/Salyut 7 (1)
	1984 Back-up Cdr	Soyuz-T 10/Salyut 7
	1984 Back-up Cdr	Soyuz-T 12/Salyut 7
	1985 Cdr	Soyuz-T 13/Salyut 7 (2)
	1985 Cdr	Soyuz-T 14/Salyut 7

(1) Replaced Romanenko.
(2) Originally, Soyuz-T 13 was to be launched with Vasyutin, Savinykh and A. A. Volkov, but the crew was split following the in-orbit failure of Salyut 7. Vasyutin and Volkov were re-assigned to Soyuz-T 14.

VAVKIN, Yuri

Date of Birth:	?	
Selection Group:	1965	Military Engineer

Filmed in 1968 by an NBC team; identified as a trainee cosmonaut.

VETROV, Vladimir

Date of Birth:	?	
Selection Group:	1970 ?	Engineer

Identified as a cosmonaut in Soviet literature, but no further details given.

VIKTORENKO, Alexander Stepanovich

Date of Birth:	29 March 1947	
Selection Group:	1978	Mllitary Pilot
Assignments:	1985 Back-up Cdr	Soyuz-T 13/Salyut 7 (1)
	1985 Back-up Cdr	Soyuz-T 14/Salyut 7
	1986 Back-up Cdr	Soyuz-T 15/Mir/Salyut 7
	1987 Cdr	Soyuz-TM 3/Mir/Syria
	1988 Back-up Cdr	Soyuz-TM 7/Mir/France

(1) Originally Soyuz-T 13 was to be a long duration mission with Viktorenko, Alexandrov and Saley as the back-up cosmonauts. Following the Salyut 7 failure, Alexandrov was re-assigned to the revised Soyuz-T 13 rescue mission while Viktorenko and Saley were assigned to the long duration Soyuz-T 14 mission.

VINOGRADOV

Date of Birth:	?	
Selection Group:	?	Military Pilot

Rumoured to be training for a visit to Salyut in 1972, together with Voronov, Sevastyankov and Kubasov. The Salyut failed to reach orbit in July and neither Vinogradov nor Voronin flew subsequently.

VOLK, Igor Petrovich

Date of Birth:	12 April 1937	
Selection Group:	1978	Shuttle Pilot
Assignments:	1984 RE	Soyuz-T 12/Salyut 7

Conducted shuttle orbiter landing tests during 1987.

VOLKOV, Alexander Alexandrovich

Date of Birth:	27 May 1948	
Selection Group:	1976 February	Military Pilot
Assignments:	1985 RE	Soyuz-T 13/Salyut 7 (1)
	1985 RE	Soyuz-T 14/Salyut 7
	1987 Back-up Cdr	Soyuz-TM 4/Mir
	1988 Cdr	Soyuz-TM 7/Mir/France

(1) Originally, Soyuz-T 13 would have carried Vasyutin, Savinykh and Volkov, but the in-orbit failure of Salyut 7 resulted in the crew being split. Savinykh was launched on the Soyuz-T 13 rescue mission, while Vasyutin and Volkov flew the Soyuz-T 14 long duration mission.

VOLKOV, Vladislav Nikolayevich

Date of Birth:	23 November 1935	
Date of Death:	29 June 1971	
Selection Group:	1966 September	Engineer
Assignments:	1969 FE	Soyuz 7
	1971 Back-up FE	Soyuz 10/Salyut 1
	1971 (Back-up) FE	Soyuz 11/Salyut 1 (1)

(1) Originally on the back-up crew for Soyuz 11, but when Kubasov fell ill Volkov was drafted into the prime crew as his replacement; subsequently, the whole of the Soyuz 11 back-up crew was assigned as prime crewmen for Soyuz 11. Died when pressure was lost during the Soyuz 11 re-entry (30 June Moscow Time).

VOLYNOV, Boris Valentinovich

Date of Birth:	18 December 1934	
Selection Group:	1960 March	Military Pilot
Assignments:	1962 Back-up Pilot Vostok 4 (1)	
	1962 Back-up Pilot Vostok 5	
	1964 Back-up Cdr Voskhod 1	
	1966 Cdr Voskhod 3 (2)	
	1968 Back-up Cdr Soyuz 3 (3)	
	1969 Cdr Soyuz 5	
	1974 Back-up Cdr Soyuz 14/Salyut 3	
	1974 Back-up Cdr Soyuz 15/Salyut 3	
	1976 Cdr Soyuz 21/Salyut 5	

(1) Replaced Komarov.
(2) Mission cancelled.
(3) Shatalov also trained as Soyuz 3 back-up commander.
It was reported that Volynov was training for a visiting mission to Salyut 7 in 1985, but the station's failure resulted in the mission's cancellation.

VORONOV

Date of Birth:	?	
Selection Group:	?	Military Pilot

Rumoured to be training in 1972 for a Salyut flight with Vinogradov, Sevastyanov and Kubasov. The Salyut failed to reach orbit, and neither Vinogradov nor Voronin subsequently made spaceflights.

YEGOROV, Boris Borisovich

Date of Birth:	26 November 1937	
Selection Group:	1963	Vostok 7 Pilot
	1964 February	Voskhod Doctor
Assignments:	1964	Pilot Vostok 7 (1)
	1964	Dr. Voskhod 1

(1) Yegorov was selected in 1963 with another cosmonaut (thought to have been Lazarev) to train for a Vostok 7 mission in 1964, but the flight was cancelled in 1963 and the cosmonauts stood down. They were re-selected for the Voskhod 1 mission approximately six months later.

YELISEYEV, Alexei Stanislavovich

Date of Birth:	13 July 1934	
Selection Group:	1966 September	Engineer
Assignments:	1967 FE	Soyuz 2 (1)
	1969 FE	Soyuz 5
	1971 FE	Soyuz 10/Salyut 1
	1971 Back-up FE	Soyuz 11/Salyut 1 (2)

(1) Mission cancelled when Soyuz 1 developed problems in orbit; the name was later given to an unmanned mission.
(2) When the original back-up crew for Soyuz 11 was re-assigned to fly the mission, the Soyuz 10 cosmonauts were re-cycled to act as back-ups.

YORKINA, Zhanna Dmitryevna

Date of Birth:	?	
Selection Group:	1962 March	Vostok 6

No assignments.

ZAIKIN, Dmitri Alekseyevich

Date of Birth:	?	
Selection Group:	1960 March	Military Pilot
Assignments:	1965 Back-up Cdr Voskhod 2 (1)	

(1) Replaced Gorbatko, who withdrew because of medical problems. Left the cosmonaut training group in April 1968, due to ill-health while training for a Soyuz mission.

ZAITSEV, Andrei Yevgenevich

Date of Birth:	5 August 1957	
Selection Group:	1987 February	Engineer
Assignments:	1988 Back-up FE	Soyuz-TM 5/Mir/Bulgaria (1)

(1) Replaced by Serebrov.

ZHOLOBOV, Vitaly Mikhailovich

Date of Birth:	18 June 1937	
Selection Group:	1963 January	Military Engineer
Assignments:	1974 Back-up FE	Soyuz 14/Salyut 3
	1974 Back-up FE	Soyuz 15/Salyut 3
	1976 FE	Soyuz 21/Salyut 5

ZUDOV, Vyacheslav Dmitrievich

Date of Birth:	8 January 1942	
Selection Group:	1965 October	Military Pilot
Assignments:	1976 Back-up Cdr	Soyuz 21/Salyut 5
	1976 Cdr	Soyuz 23/Salyut 5
	1980 Back-up Cdr	Soyuz 35/Salyut 6
	1981 Back-up Cdr	Soyuz-T 4/Salyut 6

ZYKOV, Anatoli

Date of Birth:	?	
Selection Group:	1966	Engineer

Applied for cosmonaut training in June 1966, and apparently passed all of the physical and medical exercises, but was rejected from the training group because he was too tall. Therefore, Zykov was never a trainee cosmonaut.

FURTHER READING

Probably the two leading authorities on the cosmonaut team are Rex Hall and Gordon Hooper and the works by them which are listed under ''The Cosmonauts'' in the Bibliography have proved essential in compiling this summary of the cosmonauts.

The research in this field by James E. Oberg is also acknowledged.

Appendix 4: **The Men Behind the Space Programme**

(The information for this Appendix was supplied by Rex Hall).

Appendix 3 gave brief details of the cosmonauts who have flown Soviet spacecraft, but details have not previously been given in this book about the men who designed the spacecraft and rocket engines, or who were involved with the administration of the programme. This will now be rectified. There is not the space available to discuss all of the designers who have been identified by the Soviet Union, but there follows details of the leading men in this normally overlooked area.

Georgi Nikolaevich BABAKIN
(31 Oct 1914 – 3 Aug 1971)

Babakin was an outstanding designer in the field of space engineering, working in the Korolyov bureau developing automatic space vehicles for lunar and planetary missions. He started his own design bureau in 1965, and was responsible for the three generations of Luna probes (Lunas 1-3, Lunas 4-14, and Lunas 15-24), as well as being involved in the Mars and Venera missions. He became a corresponding member of the USSR Academy of Sciences in 1970.

Vladimir Pavlovich BARMIN
(born 1909)

Barmin was a graduate of the Bauman Higher Technical School, and was involved in the design of mobile launchers used in World War II. In 1945 he met Korolyov when they were both sent to Germany to evaluate German missile technology. In 1946 he was appointed the Chief Designer of Launch Complexes, and in this role he was responsible for the design of the Tyuratam (Baikonur), Kapustin Yar and Plesetsk cosmodromes. He is an Academician of the USSR Academy of Sciences, and became a General Designer of Ground Support Equipment for space launches.

Konstantin Davidovich BUSHUYEV
(23 May 1914 – 26 Oct 1978)

Bushuyev was a graduate of the Moscow Aviation Institute and after graduating he initially went to work in an aircraft design bureau. In 1947 he joined the Korolyov design bureau and in 1954 he became the deputy head of construction work, with his responsibilities focusing on applied dynamics and the durability of rockets. He worked on the development of IRBMs and ICBMs. He was named as the Soviet director of ASTP. After ASTP, he joined the Chelomei bureau as a chief designer, working on the Salyut programme. He became a corresponding member of the USSR Academy of Sciences in 1960.

Vladimir Nikolayevich CHELOMEI
(30 June 1914 – 8 Dec 1984)

During World War II (called by the Soviets the Great Patriotic War), Chelomei worked on pulse-jet engines. From 1944 to 1954 he was the chief designer working on military missiles, and most of his work was in this area of development. In 1959 he was named a Designer General. The Chelomei bureau was responsible for the design of the Proton launch vehicle and the SS-9 Scarp missile (western missile identifications) which was used as the basis for the Tsyklon booster (SL-11 and SL-14 variants). The same bureau designed Salyut 2, Salyut 3 and Salyut 5 (the "military Salyuts"), which directly led to the design of the Heavy ·Cosmos satellites (Cosmos 929, etc.)which docked with the Salyut and Mir orbital stations. Additionally, Chelomei's bureau produced the Polyot satellites, launched in 1963 and 1964.

Chelomei·became a corresponding member of the USSR Academy of Sciences in 1958 and an Academician in 1962. He was a professor at the Bauman Higher Technical School.

Boris Yevseyevich CHERTOK

Chertok was a leading member of the Korolyov design bureau, specialising in automatic control systems. He worked on the Vostok booster with Pilyugin. Later, Chertok took control of a group to develop the Molniya-1 communications satellites. He is a corresponding member of the USSR Academy of Sciences.

Oleg G. GAZENKO

Gazenko is a biologist who has been involved in the medical support of the space programme since the early days of the Vostok programme. He is the current director of the Institute of Bio-Medical Problems in Moscow. The largest institute of its type in the world, it is responsible for the support and the monitoring of medical results from manned and unmanned missions. Gazenko is a leading authority in the field of space biology and medicine. He is an Academician in the USSR Academy of Sciences.

Valentin Petrovich GLUSHKO
(born 20 Aug 1906)

Glushko is a founder of rocket engine design in the Soviet Union. In 1929 he founded the Gas Dynamics Laboratory design bureau (GDL-OKB). This bureau later became responsible for the design of many engines used to power the first stages of Soviet launch vehicles and missiles. These included the RD-107 and the RD-108 used on the Sputnik launch vehicle and all other SS-6 derived boosters and the RD-253 used on the first stage of the Proton booster.

Glushko became a corresponding member of the USSR Academy of Sciences in 1953 and an Academician in 1958. He was known only as the "Chief Designer of Rocket Engines" in the early years of the space programme. He was a member of the State Commission and worked along side Korolyov for many years. Some other designers from his bureau became chief designers in their own right. His current deputy is **D. Sevruk**.

Alexei Mikhailovich ISAYEV
(24 Oct 1908 – 10 Jun 1971)

Isayev was a member of the Glushko design bureau but in 1944 he left to form his own design bureau to develop liquid propellant engines. Initially, the engines were used on aircraft produced by the Mikoyan and Tupolev bureaus. In 1959 he developed the engines to be used on automatic interplanetary craft and manned spacecraft; he was in charge of retro-rockets and manoeuvring units. These were used on the Vostok, Voskhod, Soyuz, Polyot, Luna, Venera, Mars, Zond and Molniya-1 spacecraft. Isayev was a corresponding member of the USSR Academy of Sciences.

Mstislav Vsevolodovich KELDYSH
(10 Feb 1911 – 24 Jun 1978)

In the early years of the Soviet space programme Keldysh was identified as the "Chief Theoretician of Cosmonautics", although he was widely known as the head of the USSR Academy of Sciences between 1961 and 1975.

Keldysh became an Academician in 1946 and was a mathematician·who·worked·in the areas of aerodynamics and applied mathematics. He worked closely with Korolyov and was a member of the famed State Commission until the launch of Vostok 1. Together with his team of mathematicians and other specialists, he did many of the calculations relating to space missions.

Lt-Col Kerim Aliyevich KERIMOV
(born 14 Nov 1917)

Kerimov was a specialist in military artillery and he worked in the Ministry of Defence during 1943-1965. In 1946 he worked with Korolyov in the field of rocketry and space technology. In 1957 he took part in the tests of a military rocket at the proving facility of Kapustin Yar (later to become a cosmodrome). On Korolyov's recommendation, Kerimov became the Chairman of the State Commission for the flight testing of manned space complexes and unmanned interplanetary stations in 1965. In this role, he supervised every stage of development and operation of manned space complexes. Kerimov still occupied that position in 1988.

Sergei Pavlovich KOROLYOV
(30 Dec 1907 – 14 Jan 1966)

During his lifetime Korolyov was not identified with the space programme by the Soviet authorities; he was referred to as being the "Chief Designer Of Rocket-Cosmic Systems".

Korolyov was born at Zhitomin in the Ukraine. A graduate of the Bauman Higher Technical School, he met the father of cosmonautics, Konstantin Tsiolkovsky and became a follower of his ideas. In 1933 Korolyov was the founder member of a now-famous group of rocket engineers. In 1938 he was arrested and imprisoned for six years; he was not finally cleared until 1957. Korolyov led a team in 1945 which was charged with developing advanced versions of the German V-2 rocket weapon. For the next few years he worked in the Military Industries to develop the Soviet Union's first ballistic missile.

His main aim was to promote Tsiolkovsky's ideas and to get man into space. Once the missile had flown, he began work on the Sputnik programme, which later led to the development of the Vostok, Voskhod and Soyuz manned spacecraft, as well as many unmanned spacecraft including planetary craft and the Molniya communications satellites. His R-7 missile (designated in the West the SS-6) was converted and is still used today as the main vehicle to launch manned and unmanned spacecraft (the Sputnik, Vostok, Soyuz and Molniya boosters are based upon the R-7 missile).

In addition, Korolyov built up a large design team which included men who could design craft, as well as some cosmonaut-engineers. The designers included Mishin, Bushuyev and Babakin, while the cosmonauts included Feoktistov, Grechko, Kubasov, Yeliseyev, Volkov, Sevastyanov, Makarov and Aksyonov.

Korolyov was plagued with ill-health due to his imprisonment during the 1930s and 1940s and he died aged 58 in January 1966. His death was a major blow to the space programme because he was a great leader and manager of the space programmes; his ideas are still current in the programme management. He became a corresponding member of the USSR Academy of Sciences in 1953 and elected a full Academician in 1958. His remains are buried in the Kremlin Wall.

Semyin Arievich KOSBERG
(1 Oct 1903 – 3 Jan 1965)

Kosberg was a graduate of the Moscow Aviation Institute and later became a member of the GDL-OKB. In 1946 he became a chief designer working on liquid rocket engines. From 1954 they developed a number of engines for the upper stages for a variety of launch vehicles, including the second and third stages of the Proton booster and the third stages for the SL-3 and SL-4 boosters. His successor was **Alexandr Konopatov.**

Vyacheslav Mikhailovich KOVTUNENKO

Kovtunenko has been a corresponding member of the USSR Academy of Sciences since 1984. He is the leading designer of automatic spacecraft, and was the project director of the VEGA (Venus-Comet Halley) and Phobos (Mars satellite) missions. He took over this role in 1978 as a direct successor to Babakin, and he was transferred from the Yangel design bureau.

Vasili Pavlovich MISHIN
(born 5 Jan 1917)

Mishin is a specialist in control processes. He became a corresponding member of the USSR Academy of Sciences in 1958 and a full Academician in 1966. He was the deputy to Korolyov in his design bureau and succeeded Korolyov after his death in 1966. Mishin headed the bureau from 1966 to 1974, when it was extensively re-organised. He is described as a general constructor of space rocket systems.

Alexandr Davydovich NADIRADZE
(20 Aug 1914 – 3 Sep 1987)

Nadiradze was a veteran of the early days of rocket design. He was responsible for the designs of various ICBMs including the SS-13 and the SS-16 (western designations). He was an Academician of the USSR Academy of Sciences.

Sergei Osipovich OKHAPKIN
(1910 – 1980)

Okhapkin was a deputy chief designer with the Korolyov design bureau. He worked as a professor at the Moscow Aviation Institute, and was involved in the Vostok and Voskhod projects.

Boris Nikolayevich PETROV
(11 Mar 1913 – 23 Aug 1980)

Petrov was an expert on automation and telemechanics. He became a corresponding member of the USSR Academy of Sciences in 1953 and an Academician in 1960. In 1966 he became Chairman of the Intercosmos Council and thus headed the development of international co-operation within the Soviet space programme. This included ASTP and the flights of guest cosmonauts to the Salyut stations. Petrov was also the Vice-President of the USSR Academy of Sciences responsible for space exploration.

Nikolai Alekseevich PILYUGIN
(18 May 1908 – 2 Aug 1982)

Pilyugin was a graduate of the Bauman Higher Technical School and was the chief designer of autonomous guidance systems. These saw applications on Soviet spacecraft and rockets. He was a close collaborator with Korolyov from the end of World War II and headed the Institute of Mechanics for 20 years. Pilyugin was a member of the State Commission and he became an Academician in 1966.

Konstantin Nikolaevich RUDNEV
(1911 – 13 Aug 1980)

Rudnev was a top administrator in the defence industries as well as a top Communist Party official. In 1952 he was the Deputy Minister for Armaments and then a Deputy Minister of the Defence Industry. During the period 1961-1965 he was Vice Premier and was the Chairman of the State Commission which played an important role in the development of rocketry. He was succeeded by **Lt-Gen Kerimov.**

Mikhail Sergeyevich RYAZANSKIY
(1909 – 5 Aug 1987)

A graduate of the Moscow Power Engineering Institute, Ryazanskiy was a specialist in the field of control systems and worked under Korolyov in his design bureau. He developed the radio control systems for launch vehicles and for the Vostok, Voskhod, Soyuz, Luna, Venera and Mars spacecraft. He also developed the systems for the land and sea command and measurement complexes. Ryazanskiy was a corresponding member of the USSR Academy of Sciences.

Roald Z. SAGDEYEV
(born 1933)

Sagdeyev is a physicist by training and in 1961 he helped found Akademgorodok in Siberia. He became the director of the Space Research Institute (IKI) in 1973. Sagdeyev now heads one of the major institutes for the planning and design of space research in the World. The Institute has a number of special design bureaus responsible for the development of space equipment. He is an Academician of the USSR Academy of Sciences.

Yuri Pavlovich SEMENOV

Semenov became a corresponding member of the USSR Academy of Sciences in 1987. He is currently the project director for the development of manned flight apparatus for international programmes within the USSR main Space Directorate (Glavcosmos). He is the designer of Salyut 6 and Salyut 7, as well as the Mir complex. Since 1978 Semenov has been the chief designer of manned spacecraft.

Boris Sergeyich STECHKIN
(1891 – 1969)

Stechkin was the chief consultant at the Korolyov design bureau for future space engines from 1963 until his death. He was an Academician with the USSR Academy of Sciences.

Mikhail Klavdievich TIKHONRAVOV
(1900 – 4 Mar 1974)

Tikhonravov worked closely with Korolyov from the mid-1930s in the original GIRD organisation. He was a leading theoretician and designer in the Korolyov bureau and led his own design bureau which worked on the original studies of Vostok. Tikhonravov was a member of the State Commission and an Academician of the USSR Academy of Sciences.

Leonid Aleksandrovich VOSKRESENSKY
(1913 – 1965)

Voskresensky was the deputy chief designer for the Korolyov design bureau. He was in charge of the Vostok flight tests and the spacecraft testing department. He was a close colleague of Korolyov for many years.

Mikhail Kuzmich YANGEL
(25 Oct 1911 – 25 Oct 1971)

Yangel was a graduate of the Moscow Aviation Institute, and worked closely with Korolyov in his design bureau for many years. In 1954 he was appointed a General Designer and given his own design bureau in Dnepropetrovsk. He developed silo-launched ballistic missiles. The Yangel bureau was responsible for the launching of small Cosmos and Intercosmos satellites. Yangel also designed military missiles, including the SS-18. He was a member of the State Commission and was appointed an Academician of the USSR Academy of Sciences in 1966. His successor was **Academician V. F. Utkin.**

Bibliography

A brief bibliography has been given at the end of each chapter and appendix, but these reflect only the information contained in each section of the book. Here, a list is given of works which have been used in preparation of this volume. In many cases the works have been referenced in the chapter and appendix bibliographies; however, other books have not been referenced earlier. The latter books are of a more general nature in their coverage of the Soviet space programme (eg, reviewing the whole programme, rather than simply the manned aspect). Inclusion of a book in this bibliography does not indicate that the writer of this volume agrees with all (or even part) of the book's content.

HISTORY OF THE SOVIET MANNED PROGRAMME

The first paper which gave details of Korolyov's design studies was:

Claude Wachtel, "Design Studies Of The Vostok-J And Soyuz Spacecraft" in *JBIS*, vol 35 (1982), pp92-94.

The implications of this paper formed the basis of the following pair of papers:

Phillip S. Clark and Ralph F. Gibbons, "The Evolution of the Soyuz Programme" in *JBIS*, vol 36 (1983), pp434-452.

Ralph F. Gibbons and Phillip S. Clark, "The Evolution of the Vostok and Voskhod Programmes" in *JBIS*, vol 38 (1985), pp3-10.

It was from the latter two papers that the present book evolved.

THE COSMONAUTS

Astronauts and Cosmonauts Biographical and Statistical Data (revised 28 June 1985), US Government Printing Office, December 1985.

Rex Hall, "The Soviet Cosmonaut Team, 1960-1971" in *JBIS*, vol 36 (1983), pp468-473.

Rex Hall, "The Soviet Cosmonaut Team, 1971-1983" in *JBIS*, vol 36 (1983), pp474-480.

Rex Hall, "The Soviet Cosmonaut Team, 1956-1967" in *JBIS*, vol 41 (1988), pp107-110.

Rex Hall, "The Soviet Cosmonaut Team, 1978-1987" in *JBIS*, vol 41 (1988), pp111-116.

Gordon R. Hooper, *The Soviet Cosmonaut Team*, GRH Publications Woodbridge, England, 1986.

WESTERN REVIEWS OF THE SOVIET MANNED PROGRAMME

Phillip S. Clark, *Soviet Launch Failures, 1957-1985*, privately published report, 1986.

Nicholas Daniloff, *The Kremlin and The Cosmos*, Alfred A. Knopf, New York, 1972.

Brian Harvey, *Race Into Space: The Soviet Space Programme*, Ellis Horwood Ltd, Chichester, 1988, and John Wiley & Sons, New York, 1988.

Peter N. James, *Soviet Conquest From Space*, Arlington House, New Rochelle, 1974.

Nicholas L. Johnson, *Handbook of Soviet Manned Space Flight*, Univelt Inc, San Diego, 1980.

James E. Oberg, *Red Star In Orbit: The Inside Story of the Soviet Space Programme*, Random House, New York, and Harrap Ltd, London, 1981.

Charles S. Sheldon II, *Review of the Soviet Space Program, with Comparative United States Data*, McGraw-Hill Book Company, New York, 1968.

William Shelton, *Soviet Space Exploration – The First Decade*, Arthur Barker Ltd, London, 1969.

Leonid Vladimirov, *The Russian Space Bluff*, Tom Stacey Ltd, London, 1971.

SOVIET BOOKS ABOUT THEIR SPACE PROGRAMME

Where a book is only available in Russian, the Russian title is given, followed by a translation into English; where the book has been published into English, only the English title is given.

V. A. Alexandrov et al, *Raketa Nosetyele* ("Carrier Rockets"), Moscow, 1981.

K. P. Feoktistov et al, *Comisheskiya Apparata* ("Cosmic Apparatus"), Moscow, 1983.

V. P. Glushko, *Kosmonautica Entsyklopediya* ("Cosmonautics Encyclopedia"), Moscow, 1985.

M. V. Keldysh, *Tvorcheskoye Nasledia Academika Sergei Pavlovicha Korolyova* ("The Creative Legacy Of Sergei Pavlovich Korolyov"), Moscow, 1980.

L. Lebedev et al, *Sons of the Blue Planet*, Amerind Publishing Co, New Delhi, 1973: Soviet edition, 1971.

Y. A Mozzhoriya, *Kosmonautica CCCP* ("Cosmonautics USSR"), Moscow, 1986.

G. I. Petrov (ed), *Conquest of Outer Space in the USSR, 1967-1970*, Amerind Publishing Co, New Delhi, 1973; Soviet edition, 1971.

Evgeny Riabchikov, *Russians In Space*, Doubleday & Co, Garden City, NY, 1971, and Weidenfeld & Nicolson, London, 1971. Note that the American edition has far more photographs than the English edition.

A. P. Voltskiya et al, *Kosmodrom* ("Cosmodrome"), Moscow, 1977.

US LIBRARY OF CONGRESS STUDIES

A series of studies has been prepared by the Congressional Research Service of the US Library of Congress, and these are detailed below. They were all published by the US Government Printing Office on the dates indicated:

Soviet Space Programs (31 May 1962).

Soviet Space Programs, 1962-1965 (30 December 1966).

Soviet Space Programs, 1966-1970 (9 December 1971).

Soviet Space Programs, 1971, Supplement to the 1966-1970 study (April 1972).

Soviet Space Programs, 1971-1975, Vol 1 (30 August 1976).

Soviet Space Programs, 1971-1975, Vol 2 (30 August 1976).

Soviet Space Programs, 1976-1980, Vol 1 (December 1982).

Soviet Space Programs, 1976-1980, Vol 2 (October 1984).

Soviet Space Programs, 1976-1980, Vol 3 (May 1985).

Soviet Space Programs, 1981-1987, Vol 1 (May 1988).

Soviet Space Programs, 1981-1987, Vol 2 (Winter 1988).

ANNUAL REVIEWS OF SOVIET SPACE ACTIVITY

The following articles and reports provide annual reviews of all of the Soviet space activity for the years shown.

Phillip S. Clark, "The Soviet Space Year Of 1981" in *JBIS*, vol 35 (1982), pp261-271.

Phillip S. Clark, "The Soviet Space Year Of 1982" in *JBIS*, vol 36 (1983), pp249-262.

Phillip S. Clark, "The Soviet Space Year Of 1983" in *JBIS*, vol 38 (1985), pp31-46, 48.

Phillip S. Clark, "The Soviet Space Year Of 1984" in *JBIS*, vol 38 (1985), pp339-353.

Phillip S. Clark, "Soviet Space Activity, 1985-1986" in *JBIS*, vol 40 (1987), pp203-221.

Nicholas L. Johnson, *The Soviet Year In Space, 1981*, Teledyne Brown Engineering, Colorado Springs, 1982.

Nicholas L. Johnson, *The Soviet Year In Space, 1982*, Teledyne Brown Engineering, Colorado Springs, 1983.

Nicholas L. Johnson, *The Soviet Year In Space, 1983*, Teledyne Brown Engineering, Colorado Springs, 1984.

Nicholas L. Johnson, *The Soviet Year In Space, 1984*, Teledyne Brown Engineering, Colorado Springs, 1985.

Nicholas L. Johnson, *The Soviet Year In Space, 1985*, Teledyne Brown Engineering, Colorado Springs, 1986.

Nicholas L. Johnson, *The Soviet Year In Space, 1986*, Teledyne Brown Engineering, Colorado Springs, 1987.

Nicholas L. Johnson, *The Soviet Year In Space, 1987*, Teledyne Brown Engineering, Colorado Springs, 1988.

Index

Notes: Page references to primary subjects are in **bold**, references occurring in picture captions are in *italic*.

Picture Credits

The author and publishers would like to thank everyone, as credited here, who supplied pictures for this book. Particular thanks go to Theo Pirard of the Space Information Center who supplied many valuable photographs from his own collection.

ADN-ZB (via SIC-Theo Pirard): 111 right
Phillip Clark Collection: 8 upper, 9 lower, 10, 11 right, 14, 15, 17 lower, 19 lower left and right, 20 lower left, 21 middle and lower, 22 both, 26, 27 right, 28 both, 29, 31 lower, 37, 38, 46, 47 upper and lower left, 49 both, 50, 51 both, 52 both, 53 both, 54, 55 upper, 56, 57 both, 58 right, 60, 61 lower, 64 lower, 67 upper right, 68, 69 upper, 73 both, 79 both, 80/1, 81 upper, 82, 87 upper, 105 lower, 107 lower, 110 lower, 112, 114, 115 upper left, 119 upper and middle right, 124 top, 134 lower, 136 upper, 137 right, 140 lower, 145 lower, 146 lower
CNES: 160
CNES/Intercosmos (via SIC-Theo Pirard): 6, 35, 36, 130, 131 all
Department of Defence, Australia: 164
DoD, USA: 165
Glavcosmos (via SIC-Theo Pirard): 40 upper, 129, 149, 150, 157, 163, 166, 176
Hasselblad: 45
Intercosmos (via SIC-Theo Pirard): 108 lower, 115 upper right, 118, 122 upper, 139 right, 140 upper

Neville Kidger: 45
NASA: 25, 88, 89, 90 both, 91 both, 92 top, 169 upper, 173
Novosti Press Agency, London: 11 middle, 40 lower, 43, 86
Novosti Press Agency, Moscow (via SIC-Theo Pirard): Jacket back (top left) 4/5, 8 lower, 16, 17 upper, 18, 19 upper 21 upper, 27 left, 31 upper, 33 upper, 55 lower, 62, 63, 64 upper, 65 lower, 67 upper left and lower, 69 lower, 74 both, 75 99 both, 102 both, 103 both, 104 113, lower, 115 lower, 117, 125 bottom, 125, 133, 136 lower, 146 upper, 151, 155 upper, 175
Curtis Peebles: 9 upper
Planeta Publishers, Moscow: 23, 47 lower right, 58 left, 61 upper, 65 upper, 78, 81 lower, 83, 87 lower left, 92 bottom left and right, 107 upper, 110 upper, 116, 119 upper left, 120 both, 121 122 lower, 123, 124 middle, 127 top left and right, 143, left, 144 both, 145 lower, 169 lower
Rockwell International: 84
SIC-Theo Pirard: 13, 20 lower right, 33 lower, 134 upper
A. Sokolov (via SIC-Theo Pirard): 45
Tass, London: Jacket front, Jacket back (top right), 2, 7, 126/7, 137 left, 138, 139 left, 141 152, 153 both, 154 both, 155 lower, 156, 158, 159, 160 middle and lower, 167
Tass, Moscow (via SIC-Theo Pirard): 71, 87 lower right, 95, 97, 98, 105 upper, 106, 108 upper, 109, 111 left, 113 upper, 132, 135, 143 right, 181
Carl Zeiss Jena: 93 both

Glossary of Terms

A number of terms which are specialist to astronomy and rocketry are used in this book, and brief definitions and/or explanations are given here.

Apogee The maximum distance of a satellite in its orbit above the surface of the Earth.

Delta-V Velocity change. In rocketry, the velocity change performed by an engine is given by:

Delta-V = Exhaust velocity × $\log_e(R)$

where the exhaust velocity is defined below and "R" is the ratio of the pre-manoeuvre and post-manoeuvre spacecraft masses.

Escape Velocity The minimum velocity required to escape from a gravitational field (eg, that of the Earth). A spacecraft launched towards another planet requires a velocity greater than escape velocity, while a lunar mission requires a velocity slightly less than escape velocity to leave the Earth.

Exhaust velocity A measurement of rocket engine performance: the exhaust velocity (in metres/sec) and the specific impulse (sec, qv) are connected by:

exhaust velocity = 9.806 × specific impulse

Inclination (or Orbital Inclination) The angle between the plane of the orbit and the equator (of the Earth for Earth orbits or the Moon's equator for lunar orbits, etc). This can range from 0° to 180° (orbits with inclinations greater than 90° are called **retrograde** orbits).

Perigee The minimum distance of a satellite in its orbit above the Earth.

Period (or Orbital Period) The time which a satellite takes to complete one circuit of the Earth (or Moon, etc). In more detailed orbital theory, there are different orbital periods used, depending upon the reference points used for the beginning and the end of the orbital circuit.

Specific Impulse A term used to measure the efficiency of a rocket engine: the higher the specific impulse, the more efficient the engine. The specific impulse, measured in seconds, is related to the propellant mass, the time of the rocket engine burn and the engine thrust by:

$$\text{specific impulse} = \frac{\text{thrust} \times \text{burn time}}{\text{propellant mass}}$$

where the thrust and propellant mass are in the same units (tonnes, kilogrammes, etc) and the burn time is in seconds.